Aghalaya S. Vatsala
April 29, 2013

Theory of Differential Equation in Cones

Theory of Differential Equations in Cones

by

V. Lakshmikantham
Florida Institute of Technology
Division of Applied Mathematics
Melbourne, Florida, USA

S. Leela
Professor Emeritus
State University of New York, Geneseo
New York, USA

A. S. Vatsala
University of Louisiana at Lafayette
College of Sciences
Department of Mathematics
Lafayette, Louisiana, USA

Cambridge Scientific Publishers

Cover design: Clare Turner

British Library Cataloguing in Publication Data
A catalogue record for this book has been requested

Library of Congress Cataloguing in Publication Data
A catalogue record has been requested

ISBN 978-1-904868-96-5 Hardback

Printed and bound in the UK by
Berforts Group Ltd, Stevenage

Cambridge Scientific Publishers Ltd
P.O.Box 806Cottenham, Cambridge CB24 8RT
UK
www.cambridgescientificpublishers.com

Theory of Differential Equations in Cones

Preface

Many dynamic systems that arise in modeling chemical reactors, neutron transport, population biology, spread of infectious diseases, as well as problems in economics, engineering, sociology and other systems demand the existence of nonnegative solutions with certain desired qualitative properties. We normally understand that non-negativity can be expressed in terms of arbitrary cones, that is, closed, convex subsets of the space under consideration. The cones define an order relation by means of which different elements can be compared in a way better than using estimates in terms of a norm. Furthermore, one way to extend the concept of monotonicity to higher dimensions is to utilize the order relation induced by a cone. Also, in the study of large-scale dynamic systems by the method of vector Lyapunov functions, employing arbitrary cones offers better results than using component-wise inequalities. Thus, it is clear that investigating nonlinear problems through abstract cones is an important branch of nonlinear analysis.

An extremely interesting and important area of study is the theory of differential equations in abstract spaces. The reason for this stems from the fact that many problems in Partial differential equations (PDEs) which arise from physical models have the form of evolution equations, which can often be considered as ordinary differential equations (ODEs) in suitable infinite dimensional spaces. Although, the theories of ODE and PDE have a striking difference, the point of view of imbedding PDEs in general evolution equations renders it possible to make use of the elegant theories and powerful techniques of ODEs and nonlinear functional analysis. The theory of nonlinear ODEs in an arbitrary Banach space has developed significantly and there are several books dealing with the area.

It is well known that Lyapunov's second method has given decisive impetus for the modern development of stability theory and that employing several Lyapunov functions instead of single one is more useful. The corresponding theory known as the method of vector Lyapunov functions offers a flexible mechanism by weakening the requirements of the original Lyapunov method. Moreover, we find that several Lyapunov functions arise naturally in some situations, such as the method of large-scale dynamic systems. It is also being realized that the same versatile tools are adaptable to investigate entirely different nonlinear systems and these effective methods offer an exciting prospect for further advancement.

This book essentially deals with the theory of differential equations in cones since this approach unifies the Lyapunov method along with obtaining nonnegative solutions of nonlinear dynamic systems. Chapter 1 is dedicated to the preliminary material required for the development of the rest of the book. In Chapter 2, we consider the theory of existence, local and global, extremal solutions and flow invariance in closed convex sets. Since cones are special cases of closed convex sets, the results provide the basic theory in cones. Chapter 3, is concerned with theoretical approximation methods, namely, monotone iterative technique and the method of quasilinearization and the corresponding abstractization of those results. In Chapter 4, we treat the Lyapunov method in cones, providing various typical results, including cone-valued Lyapunov functions and stability theory in terms two different measures. Finally, in Chapter 5, extensions to various other dynamic systems are indicated so that further progress in these areas can be entertained, employing the theory of cones.

Some of the important features of the monograph are as follows: This is the first book that

(i) treats the theory of differential equations in arbitrary cones;

(ii) demonstrates that developing Lyapunov method in cones significantly enhances the power of that method for the study of largescale systems;

(iii) presents theoretical approximation methods in terms of order relations induced by cones;

(iv) shows that utilizing arbitrary cones in investigating the theory of several dynamic systems is significantly more useful.

We offer our heartfelt thanks to Ms. Janie Wardle for her interest and cooperation in our project.

Chapter 1

Preliminaries

1.1 Introduction

This chapter is essentially introductory in nature. Its main purpose is to assemble relevant concepts and present certain results that are not so well known. Section 1.2 contains preliminary analysis most of which is taken from standard books on functional analysis. Section 1.3 introduces the basic properties of cones that form the basis for the entire book. Section 1.4 deals with the properties of directional derivatives that play a prominent role in the subsequent development of the book. Here, the notion of duality maps, generalized inner products and Lyapunov-like functions are introduced. In Section 1.5, we present needed mean value theorems. Section 1.6 contains the discussion of measures of strong and weak noncompactness and their properties. The relevant comparison results are listed in Section 1.7. Finally 1.8 deals with notes and comments.

1.2 Basic Known Results

Let E be a linear space (vector space) over the field ϕ (of real numbers or complex numbers) and let p be a function from E into R_+ the set of non-negative real numbers. The p is a norm on E if

(i) $p(ax) = |a|p(x)$ for all $x \in E$ and $a \in \phi$;

(ii) $p(x+y) \leq p(x) + p(y)$ for all $x, y \in E$;

(iii) $p(x) = 0$ if and only if $x = \theta$ where θ is the null element of E.

In this case, we write $p(\cdot) = \|\cdot\|$ and say that $(E, \|\cdot\|)$ is a normed linear space over ϕ. It is easily seen that $\rho(x, y) = \|x - y\|$, $x, y \in E$, defines a metric on E and that a sequence $\{x_n\} \subset E$ converges in this metric topology to $x \in E$ if and only if $\lim_{n\to\infty} \|x_n - x\| = 0$. We shall generally refer to this metric topology as the norm (uniform or strong) topology. A normed linear space E over the field of real numbers is said to be a real *Banach space* if it is a complete metric space when equipped with the metric $\rho(x,y)$. We shall be working mostly with real Banach space in this book.

A mapping $f: E \to R$ is said to be a *linear functional* on E if $f(x+y) = f(x) + f(y)$ and $f(ax) = af(x)$ for all $x, y \in E$ and $a \in R$, R being the field of real numbers. A linear functional f on E is bounded if there is a $M > 0$, such that $|f(x)| \leq M$ for all $x \in E$ with $\|x\| \leq 1$. The *dual space* E^* of E is defined to be the class of all continuous linear functionals on E and for each $x^* \in E^*$, we define

$$\|x^*\| = \sup\{|x^*| : x \in E, \|x\| \leq 1\}.$$

It is easy to show that this is a norm on E^* and with this norm, it is clear that E^* is a normed linear space. Also, since E is complete, it follows that E^* is complete and thus is a Banach space. Note that continuity and boundedness of a linear functional are equivalent concepts, i.e., a linear functional $f \in E^*$ if and only if it is bounded on E.

One of the fundamental results in functional analysis is the following theorem due to Hahn-Banach which assures that a linear functional on a linear subspace of a linear space that is bounded by a seminorm can always be extended to the entire space in such a manner that its seminorm boundedness is preserved. However, such an extension in general is not unique.

Theorem 1.2.1. *(Hahn-Banach) Suppose that E is linear space over R and that $p: E \to R$ is such that $p(\lambda x) = \lambda p(x)$, $p(x+y) \leq p(x) + p(y)$ for all $\lambda \geq 0$ and $x, y \in E$. Assume that Ω is a linear subspace of E and f is a linear functional from Ω into R such that $f(x) \leq p(x)$ for all $x \in \Omega$. Then there is linear functional $g: E \to R$ such that $g(x) = f(x)$for all $x \in \Omega$ and $g(x) \leq p(x)$ for all $x \in E$.*

If E is a normed linear space, we have the following version of Theorem 1.2.1.

Theorem 1.2.2. *Let Ω be a linear subspace of a normed linear space E over R and let $y^* \in \Omega^*$. Then there is an $x^* \in E^*$ such that $|x^*| = |y^*|$ and $x^*(x) = y^*(x)$ for all $x \in \Omega$.*

Some important consequences of Hahn-Banach theorem are as follows.

Corollary 1.2.1. *If $x \in E$ and $x \neq \theta$, then there is an $x^* \in E^*$ such that $|x^*| = 1$ and $x^*(x) = \|x\|$.*

Corollary 1.2.2. *Let Ω be a subspace of E and $x \in E$ with $d(x,\Omega) = d > 0$, where $d(x,\Omega) = \inf_{y\in\Omega} \|x-y\|$. Then there is an $x^* \in E^*$ with $d\|x^*\| = 1$, $x^*(x) = 1$ and $x^*(y) = 0$ for all $y \in \Omega$.*

Now, let $x \in E$, E a normed linear space, and define the mapping $\tilde{x} : E^* \to R$ by

$$\tilde{x}(x^*) = x^*(x), \quad x^* \in E^*.$$

Clearly $\tilde{x}$ is linear and $|\tilde{x}(x^*)| \leq \|x^*\|\|x\|$ for all $x^* \in E^*$. Hence $\tilde{x}$ is a bounded linear functional on E^* and $\tilde{x}$ is a member of the dual space of E^*, which is denoted by E^{**} and called the *bidual* or *second dual* of E. By Corollary 1.2.1, it follows that $\|\tilde{x}\| = \|x\|$ and the mapping $\tau : E \to E^{**}$ defined by $\tau x = \tilde{x}$ is linear and norm preserving. This mapping τ is called the *canonical embedding* of E into E^{**}. A Banach space E is said to be reflexive if $E^{**} = \{\tau x : x \in E\}$. Note that any finite dimensional Banach space is reflexive.

Let E_ω denote the space E when endowed with the weak topology generated by the continuous linear functionals on E, that is a sequence $\{x_\alpha\} \in E$ converges to $x \in E$ in this topology if and only if $\lim\limits_{\alpha\to\infty} x^*(x_\alpha) = x^*(x)$ for all $x^* \in E^*$. In this case, the sequence $\{x_\alpha\}$ is said to *converge weakly* to x and we write $x = \omega - \lim\limits_{\alpha\to\infty} x_{\alpha.}$ A sub-base E_ω is

$$B = \{S\{x, x^*, r\} : x \in E, x^* \in E^*, r > 0\}$$

where

$$S\{x, x^*, r\} = \{y \in E : x^*(x - y) < r\}.$$

It should be noted that

$$\rho_{x^*}(x - y) = |x^*(x - y)|$$

is a pseudo-metric so that $\{\rho_{x^*} : x^* \in E^*\}$ generates a uniformity on E. The weak topology is the same as the uniform topology generated by $\{\rho_{x^*} : x^* \in E^*\}$. In a reflexive Banach space, the following result due to weak convergence is true.

Theorem 1.2.3. *If E is a reflexive Banach space, any norm bounded sequence in E has a weakly convergent subsequence.*

If $x^* \in E^*$ and $\{x_n^*\}$ is a sequence in E^* then $\{x_n^*\}$ is said to be weak* convergent to x^* if

$$\lim x_n^*(x) = x^*(x), \quad \text{for all} \quad x \in E$$

In this case, we say that $\{x_n^*\}$ converges to x^* in the weak topology.
The next result concerning weak* topology is also valid.

Theorem 1.2.4. *Let E be a Banach space and $\{x_n^*\}$ be a bounded sequence in E^*. Then $\{x_n^*\}$ has a weak*−convergent sub-sequence.*

We list below definitions of some useful concepts in terms of weak topology.

Definition 1.2.1. *(i) A subset A of E_ω is totally bounded if and only if for all $x^* \in E^*$ and $\epsilon > 0$, the set A can be covered by a finite number of x^*−balls of radius ϵ;*

(ii) *If* $\{y_n\}$ *is a sequence in* E, *then* $\{y_n\}$ *is weakly Cauchy if given* $\epsilon > 0$, $x^* \in E^*$ *there exists an* $N = N(x^*, \epsilon)$ *such that* $n, m \geqslant N$ *implies* $|x^*(y_n - y_m)| < \epsilon$;

(iii) E *is weakly complete if every weakly Cauchy sequence converges weakly to a point in* E. *indexweakly complete*

With these definitions, we can now state that a set A is compact in the weak topology (or *weakly compact*) if and only if it is weakly complete and totally bounded.
Let $x(t)$ be a function mapping some interval $I \subset R$ into E. The function $x(t)$ is said to be *strongly continuous* at $t_0 \in I$ if

$$\lim_{t \to t_0} \|x(t) - x(t_0)\| = 0$$

that is, the convergence of $x(t)$ to $x(t_0)$ as $t \to t_0$ is in the norm topology on E. If $x : I \to E$ is continuous at each point of I then we say that x is continuous on I and write $x \in C[I, E]$. $x'(t)$ is said to be the *strong derivative* of $x(t)$ if

$$\lim_{h \to 0} \|\frac{1}{h}[x(t+h) - x(t)] - x'(t)\| = 0.$$

The Riemann integral of $x(t)$ can be similarly defined. Some useful properties of the integral are given in the following lemmas.

Lemma 1.2.1. *Let* E *be a real Banach space and* $x(t)$ *be an integrable function from* I *into* E. *Then*

$$(\frac{1}{b-a}) \int_a^b x(s)\mathrm{d}s \in \bar{co}(\{x'(s) : s \in [a, b]\})$$

for all $a, b \in I$ *with* $a < b$, *where* $\bar{co}(A)$ *is the closed convex hull of* A.

Lemma 1.2.2. *If* $\{x_n\}$, $n = 1, 2, ...$, *is a sequence of continuous functions from* I *into* E *such that* $\lim_{n \to \infty} x_n(t) = x(t)$ *uniformly for* $t \in I$, *then*

$$\lim_{n \to \infty} \int_a^b x_n(s)\mathrm{d}s = \int_a^b x(s)\mathrm{d}s \text{ for all } [a, b] \subset I.$$

Definition 1.2.2. *Let* $x(t)$ *be a function from* I *into* E. *Then*

(i) $x(t)$ *is weakly continuous at* t_0 *if for every* $x^* \in E^*$, *the scalar function* $x^*(x(t))$ *is continuous at* t_0;

(ii) $x(t)$ *is weakly differentiable at* t_0 *if there exists a point in* E, *denoted by* $x'(t_0)$ *such that* $x^*(x'(t_0)) = (x^*(x))'(t_0)$ *for each* $x^* \in E^*$;

(iii) $x(t)$ *is weakly (Riemann) integrable on* I *if there is an* $x_0 \in E$ *such that for each* $x^* \in E^*$

$$x^*(x_0) = \int_I x^*(x(s))\mathrm{d}s;$$

we will then write $x_0 = (\omega \int_I x(s)\mathrm{d}s$.

The following facts result from the above definitions:

(1) if $x(t)$ is weakly continuous and E_ω is weakly complete, then $x(t)$ is weakly integrable;

(2) if $x(t)$ is weakly differentiable, then $x(t)$ is weakly continuous;

(3) if $x(t)$ is weakly continuous and $F(t) = (\omega)\int^t x(s)\mathrm{d}s$, $s \in I$ then $F'(t) = x(t)$, where $F'(t)$ is the weak derivative of F.

Let $x_n(t)$ be a sequence of functions from I into E. Then $\{x_n(t)\}$ converges weakly uniformly to $x(t)$, where $x : I \to E$, if for any $\epsilon > 0$, $x^* \in E^*$ there exists $N = N(x^*, \epsilon)$ such that $n > N$ implies $|x^*(x_n(t) - x(t)| < \epsilon$ for all $t \in I$. The family $x_n(t)$ is said to be *weakly equicontinuous* if given $\epsilon > 0$, $x^* \in E^*$, there exists a $\delta = \delta(x^*, \epsilon)$ such that

$$|x^*(x_n(t) - x_n(s))| < \epsilon$$

whenever $|t - s| < \delta$ and for any n.
The following form of Ascoli-Arzela theorem for a family of functions from I into E is very useful.

Theorem 1.2.5. *(Ascoli-Arzela) Let F be an equicontinuous family of functions from I into E. Let $\{x_n(t)\}$ be a sequence in F such that, for each $t \in I$, the set $\{x_n(t) : n \geq 1\}$ is relatively compact in E,i.e., the closure of the set $\{x_n(t) : n \geq 1\}$ is compact. Then there is a subsequence $\{x_{n_k}(t)\}$ which converges uniformly on I to a continuous function $x(t)$.*

The analogue of Theorem 1.2.5 in terms of weak topology is as follows:

Theorem 1.2.6. *Let F be a weakly equicontinuous family of functions from I into E. Let $\{x_n(t)\}$ be a sequence in F such that, for each $t \in I$, the set $\{x_n(t) : n \geq 1\}$ is weakly relatively compact. Then there is a subsequence $\{x_{n_k}(t)\}$ which converges weakly uniformly on I to a weakly continuous function.*

Definition 1.2.3. *Let $f : I \times A \subset R \times E \to E$. Then the function $f(t, x)$ is said to be*

(i) weakly weakly continuous at (t_0, x_0) if given $\epsilon > 0, x^* \in E^*$, there exists $\delta = \delta(x^*, \epsilon)$ and a weakly open set $\Omega = \Omega(x^*, \epsilon)$ containing x_0 such that

$$|x^*(f(t, x) - f(t_0, x_0))| < \epsilon$$

whenever $|t - t_0| < \delta$ and $x \in \Omega$;

(ii) weakly weakly uniformly continuous if given $\epsilon > 0$, and $x^* \in E^*$, there exists $\delta = \delta(x^*, \epsilon)$ and$\{x_i^* : x_i^* \in E^*, i = 1, 2, ...n = n(x^*, \epsilon)\}$ such that

$$|x^*(f(t, x) - f(s, y))| < \epsilon$$

whenever $|t - s| < \delta$ and $|x_i^*(x - y)| < \delta$, $i = 1, 2, \ldots, n$.

We shall next consider a class of sets, called convex sets which play an important role in many results.
A subset K of a linear space E is said to be convex if $\lambda x + (1-\lambda)y \in E$ whenever $x, y \in K$ and $0 \leq \lambda \leq 1$. The convex hull of K, denoted by $co(K)$, is the intersection of all convex subsets of E which contains K. The convex hull of K can also be characterized as the set of all elements of E which are convex combinations of K, that is, $x \in co(K)$ only if $x = \sum_{i=1}^{n} \lambda_i x_i$ where $x_i \in K$, and $\lambda_i \geq 0$ are such that $\sum_{i=1}^{n} \lambda_i = 1$. It follows that $co(K)$ is the smallest convex set of E which contains K and $co(K) = K$ if and only if K is convex. If K is a subset of a normed linear space, then the closure $\bar{K}$ and the interior K^0 are also convex. Moreover, the intersection of all closed, convex subsets of E containing K is called the closed convex hull of K and is denoted by $\bar{co}(K)$. It is easy to show that $\overline{co}(K) = \overline{co(K)}$.
The important property of convex sets is given in the following.

Theorem 1.2.7. *Let E be Banach space and let $\{x_n\}$ be a sequence in E which converges to $x \in E$. Then there is a sequence of convex combination of $\{x_n : n = 1, 2, ...\}$ which converges strongly to x.*

An immediate consequence of Theorem 1.2.7 is

Corollary 1.2.3. *Let E be Banach space and $M \subset E$. If $\{x_n\}$ is a sequence in M which converges weakly to $x \in E$, then $x \in \bar{co}(M)$.*

Theorem 1.2.8. *(Dugundji). Suppose that E_1 and E_2 are two Banach spaces, $\Omega \subset E_1$ and $f : \Omega \to E_2$ is a continuous mapping. Then there is a continuous extension $\tilde{f} : E_1 \to E_2$ of f such that $\tilde{f}(E_1) \subset co(f(\Omega))$.*

Let K and L be two convex subsets of a Banach space E. A member $x^* \in E^*$ is said to separate K and L if there is a number $\lambda \in R$ such that $x^*(x) \leq \lambda$ for $x \in K$ and $x^*(x) \geq \lambda$ for $x \in L$. A fundamental result on seperation of convex sets is the following.

Theorem 1.2.9. *Let K and L be disjoint convex subsets of the normed linear space E. Suppose the interior of K is nonempty. Then there exists a nonzero $x^* \in E^*$ which separates K and L.*

Some other fundamental results of functional analysis which we need in the later chapters are:

(i) the Uniform Boundedness Theorem which asserts that a family of continuous linear transformations between two Banach spaces that is pointwise bounded is actually uniformly bounded;

(ii) the Krein-$\tilde{S}$mulian Theorem which says that a convex set in the dual space of a Banach space is weak*−closed if and only if its intersection with every norm closed unit ball about the origin is weak*−closed.

(iii) the Eberlein-$\tilde{S}$mulian Theorem which shows that a weakly closed set in a Banach space is weakly compact if and only if it is weakly sequentially compact.

For the sake of completeness we state these three results in the following three theorems.

Theorem 1.2.10. *(Uniform Boundedness Theorem) Let E_1 and E_2 be two normed linear spaces over the field Φ with $\|\cdot\|_1$ and $\|\cdot\|_2$ respectively. Let $\{T_\alpha\}$, $\alpha \in \bigwedge$, be a family of continuous linear transformations from E_1 to E_2 where $\bigwedge$ is some index set. If $F \subset E_1$ is a set of category II such that, for some $x \in F$, there exists some constant $M_x > 0$ for which*

$$\sup_{\alpha\in\Lambda} \|T_\alpha(x)\|_2 \leq M_x.$$

Then there exists some $M > 0$ such that

$$\sup_{\alpha\in\Lambda} \|T_\alpha\|_2 \leq M.$$

Note that for bounded linear transformation,

$$\|T\|_2 = \sup_{\substack{x\in E_1 \\ x\neq 0}} \frac{\|T(x)\|_2}{\|x\|_1} = \sup_{\substack{x\in E_1 \\ \|x\|_1\leq 1}} \|T(x)\|_2 = \sup_{\substack{x\in E_1 \\ \|x\|_1=1}} \|T(x)\|_2.$$

Theorem 1.2.11. *(Krein-$\tilde{S}$mulian Theorem) Let $(E_1, \|\cdot\|)$ be a Banach space over Φ and suppose $K \in E^*$ is convex. Then the following statements are equivalent:*

(i) K is weak$^*-$ closed ;

(ii) For each $\alpha > 0$ the set $K \cap \alpha B_1^*$ is weak$^*-$closed where $B_1^* = \{x^* \in E^* : \|x^*\| \leq 1\}$.

Theorem 1.2.12. *(Eberlein-$\tilde{S}$mulian Theorem) Let $(E_1, \|\cdot\|)$ be a Banach space over Φ and suppose $K \in E$ is weakly closed in E. Then the following statements are equivalent.*

(i) K is weakly compact;

(ii) K is weakly sequentially compact.

1.3 Basic Properties of Cones

This section introduces the basic properties of cones, and therefore forms a basis for the remaining chapters.

Definition 1.3.1. *Let E be a real Banach space. A nonempty convex closed set $P \subset E$ is called a cone if it satisfies the following two conditions*

(i) $x \in P$, $\lambda \geq 0$ implies $\lambda x \in P$;

(ii) $x \in P$, $-x \in P$ implies $x = \theta$, where θ denotes the zero element E.

A cone is said to be solid if it contains interior points, i.e., $P^0 \neq \phi$(empty set). A cone P is said to be generating if $E = P - P$, i.e., every element $x \in E$ can be represented in the form $x = u - v$, where $u, v \in P$.
Every cone $P \in E$ defines a partial ordering in E given by

$$x \leq y \text{ iff } y - x \in P.$$

If $x \leq y$ and $x \neq y$, we write $x < y$; if cone P is solid and $y - x \in P^0$, we write $x << y$.

Definition 1.3.2. *A cone $P \in E$ is said to be normal if there exists a positive constant δ such that $\|x + y\| \geq \delta$, for all $x, y \in P$, $\|x\| = 1, \|y\| = 1$. Geometrically, normality means that the angle between two positive unit vectors has to be bounded away from π. In other words, a normal cone cannot be too large.*

Theorem 1.3.1. *Let P be a cone in E. Then the following assertions are equivalent:*

(i) P is normal;

(ii) there exists a $\gamma > 0$ such that

$$\|x + y\| \geq \gamma \max\{\|x\|, \|y\|\} \quad \textit{for all} \quad x, y \in P;$$

(iii) there exists a constant $N > 0$ such that $\theta \leq x \leq y$ implies $\|x\| \leq N\|y\|$, i.e., the $\|\cdot\|$ is semi-monotone;

(iv) there exists an equivalent norm $\|\cdot\|_1$ on E such that $\theta \leq x \leq y$ implies $\|x\|_1 \leq \|y\|_1$, i.e., $\|\cdot\|_1$ is monotone;

(v) $x_n \leq z_n \leq y_n$ $(n = 1, 2, 3, ...)$ and $\|x_n - x\| \to 0$, $\|y_n - x\| \to 0$ imply $\|z_n - x\| \to 0$;

(vi) set $(B + P) \bigcap (B - P)$ is bounded, where

$$B = \{x \in E \mid \|x\| \leq 1\};$$

(vii) every order interval $[x, y] = \{x \in E \mid x \leq a \leq y\}$ is bounded.

Proof. $(i) \Rightarrow (ii)$: We assume $\|x\| = 1$ and $\|y\| \leq 1$, hence

$$1 = \|x\| \leq \|x + y\| + \| - y\| = \|x + y\| + \|y\|.$$

On the other hand,

$$\|x+y\| = \|x + \frac{y}{\|y\|} - \frac{1-\|y\|}{\|y\|}y\| \geq \|x + \frac{y}{\|y\|}\| - 1 + \|y\| \geq \delta - 1 + \|y\|.$$

It follows therefore $\|x+y\| \geq \gamma$ where $\gamma = \delta/2$.
(ii) $\Rightarrow$ (iii): Suppose (iii) is not true. Then there exists $x_n, y_n \in P$ such that $\theta \leq x_n \leq y_n$ and $\|x_n\| > n\|y_n\|$ $(n = 1, 2, 3....)$. Let

$$u_n = \frac{x_n}{\|x_n\|} + \frac{y_n}{n\|y_n\|}, \quad v_n = -\frac{x_n}{\|x_n\|} + \frac{y_n}{n\|y_n\|}, \quad (n = 1, 2, 3, \dots).$$

It is easy to see that $u_n, v_n \in P$ and

$$\|u_n\| \geq 1 - \frac{1}{n}, \quad \|v_n\| \geq 1 - \frac{1}{n}.$$

Hence, by virtue of (ii)

$$\frac{2}{n} = \|u_n + v_n\| \geq \gamma(1 - \frac{1}{n}), \quad (n = 1, 2, 3, ...),$$

which is evidently impossible.
(iii)$\Rightarrow$ (iv): Let

$$\|x\|_1 = \inf_{u \leq x} \|u\| + \inf_{v \leq x} \|v\|.$$

We prove $\|\cdot\|_1$ is a monotone norm on E. In fact, $\|\theta\|_1 = 0$ is clear. Suppose $\|x\|_1 = 0$. For any $\epsilon > 0$ there exist $u, v \in E$ such that $u \leq x \leq v$ and $\|u\| < \epsilon$, $\|v\| < \epsilon$; hence, by (iii), $\|x\| \leq \|x - u\| + \|u\| \leq N\|v - u\| + \|u\| < (2N+1)\epsilon$. Since $\epsilon > 0$ is arbitrary, we get $x = \theta$. The inequality $\|\lambda x\|_1 = |\lambda| \cdot \|x\|_1$ obviously holds for any $x \in E$ and $\lambda \in R$. Now, suppose $x, y \in E$. For any $\epsilon > 0$, there exist $u_1, v_1, u_2, v_2 \in E$ such that $u_1 \leq x \leq v_1$, $u_2 \leq y \leq v_2$ and $\|u_1\| + \|v_1\| < \|x\|_1 + \epsilon$, $\|u_2\| + \|v_2\| < \|y\|_1 + \epsilon$. From $u_1 + u_2 \leq x + y \leq v_1 + v_2$, we know $\|u_1 + u_2\| + \|v_1 + v_2\| \geq \|x + y\|_1$, and so we get

$$\|x + y\|_1 \leq \|u_1\| + \|u_2\| + \|v_1\| + \|v_2\| \leq \|x\|_1 + \|y\|_1 + 2\epsilon.$$

Since $\epsilon > 0$ is arbitrary, this implies $\|x + y\|_1 \leq \|x\|_1 + \|y\|_1$.

It is easy to see that $\|.\|_1$ is monotone, since $\theta \leqslant x \leqslant y$ implies $\inf\limits_{u \leqslant x} \|u\| = \inf\limits_{u \leqslant y} \|u\| = 0$ and therefore

$$\|x\|_1 = \inf_{v \geq x} \|v\| \leqslant \inf_{v \geq y} \|v\| = \|y\|_1.$$

Finally, we prove that $\|.\|$ and $\|.\|_1$ are equivalent. Obviously, $\|x\|_1 \leqslant 2\|x\|$. On the other hand, for any $u \leqslant x \leqslant v$, we have $\|x\| \leqslant \|x - u\| + \|u\| \leqslant N\|v - u\| + \|u\| \leqslant (N+1)(\|u\| + \|v\|)$, hence $\|x\| \leqslant (N+1)\|x\|_1$.

(iv) $\Rightarrow$ (v): From $\theta \leqslant z_n - x_n \leqslant y_n - x_n$ we find $\|z_n - x_n\|_1 \leqslant \|y_n - x_n\|_1$. Since $m\|x\| \leqslant \|x\|_1 \leqslant M\|x\|$ for any $x \in E$, where $M > m > 0$ are two constants, it follows

$$\|z_n - x_n\| \leqslant \frac{1}{m}\|z_n - x_n\|_1 \leqslant \frac{1}{m}\|y_n - x_n\|_1 \leqslant \frac{M}{m}\|y_n - x_n\| \to 0,$$

and hence

$$\|z_n - x\| \leqslant \|z_n - x_n\| + \|x_n - x\| \to 0.$$

(v)⇒ (vi): If $(B + P) \cap (B - P)$ is unbounded, then there exists $\{z_n\} \subset (B+P)\cap(B-P)$ such that $\|z_n\| \to +\infty$, Hence $x_n \leqslant z_n \leqslant y_n$, where $x_n, y_n \in B$. Setting $u_n = x_n/\|z_n\|$, $v_n = y_n/\|z_n\|$ and $w_n = z_n/\|z_n\|$, we get $u_n \leqslant w_n \leqslant v_n$ and $u_n \to \theta$, $v_n \to \theta$, but $w_n \nrightarrow \theta$ (since $\|w_n\| = 1$), which contradicts (v).

(vi)⇒(vii): Suppose $(B + P) \cap (B - P) \subset \rho B$, where $\rho > 0$. Putting $r = \max\{\|x\|, \|y\|\}$, it is easy to show that $z/r \in (B+P)\cap(B-P)$ for any $z \in [x, y]$; hence $[x, y] \subset r\rho B$.

(vii)⇒(i): Suppose (i) is not true. Then there exist $\{x_n\} \subset P$ and $\{y_n\} \subset P$ such that $\|x_n\| = \|y_n\| = 1$ and $\|x_n + y_n\| < \dfrac{1}{4^n}$ $(n = 1, 2, \dots)$.

Letting

$$u_n = \frac{x_n}{\sqrt{\|x_n + y_n\|}}, \quad v_n = \frac{x_n + y_n}{\sqrt{\|x_n + y_n\|}} \quad (n = 1, 2, 3, \dots),$$

we have $\theta \leq u_n \leq v_n$ and, since

$$\sum_{n=1}^{\infty} \|v_n\| < \sum_{n=1}^{\infty} \frac{1}{2^n} < +\infty,$$

the series $\sum_{n=1}^{\infty} v_n$ is convergent to some element $v \in E$. Evidently, $\theta \leq u_n \leq v_n \leq v$ and

$$\|u_n\| = \frac{1}{\sqrt{\|x_n + y_n\|}} > 2^n \quad (n = 1, 2, 3, \dots);$$

hence, the order interval $[\theta, v]$ is unbounded, which contradicts (vi).

Some authors use the assertion (iii) as the definition of normality of a cone P and call the smallest number N the normal constant of P.

Example 1.3.1. *Let $E = R^n$, the Euclidean space, and $P_1 = \{x = (x_1, x_2, \dots, x_n) | x_i \geq 0, \quad i = 1, 2, \dots, n\}$. It is clear that P_1 is a cone in R^n and P_1 is solid and generating. Since the norm in R^n is monotone, it follows from Theorem 1.3.1 that P_1 is normal.*

Example 1.3.2. *Let $E = C(G)$, space of continuous functions on a bounded closed set G in R^n and $P_2 = \{x(t) \in C(G) \mid x(t) \geq 0, \quad t \in G\}$. It is easy to see that P_2 is a solid and generating normal cone in $C(G)$. Later we shall use other cones in $C(G)$, such as*

$$P_3 = \{x(t) \in C(G) \mid x(t) \geq 0 \quad \textit{and} \quad \int_{G_0} x(t)\mathrm{d}t \geq \epsilon_0 \|x\|_c\},$$

$$P_4 = \{x(t) \in C(G) \mid x(t) \geq 0 \quad \text{and} \quad \min_{t \in G_0} x(t) \geq \epsilon_0 \|x\|_c\},$$

where G_0 is a closed subset of G and ϵ_0 is a given number satisfying $0 < \epsilon_0 < 1$. It is easy to show that P_3 and P_4 are solid normal cones in $C(G)$.

Example 1.3.3. *Let $E = L^P(\Omega)$. the space of Lebesgue measurable functions which are pth power summable on $\Omega \subset R^n$, where $p \geq 1$ and $0 < \text{mes}\,\Omega < +\infty$. Let $P_5 = \{x(t) \in L^P(\Omega)|x(t) \geq 0\}$. It is easy to show that P_5^0 is a generating normal cone in $L^p(\Omega)$, but P_5 is not solid, i.e., $P_5^0 = \phi$.*

Example 1.3.4. *Let $E = C^1[0, 2\pi]$, the space of continuously differentiable functions on $[0, 2\pi]$ with the norm*

$$\|x\| = \max_{0 \leq t \leq 2\pi} |x(t)| + \max_{0 \leq t \leq 2\pi} |x'(t)|,$$

and let $P_6 = \{x(t) \in C^1[0, 2\pi] \mid x(t) \geq 0, \quad 0 \leq t \leq 2\pi\}$. Clearly P_6 is a solid cone in $C^1[0, 2\pi]$. Also, P_6 is generating, since every $x(t) \in C^1[0, 2\pi]$ can be expressed in the form

$$x(t) = y(t) - z(t),$$

where $y(t) \equiv M > 0$ and $z(t) = M - x(t)$, $M > \max_{0 \leq t \leq 2\pi} x(t)$, $y(t)$, $z(t) \in P_6$. P_6 is not normal. In fact, if P_6 is normal, then, by Theorem 1.3.1, there exists an $N > 0$ such that $\theta \leq x \leq y \quad \|x\| \leq N\|y\|$. Let $x_n(t) = 1 - cos(nt)$, $y_n(t) \equiv 2$. Then we have $\theta \leq x_n \leq y_n$, $\|x_n\| = 2 + n$, and $\|y_n\| - 2$. Consequently, $2 + n \leq 2N \quad (n = 1, 2, 3, \dots)$, which is impossible.

Definition 1.3.3. *A cone $P \subset E$ is said to be regular if every increasing and bounded in order sequence in E has a limit, i.e., if $\{x_n\} \subset E$ and $y \in E$ satisfy*

$$x_1 \leq x_2 \leq \cdots \leq x_n \leq \cdots \leq y, \tag{1.3.1}$$

then there exists $x^* \in E$ such that $\|x_n - x^*\| \to 0$.

It is evident that cone P is regular if and only if every decreasing and bounded in order sequence in E has a limit.

Definition 1.3.4. *A cone $P \subset E$ is said to be fully regular if every increasing and bounded in norm sequence in E has a limit, i.e., if $\{x_n\} \subset E$ satisfies*

$$x_1 \leq x_2 \leq \cdots \leq x_n \leq \dots, M = \sup_n \|x_n\| < +\infty, \tag{1.3.2}$$

then there exists $x^* \in E$ such that $\|x_n - x^*\| \to 0$.

It is evident that cone P is fully regular if and only if every decreasing and bounded in norm sequence in E has a limit.

Theorem 1.3.2. *Cone P is fully regular $\to$ P is regular $\Rightarrow$ P is normal.*

Proof. We first prove that if P is not normal, then P is neither regular nor fully regular. Suppose that P is not normal. By virtue of Theorem 1.3.1 (iii), there exist $\{x_n\} \subset P$ and $\{y_n\} \subset P$ such that $\theta \leq x_n \leq y_n$ and

$$\|x_n\| > 2^n \|y_n\| \qquad (n = 1, 2, 3, \dots). \tag{1.3.3}$$

Put $z_n = \dfrac{x_n}{\|x_n\|}$ and $v_n = \dfrac{y_n}{2^n \|y_n\|}$ $(n = 1, 2, 3, \dots)$. Then, by (1.3.3)

$$\theta < z_n \leq \frac{x_n}{2^n \|y_n\|} \leq \frac{y_n}{2^n \|y_n\|} = v_n \quad (n = 1, 2, 3, \dots) \tag{1.3.4}$$

and

$$\sum_{n=1}^{\infty} \|v_n\| = \sum_{n=1}^{\infty} \frac{1}{2^n} = 1 < +\infty. \tag{1.3.5}$$

Hence the series $\sum_{n=1}^{\infty} v_n$ converges to some $v \in E$, i.e.,

$$\sum_{n=1}^{\infty} v_n = v. \tag{1.3.6}$$

Now we define

$$\omega_n = \begin{cases} v_1 + v_2 + \cdots + v_{2m}, & \text{when} \quad n = 2m, \quad m = 1, 2, 3, \dots \\ v_1 + v_2 + \cdots + v_{2m} + z_{2m+1}, & \text{when} \quad n = 2m+1, \quad m = 1, 2, 3, \dots. \end{cases}$$

It is easy to show from (1.3.4),(1.3.5), and (1.3.6) that

$$\theta < \omega_2 \leq \omega_3 \leq \omega_4 \leq \omega_5 \leq \omega_6 \leq \cdots \leq v \tag{1.3.7}$$

and

$$\sup_n \|\omega_n\| \leq 2 < +\infty. \tag{1.3.8}$$

But the sequence ω_n does not converge, since $\|\omega_{2m+1} - \omega_{2m}\| = \|z_{2m+1}\| = 1$. Hence P is neither regular nor fully regular.

Finally, we need only to prove that the full regularity of P implies the regularity of P. Suppose P is fully regular and (1.3.1) is satisfied. Then, by the conclusion mentioned above, P is normal. It follows therefore from Theorem 1.3.1 and $\theta \leq y - x_n \leq y - x_1 \quad (n = 1, 2, 3, \dots)$ that

$$\|y - x_n\| \leq N \|y - x_1\| \quad (n = 1, 2, 3, \dots).$$

Hence$\{\|x_n\|\}$ is bounded and x_n converges to some $x^* \in E$.

Example 1.3.5. *We prove the cone P_5 in Example 1.3.3 is fully regular (hence, by Theorem 1.3.1, it is regular). Suppose* (1.3.2) *is satisfied, i.e.,*

$$x_1(t) \leq x_2(t) \leq \cdots \leq x_n(t) \leq \dots, \int_\Omega \|x_n(t)\|^p \mathrm{d}t \leq M^p \quad (n = 1, 2, 3, \dots).$$

Hence, by Fatou's Lemma, $\int_\Omega \|x^*(t)\|^p \mathrm{d}t \leq M^p$, *i.e.,* $x^* \in L^p(\Omega)$, *where* $x^*(t) = \lim_{n\to\infty} x_n(t)$. *It follows easily from*

$$0 \leq x^*(t) - x_n(t) \leq x^*(t) - x(t) \quad (n = 1, 2, 3, \dots)$$

and the Vitali convergence theorem that

$$\|x^* - x_n\|^p = \int_\Omega |x^*(t) - x_n(t)|^p \mathrm{d}t \to 0 \quad (n \to \infty).$$

The full regularity of P_5 *is proved.*

Example 1.3.6. *It is easy to prove that the cone* P_2 *in Example 1.3.2 is not regular (even if it is normal). For simplicity we consider the case* $G = [0, 1]$. *Let* $x_n(t) = 1 - t^n \quad (n = 1, 2, 3, \dots)$, *then*

$$x_1(t) \leq x_2(t) \leq \cdots \leq x_n(t) \leq \cdots \leq y(t), \quad y(t) \equiv 1,$$

but x_n *does not converge in* $C[0, 1]$, *since* $x_n(t)$ *does not converge uniformly on* $[0, 1]$.

Example 1.3.7. *Let* E *be the real Banach space* $C_0 = \{x = (x_1, x_2, \dots, x_k, \dots) \mid x_k \to 0\}$ *with the norm* $\|x\| = \sup_k |x_k|$ *and* $P_7 = \{x = (x_1, x_2, \dots, x_k, \dots) \in C_0 \mid x_k \geq 0, \quad k = 1, 2, 3, \dots\}$. *If* $x^{(n)} = (x_1^{(n)}, x_2^{(n)}, \dots, x_k^{(n)}, \dots) \in C_0$, $y = (y_1, y_2, \dots, y_k, \dots) \in C_0$ *such that*

$$x^{(1)} \leq x^{(2)} \leq \cdots \leq x^{(n)} \leq \cdots \leq y,$$

it is easy to see that $\|x^{(n)} - x^*\| \to 0 \quad (n \to \infty)$, *where*

$$x^* = (x_1^*, x_2^*, \dots, x_k^*, \dots), \quad x_k^* = \lim_{n\to\infty} x_k^{(n)}.$$

Hence P_7 *is regular. On the other hand, putting* $z^{(n)} = (z_1^{(n)}, z_2^{(n)}, \dots, z_k^{(n)}, \dots)$, *where*

$$z_k^{(n)} = \begin{cases} 1, & k \leq n; \\ 0, & k > n. \end{cases}$$

we see $z^{(n)} \in C_0$, $z^{(1)} \leq z^{(2)} \leq \cdots \leq z^{(n)} \leq \dots$ *and* $\|z^{(n)}\| = 1 \quad (n = 1, 2, 3, \dots)$, *but* $\{z^{(n)}\}$ *does not converge in* C_0 *and hence* P_7 *is not fully regular.*

Similarly, we can prove that the cone $P_8 = \{x(t) \in C_0[a, +\infty) \mid x(t) \geq 0\}$ in the real Banach space $C_0[a, +\infty) = \{x = x(t) \mid x(t)$ is continuous on $a \leq t \leq +\infty$ and $x(t) \to 0$ as $t \to +\infty\}$ with norm

$$\|x\| = \sup_{a \leq t \leq +\infty} |x(t)|$$

is regular, but is not fully regular.

Theorem 1.3.3. *If E is reflexive and P is a cone in E, then the following assertions are equivalent:*

(i) P is normal;

(ii) P is regular;

(iii) P is fully regular;

Proof. By virtue of Theorem 1.3.1, we only need to prove that the normality of P implies the full regularity of P. Suppose that (1.3.2) is satisfied. We divide the proof into three steps:

(a) It is sufficient to prove that $\{x_n\}$ contains a convergent subsequence $\{x_{n_k}\}$. In fact, if $x_{n_k} \to x^* \in E$ (i.e., $\|x_{n_k} - x^*\| \to 0$), then, taking $k \to \infty$ in the inequality $x_{n_k} \geqslant x_m$, where m is any given positive integer and k is sufficiently large, we get $x^* \geqslant x_m$, and hence

$$x^* \geqslant x_n \qquad (n = 1, 2, 3, \dots).$$

By the normality of P, it follows from $\theta \leqslant x^* - x_n \leqslant x^* - x_{n_k}$ for $n > n_k$ that $\|x^* - x_n\| \leqslant N\|x^* - x_{n_k}\|$, and therefore from $\|x^* - x_{n_k}\| \to 0 \quad (k \to \infty)$ we know $\|x^* - x_n\| \to 0 \quad (n \to \infty)$.

(b) Since E is reflexive and $\{x_n\}$ is bounded, $\{x_n\}$ contains a subsequence $\{x_{n_i}\}$, which converges weakly to $x^* \in E$, i.e., $f(x_{n_i}) \to f(x^*)$ for any $f \in E^*$, where E^* denotes the dual space of E. It must be $x_{n_i} \leqslant x^* \quad (i = 1, 2, 3, \dots)$, since otherwise there exists $x_{n_{i_0}}$ such that $x_{n_{i_0}} \nleqslant x^*$, i.e., $x^* - x_{n_{i_0}} \in P$. By virtue of the second separation theorem of convex sets, there exists $f \in E^*$ such that $f(x^* - x_{n_{i_0}}) < c$ and $f(x) > c$ for any $x \in P$, where c is a real number. Consequently,

$$f(x^*) < f(x_{n_{i_0}}) + c \tag{1.3.9}$$

and

$$f(x_{n_i}) > f(x_{n_{i_0}}) + c, \qquad i > i_0. \tag{1.3.10}$$

Taking limit $i \to \infty$ in (1.3.10), we get

$$f(x^*) \geqslant f(x_{n_{i_0}}) + c$$

which contradicts (1.3.9).

(c) Finally, we prove the sequence $\{x_{n_i}\}$ in step (b) must contain a subsequence which converges (strongly) to x^*. If it is not the case, then there exist $\epsilon_0 > 0$ and $m > 0$ such that

$$\|x_{n_i} - x^*\| \geqslant \epsilon_0, \qquad i > m.$$

Put $M_i = \{x \in E \mid x \leqslant x_{n_i}\}$, $\quad M = \cup_{i>m} M_i$. Since M_i is convex and $M_i \subset M_{i+1} \quad (i = 1, 2, 3, \dots)$, it is easy to see that M is convex and hence

$\overline{M}$, the closure of M, is convex. We now prove $x^* \in \overline{M}$. For any $x \in M$, x belongs to some $M_i \quad (i > m)$, i.e., $x \leqslant x_{n_i}$. By step (b), we have $\theta \leqslant x^* - x_{n_i} \leqslant x^* - x$, which implies that $\epsilon_0 \leqslant \|x^* - x_{n_i}\| \leqslant N\|x^* - x\|$. It follows therefore $\|x^* - x\| \geqslant \epsilon_0/N$ for every $x \in \overline{M}$, which proves $x^* \notin \overline{M}$. Now, by the second separation theorem of convex sets, there exists $f_0 \in E^*$ and $c_0 \in R^1$ such that $f_0(x^*) < c_0$ and $f_0(x) > c_0$ for any $x \in \overline{M}$. Since $x_{n_i} \in M_i \subset M, \quad i > m$, we have $f_0(x_{n_i}) > c_0$ for $i > m$, and therefore

$$f_0(x^*) = \lim_{i\to\infty} f_0(x_{n_i}) \geqslant c_0,$$

which contradicts $f_0(x^*) < c_0$.

Evidently, the full regularity of P_5 in Example 1.3.3 can be deduced directly from Theorem 1.3.2 when $p > 1$.

Theorem 1.3.4. *Let P be a cone in E. P is regular if and only if the following condition holds:*

(H_1) *$\{x_i\} \subset P$ and $\inf_i \|x_i\| > 0$ imply that $\left\{\sum_{i=1}^{n} x_i\right\}$ is unbounded in order, i.e., there does not exist a $z \in E$ such that $\sum_{i=1}^{n} x_i \leqslant z, \quad n = 1, 2, 3, \ldots$.*

Similarly, P is fully regular if and only if the following condition holds:

(H_2) *$\{x_i\} \subset P$ and $\inf_i \|x_i\| > 0$ imply that $\left\{\sum_{i=1}^{n} x_i\right\}$ is unbounded in norm, i.e., $\sup_n \|\sum_{i=1}^{n} x_i\| = +\infty$.*

Proof. It is easy to see the regularity of P implies (H_1), since, if (H_1) does not hold, then there exists $\{x_i\} \subset P$ and $z \in E$ such that $\inf_i \|x_i\| = a > 0$ and $\sum_{i=1}^{n} x_i \leqslant z, \quad n = 1, 2, 3, \ldots$. Setting $y_n = \sum_{i=1}^{n} x_i$, we see that

$$y_1 \leqslant y_2 \leqslant \cdots \leqslant y_n \leqslant \cdots \leqslant z.$$

But $\|y_{n+1} - y_n\| = \|x_{n+1}\| \geqslant a$, and hence, y_n does not converge, in contradiction with the regularity of P. Similarly, we know the full regularity of P implies (H_2).

Now we prove, if one of the two conditions (H_1) and (H_2) is satisfied, then P must be normal. In fact, if P is not normal, then by the proof of Theorem 1.3.1, (1.3.3) to (1.3.8) all hold. Let $u_n = w_{n+1} - w_n, \quad (n = 2, 3, \ldots)$. It is easy to see

$$\{u_n\} \subset P, \qquad \inf_{n=2,3,\ldots} \|u_n\| \geqslant \frac{13}{16} > 0$$

and

$$\sum_{i=2}^{n} u_i = w_{n+1} - w_2 \leqslant \nu - w_2, \quad \|\sum_{i=2}^{n} u_i\| \leqslant 4, \quad n = 2, 3, \ldots,$$

and hence, neither (H_1) nor (H_2) can be satisfied.

Now, suppose (H_1) holds and (1.3.1) is satisfied. We need to prove $\{x_n\}$ is convergent. From the first step (a) in the proof of Theorem 1.3.2, it is sufficient to prove that $\{x_n\}$ contains a convergent subsequence. If $\{x_n\}$ does not contain any convergent subsequence, then set $\{x_1, x_2, x_3, \ldots\}$ is not relatively compact and hence there exists subsequences $\{x_{n_i}\} \subset \{x_n\}$ and $\epsilon_0 > 0$ such that $\|x_{n_i} - x_{n_j}\| \geqslant \epsilon_0 \quad (i \neq j)$.

Putting $x'_i = x_{n_{i+1}} - x_{n_i}, \quad (i = 1, 2, 3, \ldots)$, we see $\{x'_i\} \subset P, \quad \|x'_i\| \geqslant \epsilon_0 \quad (i = 1, 2, 3, \ldots)$ and

$$\sum_{i=1}^{m} x'_i = x_{n_{m+1}} - x_{n_1} \leqslant y - x_{n_1},$$

which contradicts (H_1). Similarly, we can prove (H_2) implies the full regularity of P.

An element $z \in E$ is called the least upper bound (i.e., supremum) of set $D \subset E$, if it satisfies two conditions: (i) $x \leqslant z$ for any $x \in D$; (ii) $x \leqslant y, \quad x \in D$ implies $z \leq y$. We denote the least upper bound of D by $\sup D$, i.e., $z = \sup D$. The greatest lower bound $\inf D$ is defined analogously.

A set $D \subset E$ is said to be bounded above in order if there is an element $w \in E$ such that $x \leqslant w$ for any $x \in D$. Similarly, we can define a set to be bounded below in order.

Definition 1.3.5. *Cone $P \subset E$ is said to be minihedral if* $\sup\{x, y\}$ *exists for any pair $\{x, y\}$, which is bounded above in order (i.e., $w \in E$ such that $x \leqslant w, y \leqslant w$); cone P is said to be strongly minihedral if* $\sup D$ *exists for any bounded above in order set $D \subset E$.*

It is evident, P is minihedral if and only if $\inf\{x, y\}$ exists for any pair $\{x, y\}$, which is bounded below in order. Similarly, P is strongly minihedral if and only if $\inf D$ exists for bounded below in order set $D \subset E$.

It is easy to show that if P is minihedral, then $\sup D$ exists for any finite set $D = \{x_1, x_2, \ldots, x_n\} \subset E$, which is bounded above in order.

Moreover, the following equality holds:

$$\sup\{x_1, x_2, \ldots, x_n\} = \sup\{x_1, \sup\{x_2, \ldots, x_n\}\}.$$

By definition it is clear that strong minihedrality of a cone P implies the minihedrality of P.

Example 1.3.8. *The cone P_2 in Example 1.3.2 is minihedral, since for every $x(t), y(t) \in C(G)$, we have* $\sup\{x, y\} = z$, *where* $z(t) = \max\{x(t), y(t)\}\} \in C(G)$. *$P_2$ is not strongly minihedral, however, since when we let $G = [0, 2]$ and*

$D = \{x(t) \in C[0,2] | x(t) < 1$ *for* $0 < t < 1$ *and* $x(t) < 2$ *for* $1 < t < 2\}$, *it is easy to see that* $\sup D$ *does not exist.*

Example 1.3.9. *The cone* P_6 *in Example 1.3.4 is not minihedral, since* $\sup\{x,y\}$ *does not exist for*

$$x(t) = t, \quad 0 \leqslant t \leqslant 2\pi \quad \text{and} \quad y(t) = 2\pi - t, \quad 0 \leqslant t \leqslant 2\pi.$$

It is evident that $x \leqslant z, y \leqslant z$, where $z(t) = 2\pi, \quad 0 \leqslant t \leqslant 2\pi$.

On the other hand, the cone P_7 in Example 1.3.3 is strongly minihedral, since $\sup D = z \in C_0$ for every bounded above in order set $D \subset C_0$, where

$$z = (z_1, z_2, \ldots, z_k, \ldots)$$

and

$$z_k = \sup\{x_k \,|\, x = (x_1, x_2, \ldots, x_k, \ldots) \in D\}, \quad k = 1, 2, 3, \ldots \quad .$$

Theorem 1.3.5. *If* E *is separable and the cone* $P \subset E$ *is regular and minihedral, then* P *is strongly minihedral.*

Proof. Suppose $D \subset E$ and there is a $z \in E$ such that $x \leq z, \quad \forall x \in D$. Since E is separable, there exists a denumerable set $M = \{x_1, x_2, \ldots, x_n, \ldots\} \subset D$, which is dense in D. Denote $y_n = \sup\{x_1, x_2, \ldots, x_n\} \quad (n = 1, 2, 3, \ldots)$, which exist by virtue of the minihedrality of P. Evidently

$$y_1 \leqslant y_2 \leq \cdots \leq y_n \leq \cdots \leq z.$$

It follows from the regularity of P that $y_n \to x^* \in E \quad$ (i.e., $\quad \|y_n - x^*\| \to 0$).

Now, we prove $x^* = \sup D$. First, we notice

$$x_n \leq x^* \quad (n = 1, 2, 3, \ldots).$$

For any $x \in D$, there exists a sequence in M, which converges to x, and we get $x \leq x^*$, hence x^* is an upper bound of D. On the other hand, if $v \in E$ is any upper bound of D, then

$$y_n \leq v \quad (n = 1, 2, 3, \ldots)$$

and therefore $x^* \leq v$.

Corollary 1.3.1. *If* E *is separable and reflexive and the cone* $P \subset E$ *is normal and minihedral, then* P *is strongly minihedral.*

Proof. The required conclusion follows directly from Theorems 1.3.3 and 1.3.4.

Example 1.3.10. *The cone* P_5 *in Example 1.3.3 is minihedral, since* $\sup\{x,y\} = z$ *exists for every* $x(t), y(t) \in L^p(\Omega)(p \geq 1)$, *where* $z(t) = \max\{x(t), y(t)\} \in L^p(\Omega)$. *Observing that* $L^p(\Omega)$ *is separable and* P_5 *is regular, as proven in Example 1.3.1, it follows from Theorem 1.3.5 that* P_5 *is strongly minihedral.*

Example 1.3.11. *The cone P_6 in example 1.3.4 is neither minihedral nor normal. Now, we give an example of a cone which is minihedral, but is not normal. Let $E = l^2$ with norm*

$$\|x\| = \Big(\sum_{i=1}^{\infty} x_i^2\Big)^{\frac{1}{2}}, \quad x = (x_1, x_2, \ldots, x_i, \ldots).$$

Let

$$P_8 = \{x = (x_1, x_2, \ldots, x_i, \ldots) \in l^2 \mid x_1 \geq 0, \ x_i \leq x_1, \quad i = 2, 3, \ldots\}.$$

Obviously, P_8 is a cone in l^2. For every pair of elements $x = (x_1, x_2, \ldots, x_i, \ldots) \in l^2$ and $y = (y_1, y_2, \ldots, y_i, \ldots) \in l^2$, it is easy to verify that $\sup\{x, y\} = z$, where $z = (z_1, z_2, \ldots, z_i, \ldots)$,

$$z_1 = \max\{x_1, y_1\},$$
$$z_i = \min\{z_1 - x_1 + x_i, z_1 - y_1 + y_i\}, \quad i = 2, 3, \ldots \quad .$$

In case $z_1 = x_1$, we have $z_i = x_i$ or $y_i \leq z_i \leq x_i$, and in case $z_1 = y_1$, we have $z_i = y_i$ or $x_i \leq z_i \leq y_i$; hence we must have $z \in l^2$, and the minihedrality of P_8 has been proven.

Now, let $u = (1, 1/2, 1/3, \ldots, 1/i, \ldots)$, $\theta = (0, 0, \ldots, 0, \ldots)$, and $x^{(n)} = (x_1^{(n)}, x_2^{(n)}, \ldots, x_i^{(n)}, \ldots)$, where

$$x_i^{(n)} = \begin{cases} \frac{1}{2}, & \text{if } \ i \leq n; \\ 0, & \text{if } \ i > n. \end{cases}$$

It is easy to see that $\theta \leq x^{(n)} \leq u \quad (n = 1, 2, 3, \ldots)$, $\|x^{(n)}\| = \sqrt{n}/2 \to +\infty$, and hence $[\theta, u]$ is unbounded. Consequently by Theorem 1.3.1, P_8 is not normal.

We can also give a simple example of a cone which is strongly minihedral and normal, but is not regular. Let E be the space m of all bounded sequences $x = (x_1, x_2, \ldots, x_i, \ldots)$ with norm $\|x\| = \sup\limits_i |x_i|$ and let $P_9 = \{x = (x_1, x_2, \ldots, x_i, \ldots) \in m \mid x_i \geq 0, \quad i = 1, 2, 3, \ldots\}$. Since the norm is monotone, P_9 is a normal cone in m. Let D be a bounded above in order set in m. It is clear $\sup D = z \in m$, where $z = (z_1, z_2, \ldots, z_i, \ldots)$,

$$z_i = \sup\{x_i \mid x = (x_1, x_2, \ldots, x_i, \ldots) \in D\}.$$

Hence P_9 is strongly minihedral. Finally, the sequence $\{x^{(n)}\}, x^{(n)} = (x_1^{(n)}, x_2^{(n)}, \ldots, x_i^{(n)}, \ldots)$, where

$$x_i^{(n)} = \begin{cases} 1, & \text{if } \ i \leq n; \\ 0, & \text{if } \ i > n, \end{cases}$$

satisfies

$$x^{(1)} \leq x^{(2)} \leq \cdots \leq x^{(n)} \leq \cdots \leq y,$$

with $y = (1, 2, \ldots, n, \ldots)$. But $x^{(n)}$ does not converge in m and hence P_9 is not regular.

1.4 Directional Derivatives

When we use $\|x\|$ or $\|x\|^2$ as a measure in estimates later, we need to assume conditions in terms of their one-sided directional derivatives. Here, we define and list several properties of such derivatives.
Let $x, y \in E$, E being a real Banach space with $\|x\|$. Define

$$[x, y]_h = \frac{1}{h}(\|x + hy\| - \|x\|)$$

for any $h \in R$. Then we have the following result.

Lemma 1.4.1. *(i) The limits* $\lim_{h\to 0^+}[x, y]_h = [x, y]_+$ *and* $\lim_{h\to 0^-}[x, y]_h = [x, y]_-$ *exist; and (ii)* $[x, y]_+$ *is upper semi-continuous and* $[x, y]_-$ *is lower semi-continuous.*

Proof. Let us first show that $[x, y]_h$ is monotone nondecreasing in h. Suppose $0 < h_1 < h_2$ and let $\beta \in (0, 1)$ be such that $h_1 = (1 - \beta)h_2$. Since

$$x + h_1 y = x + (1 - \beta)h_2 y = \beta x + (1 - \beta)(x + h_2 y),$$

we have

$$\begin{aligned}
[x, y]_{h_1} &= \frac{1}{h_1}(\|x + h_1 y\| - \|x\|) = \frac{1}{h_1}\{\|\beta x + (1 - \beta)(x + h_2 y)\| - \|x\|\} \\
&\le \frac{1}{h_1}\{\beta\|x\| + (1 - \beta)\|(x + h_2 y)\| - \|x\|\} = \frac{\|x + h_2 y\| - \|x\|}{h_2} = [x, y]_{h_2}.
\end{aligned}$$

Similarly, one gets $[x, y]_{h1} \le [x, y]_{h_2}$ if $h_1 < h_2 < 0$. If $h_1 < 0 < h_2$, we let $h = \min(-h_1, h_2)$ and note that

$$2\|x\| = \|x + hy + x - hy\| \le \|x + hy\| + \|x - hy\|.$$

This implies that $[x, y]_{-h} \le [x, y]_h$, which in turn yields

$$[x, y]_{h_1} \le [x, y]_{-h} \le [x, y]_h \le [x, y]_{h_2},$$

proving that $[x, y]_h$ is monotone. If $-1 \le h_1 \le h_2 \le 1$, the monotone property of $[x, y]$ gives

$$[x, y]_{-1} \le [x, y]_{h_1} \le [x, y]_{h_2} \le [x, y]_1$$

and hence the limits

$$[x, y]_+ = \lim_{h\to 0^+}[x, y]_h, \text{ and}$$

$$[x, y]_- = \lim_{h\to 0^-}[x, y]_h,$$

exist. To prove that $[x, y]_+$ is upper semi continuous, let $\{x_n\}, \{y_n\}$ be two sequences in E such that $\lim_{n\to\infty} x_n = x$ and $\lim_{n\to\infty} y_n = y$. Then, for $h > 0$

$$[x_n, y_n]_+ \le \frac{1}{h}\{\|x_n + hy_n\| - \|x_n\|\} \text{ for all } n \ge 1.$$

Letting $n \to \infty$, we get

$$\limsup_{n\to\infty}[x_n, y_n]_+ \le \frac{1}{h}\{\|x + hy\| - \|x\|\}$$

for all $h > 0$. We now let $h \to 0^+$, obtaining

$$\limsup_{n\to\infty}[x_n, y_n]_+ \le [x, y]_+. \tag{1.4.1}$$

Since $[x, y]_- = -[x, -y]_+$ by definition, lower semi-continuity of $[x, y]_-$ follows from (1.4.1), thus proving the lemma.
Some properties of $[x, y]_\pm$ are listed in the following lemma.

Lemma 1.4.2. *Let $[x, y]_\pm$ be defined as in Lemma 1.4.1. Then,*

(i) $[x, y]_- \le [x, y]_+$;

(ii) $|[x, y]_\pm| \le \|y\|$;

(iii) $|[x, y]_\pm - [x, z]_\pm \le \|y - z\|$;

(iv) $[x, y]_+ = -[x, -y]_- = -[-x, y]_-$;

(v) $[sx, ry]_\pm = r[x, y]_\pm$ *for* $r, s \ge 0$;

(vi) $[x, \alpha x]_\pm = \alpha\|x\|,\ \alpha \in R$;

(vii) $[x, y + z]_+ \le [x, y]_+ + [x, z]_+$ *and* $[x, y + z]_- \ge [x, y]_- + [x, z]_-$;

(viii) $[x, y + z]_+ \ge [x, y]_+ + [x, z]_-$ *and* $[x, y + z]_- \le [x, y]_- + [x, z]_+$;

(ix) $[x, y + \alpha x]_\pm = [x, y]_\pm + \alpha\|x\|,\ \alpha \in R$;

(x) If $x \in [a, b] \to E$ *such that* $x'_\pm(t)$ *the right and left derivatives of* $x(t)$ *exists for some* $t \in (a, b)$ *and* $m(t) = \|x(t)\|$, *then* $m'_\pm(t) = [x(t), x'_\pm(t)]_\pm$.

Proof. Properties (i) to (v) follow easily from the definition. To prove (vi), note that for any $\alpha \in R$,

$$\lim_{h\to 0^\pm}(\|x + h\alpha x\| - \|x\|) = \lim_{h\to 0^\pm}\|x\|\Big(\frac{|1 + h\alpha| - 1}{h}\Big) = \alpha\|x\|.$$

Assertion (vii) follows because of the inequality

$$\|x + h(y + z)\| \le \frac{1}{2}\|x + 2hy\| + \frac{1}{2}\|x + 2hz\|.$$

Observing that

$$\|x + hy\| \le \frac{1}{2}\|x + 2h(y + z)\| + \frac{1}{2}\|x - 2hz\|,$$

we obtain

$$[x, y]_+ \leq [x, y+z]_+ + [x, -z]_+ = [x, y+z]_+ - [x, z]_-,$$

which proves the first part of (viii). The second part of (viii) follows similarly. To prove (ix), we use (vi)-(viii) to get

$$[x, y + \alpha x]_+ \leq [x, y]_+ + \alpha\|x\|$$

and

$$[x, y + \alpha x]_+ \geq [x, y]_+ + \alpha\|x\|.$$

Finally, we see that

$$\begin{aligned} &|\frac{1}{h}[m(t+h) - m(t)] - \frac{1}{h}[\|x(t) + hx'_+(t)\| - \|x(t)\|]| \\ &\qquad = \frac{1}{h}[\|x(t+h)\| - \|x(t) + hx'_+(t)\|] \\ &\qquad \leq \|\frac{1}{h}[x(t+h) - x(t)] - x'_+(t)\| \to 0 \text{ as } h \to 0^+, \end{aligned}$$

proving $m'_+(t) = [x(t), x'_+(t)]_+$. The proof of the other in (x) is similar. This completes the proof of the lemma.
A function $\phi : E \to R$ is said to be convex if $\phi(\lambda x + (1-\lambda)y) \leq \lambda\phi(x) + (1-\lambda)\phi(y)$ for all $x, y \in E$ and $0 \leq \lambda \leq 1$.Note that the function $\phi(x) = \|x\|$ and $\phi(x) = \|x\|^p, 1 \leq p < \infty$, are convex.
Suppose that $\phi \in C[E, R]$ is convex. We note that the limits

$$\lim_{h \to 0^+} \frac{1}{h}[\phi(x + hy) - \phi(x)] = D^+\phi[x, y],$$

$$\lim_{h \to 0^-} \frac{1}{h}[\phi(x + hy) - \phi(x)] = D^-\phi[x, y]$$

exist and are upper and lower semi-continuous respectively.
Let E^* be the dual space of E. Let us denote by $x^*(x)$ or (x^*, x) the value of $x^* \in E^*$ at $x \in E$. As a consequence of the Hahn Banach Theorem we have for each $x \in E$, there exists a $x^* \in E^*$ such that $\|x^*\| = 1$ and $x^*(x) = \|x\|$. Hence the set

$$F(x) = \{x^* \in E^* : x^*(x) = \|x\|^2 \text{ and } \|x^*\| = \|x\| \qquad (1.4.2)$$

is nonempty for every $x \in E$. The mapping

$$F : E \to 2^{E^*} (\text{the set of subsets of} E^*)$$

is called the duality map. By means of F, we define the semi-inner products $(\cdot, \cdot)_\pm$ which are maps from $E \times E$ into R given by

$$(x, y)_+ = \sup\{x^*(y) : x^* \in F(x)\}$$

$$(x, y)_- = \inf\{x^*(y) : x^* \in F(x)\}.$$

It turns out that these semi-inner products have some properties of an inner product. Before proceeding to list these properties we shall prove that $(x, y)_\pm = \|x\|[x, y]_\pm$. For this purpose we shall first prove a simple lemma, which shows the relation between the subdifferential of the convex function $\frac{1}{2}\|x\|^2$ and the duality mapping. Given a convex function $\phi : E \to R$, the subdifferential of ϕ at x, denote by $\partial\phi(x)$, is defined by

$$\partial\phi(x) = \{x^* \in E^* : \phi(x) \leq \phi(y) + x^*(x - y), \quad \text{for every} \quad y \in E\}. \tag{1.4.3}$$

Lemma 1.4.3. *Let $\phi(x) = \frac{1}{2}\|x\|^2$ for all $x \in E$. Then the subdifferential $\partial\phi$ is identical to the duality mapping F.*

Proof. Let $x^* \in F(x)$. Then, using (1.4.2) and the fact $x^*(y) \leq \|x\|\|y\|$, we have

$$\begin{aligned} x^*(x - y) = x^*(x) - x^*(y) &\geq \|x\|^2 - \|x\|\|y\| \geq \frac{1}{2}(\|x\|^2 - \|y\|^2) \\ &= \phi(x) - \phi(y), \quad \text{for every} \quad y \in E. \end{aligned}$$

This implies, from (1.4.3), that $x^* \in \partial\phi(x)$.
Conversely, let $x^* \in \partial\phi(x)$. Then by (1.4.3), we get

$$\|x\|^2 \leq \|y\|^2 + 2x^*(x - y),$$

for every $y \in E$, which, by taking y as $x + hy$, becomes

$$\|x\|^2 \leq \|x + hy\|^2 - 2hx^*(y) \tag{1.4.4}$$

for every $y \in E$ and $h \in R$. It is easy to see, from (1.4.4), that for $h > 0$,

$$\frac{1}{2h}\left[(\|x\| + \|x + hy\|)(\|x + hy\| - \|x\|)\right] \geq x^*(y)$$

which yields, by taking the limit as $h \to 0^+$,

$$\|x\|[x, y]_+ \geq x^*(y),$$

for every $y \in E$. On the other hand, for $h < 0$, (1.4.4) yields

$$\frac{1}{2k}[(\|x\| + \|x - ky\|)(\|x\| - \|x - ky\|) \leq x^*(y)$$

where $k = -h > 0$. This in turn shows that, for every $y \in E$,

$$\|x\|[x, y]_- \leq x^*(y).$$

Thus for any $x^* \in \partial\phi(x)$, we have

$$\|x\|[x, y]_- \leq x^*(y) \leq \|x\|[x, y]_+,$$

for every $y \in E$. Now taking $y = x$ and using the fact $[x, x]_\pm = \|x\|$, we obtain $x^*(x) = \|x\|^2$. Also $\|x^*\| = \|x\|$. Hence $x^* \in F(x)$, and the lemma is proved.
Let us define $G(x)$ by

$$G(x) = \{x^* \in E^* : \|x^*\| = 1 \quad \text{and} \quad \|x^*(x)\| = \|x\|\}.$$

Then we have immediately $F(x) = \|x\|G(x)$, where F is the duality mapping.

Theorem 1.4.1. *If λ is a real number satisfying*

$$[x,y]_- \leq \lambda \leq [x,y]_+ \tag{1.4.5}$$

then there is an $x^* \in F(x)$ such that $\|x\|\lambda = x^*(y)$. In particular $(x,y)_\pm = \|x\|[x,y]_\pm$.

Proof. Let λ be a real number such that inequality (1.4.5) holds. Let $\triangle = \{\alpha x + \beta y : \alpha, \beta \in R\}$, where we assume without loss of generality that $y \neq \alpha x$. Define the linear function $p : \triangle \to R$ by

$$p(\alpha x + \beta y) = \alpha\|x\| + \beta\lambda.$$

If $\beta \geq 0$, the $\beta\lambda \leq \beta[x,y]_+ = [x,\beta y]_+$. If $\beta < 0$, the $\beta\lambda \leq \beta[x,y]_- = -\beta[x,-y]_+ = [x,\beta y]_+$. Consequently, $\beta\lambda \leq [x,\beta y]_+$ for all $\beta \in R$. Furthermore,

$$\begin{aligned} p(\alpha x + \beta y) &= \alpha\|x\| + \beta\lambda \leq \alpha\|x\| + [x,\beta y]_+ \\ &= [x, \alpha x + \beta y]_+ \leq \|\alpha x + \beta y\|. \end{aligned}$$

Hence by the Hahn-Banach theorem, there exists a linear functional $x_0^* : E \to R$ with the properties

$$x_0^*(x) \leq \|x\|, \text{ for all } x \in E$$

and

$$x_0^*(x) = p(x) = \|x\|, \text{ for all } x \in \triangle.$$

Thus $x_0^* \in G(x)$. Moreover $x_0^*(y) = p(y) = \lambda$. Now, taking $\lambda = [x,y]_+$, we get

$$x^*(y) = \|x\|x_0^*(y) \in F,$$

using the definition of F and G, and hence

$$(x,y)_+ = \|x\|[x,y]_+.$$

Similarly, $(x,y)_- = \|x\|[x,y]_-$. The proof is complete. In view of Lemmas 1.4.1, 1.4.3 and Theorem 1.4.1, we can now list the properties of $(x,y)_\pm$.

Corollary 1.4.1. *The semi-inner products $(x,y)_\pm$ have the following properties:*

(i) $(x,y)_- \leq (x,y)_+$;

(ii) $|(x,y)_\pm| \leq \|x\|\|y\|$;

(iii) $|(x,y)_\pm - (x,z)_\pm| \leq \|x\|\|y-z\|$;

(iv) $(x,y)_+ = -(x,-y)_- = -(-x,y)_-$;

(v) $(sx,ry)_\pm = sr(x,y)_\pm$, *for* $s,r \geq 0$;

(vi) $(x,\alpha x)_\pm = \alpha\|x\|^2, \alpha \in R$;

(vii) $(x, y+z)_+ \leq (x,y)_+ + (x,z)_+$ *and* $(x, y+z)_- \geq (x,y)_- + (x,z)_-$;

(viii) $(x, y+z)_+ \geq (x,y)_+ + (x,z)_-$ *and* $(x, y+z)_- \leq (x,y)_- + (x,z)_+$;

(ix) $(x, y+\alpha x)_\pm = (x,y)_\pm + \alpha\|x\|^2, \alpha \in R$;

(x) *Suppose that* $x : [a,b] \to E$ *and that* $m(t) = \|x(t)\|^2$. *If* $x'_\pm(t)$ *exists for some* $t \in (a,b)$, *then* $m'_\pm(t) = 2(x(t), x'_\pm(t))$;

(xi) $(x,y)_+$ *is upper semicontinuous and* $(x,y)_-$ *is lower semicontinuous.*

By isolating the ideas involved in the use of $\|x\|$ or $\|x\|^2$, we can define directional derivatives for a general Lyapunov-like function which can be used as a candidate for measure. Let $V : C[\Omega, R]$ where $\Omega \subset E$. Let $M_\pm : \Omega \times E \to R$ be mappings such that

(a) $M_+(x,y)$ is upper semicontinuous and $M_-(x,y)$ is lower semicontinuous;

(b) $o(\|y\|) + M_-(x,y) \leq V(x+y) - V(x) \leq M_+(x,y) + o(\|y\|)$;

(c) $M_+(x, \lambda y) \leq \lambda M_+(x,y),\ \lambda \geq 0,$
$M_-(x, \lambda y) \geq \lambda M_-(x,y),\ \lambda \geq 0;$

(d) $M_+(x, y_1+y_2) \leq M_+(x,y_1) + N\|x\|\|y_2\|,$
$M_-(x, y_1+y_2) \geq M_-(x,y_1) - N\|x\|\|y_2\|.$

Lemma 1.4.4. *Let* $x : (a,b) \to E$ *be differentiable at* $t \in (a,b)$ *and let* $m(t) = V(x(t))$. *Then*

(i) $D^+m(t) \leq M_+(x(t), x'(t))$;

(ii) $D_-m(t) \geq M_-(x(t), x'(t))$.

Proof. We shall prove *(i)*. The proof of *(ii)* is similar. We have for $h > 0$, in view of properties of $M_+(x,y)$,

$$
\begin{aligned}
\limsup_{h\to 0^+} \frac{m(t+h)-m(t)}{h} &= D^+m(t) \\
&= \limsup_{h\to 0^+} \frac{V(x(t+h)) - V(x(t)}{h} \\
&\leq \limsup_{h\to 0^+} \frac{V(x(t) + x(t+h) - x(t)) - V(x(t)}{h} \\
&\leq \limsup_{h\to 0^+} M_+(x(t), \frac{x(t+h)-x(t)}{h}) \\
&+ \limsup_{h\to 0^+} o\Big(\frac{\|x(t+h)-x(t)\|}{h}\Big) \\
&\leq M_\pm(x(t), x'(t))
\end{aligned}
$$

Remark. If $V(x) = \|x\|$, we can take $M_{\pm}(x, y) = [x, y]_{\pm}$. Similarly, if $V(x) = \|x\|^2$, $M_{\pm}(x, y)$ can be taken to be $(x, y)_{\pm}$. In fact, whenever $V(x)$ is a convex function, we can define

$$M_{+}(x, y) = \lim_{h \to 0^{\pm}} \frac{V(x + hy) - V(x)}{h}.$$

In many situations the continuity requirement of $M_{\pm}(x, y)$ are not necessary. In which case we can utilize the generalized derivative of the function $V(x)$ defined by

$$\limsup_{h \to 0^{+}} \frac{V(x + hy) - V(x)}{h}$$

or

$$\liminf_{h \to 0^{-}} \frac{V(x + hy) - V(x)}{h}$$

and impose conditions in terms of these generalized derivatives.

1.5 Mean Value Theorems

It is a known fact that classical mean value theorem for real valued functions does not hold for vector-valued functions. For example, let $f : [0, 2\pi] \to R^2$ defined by $f(t) = (-1 + \cos t, \sin t)$, $t \in [0, 2\pi]$. Then, $f(2\pi) - f(0) = (0, 0) \neq f'(\xi)2\pi = 2\pi(-\sin\xi, \cos\xi)$ for any $\xi \in (0, 2\pi)$. In this section, we discuss some results which extend the mean value theorem to functions with values in a Banach space in terms of both strong and weak topologies.

Theorem 1.5.1. *Let $f \in C[J, E]$, where J is an interval and E is a real Banach space. Suppose that $a, b \in J$, $a < b$, and there is at most a countable subset Γ of $[a, b]$ such that $f'_{+}(t)$ exists for all $t \in [a, b] - \Gamma$. Then the following relation holds:*

$$f(b) - f(a) \in (b - a)\overline{\text{co}}\left(\{f'_{+}(t) : t \in [a, b] - \Gamma\}\right),$$

where $\overline{\text{co}}(A)$ is the closed convex hull of A.

Proof. Let $\Omega = \{f'_{+}(t) : t \in [a, b] - \Gamma\}$ and $\Gamma = \{\gamma_k : k = 1, 2, 3, ...\}$. Let $\varepsilon > 0$ and consider the set T defined by

$$T = \{t \in [a, b] : d(f(t) - f(a), (t - a)\text{co}(\Omega)) \leq (t - a)\varepsilon + \varepsilon \sum_{\gamma_k < t} 2^{-k}\}.$$

Observe that $a \in T$ and thus T is nonempty. Set $\lambda = \sup\{t : t \in T\}$ and let $\{t_n\}_1^\infty$ be a sequence in T such that $\lim_{n \to \infty} t_n = \lambda$. Then, for each $n \geq 1$, there exists a $\xi_n \in \text{co}(\Omega)$ such that

$$\begin{aligned}\|f(t_n) - f(a) - (t_n - a)\xi_n\| &\leq (t_n - a)\varepsilon + \varepsilon \sum_{\gamma_k < t_n} 2^{-k} + \frac{1}{n} \\ &\leq \varepsilon(\lambda - a) + \varepsilon \sum_{\gamma_k < \lambda} 2^{-k} + \frac{1}{n}.\end{aligned}$$

Since $\|f(t_n) - f(a)\|$ is bounded, it can be assumed that there is an $M > 0$ such that $\|\xi_n\| \leq M$ for all n. Hence,

$$\begin{aligned} d(f(\lambda) - f(a), (\lambda - a)\text{co}(\Omega)) &\leq \lim_{n\to\infty} \|f(\lambda) - f(a) - (\lambda - a)\xi_n\| \\ &= \lim_{n\to\infty} \|f(t_n) - f(a) - (t_n - a)\xi_n\| \\ &\leq \varepsilon[(\lambda - a) + \sum_{\gamma_k<\lambda} 2^{-k}] \end{aligned}$$

which shows that $\lambda \in T$.

Suppose, for contradiction, that $\lambda < b$. We shall consider the two cases: (i) $\lambda = \gamma_m \in \Gamma$; (ii) $\lambda \in [a, b] - \Gamma$. In case (i); let $\xi \in \text{co}(\Omega)$ be such that

$$\|f(\lambda) - f(a) - (\lambda - a)\xi\| \leq \varepsilon[(\lambda - a) + 2^{-m}/3 + \sum_{\gamma_k<\lambda} 2^{-k}] \tag{1.5.1}$$

and let $\delta > 0$ be such that $\lambda + \delta \leq b$ and

$$\|f(\lambda + \delta) - f(\lambda)\| \leq \varepsilon 2^{-m}/3 \text{ and } \delta\|\xi\| \leq \varepsilon 2^{-m}/3. \tag{1.5.2}$$

Then

$$\begin{aligned} d(f(\lambda + \delta - f(a), (\lambda + \delta - a)\text{co}(\Omega)) &\leq \|f(\lambda + \delta - f(a) - (\lambda + \delta - a)\xi\| \\ &\leq \|f(\lambda) - f(a) - (\lambda - a)\xi\| + \|f(\lambda + \delta) - f(\lambda)\| + \delta\|\xi\| \\ &\leq \varepsilon(\lambda + \delta - a) + \varepsilon \sum_{\gamma_k<\lambda+\delta} 2^{-k}, \end{aligned}$$

in view of (1.5.1) and (1.5.2). However, the above inequality implies that $\lambda + \delta \in T$, which contradicts the definition of λ.

In case (ii); $f'_+(\lambda)$ exists and, in particular, there exists a $\delta > 0$ such that $\lambda + \delta \leq b$ and $\|f(\lambda + \delta) - f(\lambda) - \delta f'_+(\lambda)\| \leq \delta\varepsilon/2$. Moreover, since $\lambda \in T$, there exists a $\xi \in \text{co}(\Omega)$ such that

$$\|f(\lambda) - f(a) - (\lambda - a)\xi\| \leq [(\lambda - a) + \sum_{\gamma_k<\lambda} 2^{-k} + \delta/2]\varepsilon.$$

Since $f'_+(\lambda)$, $\xi \in \text{co}(\Omega)$ we have

$$(\lambda + \delta - a)^{-1}(\delta f'_+(\lambda) + (\lambda - a)\xi) \in \text{co}(\Omega)$$

and consequently,

$$\begin{aligned} d(f(\lambda + \delta) - f(a), (\lambda + \delta - a)\text{co}(\Omega)) &\leq \|f(\lambda + \delta) - f(a) - (\delta f'_+(\lambda) + (\lambda - a)\xi\| \\ &\leq \|f(\lambda + \delta) - f(\lambda) - \delta f'_+(\lambda)\| + \|f(\lambda) - f(a) - (\lambda - a)\xi\| \\ &\leq \delta\varepsilon/2 + [(\lambda - a) + \sum_{\gamma_k<\lambda} 2^{-k} + \delta/2]\varepsilon \leq (\lambda + \delta - a)\varepsilon + \varepsilon \sum_{\gamma_k\leq\lambda+\delta} 2^{-k}. \end{aligned}$$

But this implies $\lambda + \delta \in T$, which is a contradiction. Hence $\lambda = b$ and we have

$$d(f(b) - f(a), (b-a)\text{co}(\Omega)) \leq \varepsilon(b-a) + \varepsilon,$$

for all $\varepsilon > 0$. Therefore $d(f(b) - f(a), (b-a)\text{co}(\Omega)) = 0$, and the conclusion of the theorem follows from this.

Remark 1.5.1. *Theorem 1.5.1 is not valid if we assume only that Γ has measure zero; for there is a continuous, strictly increasing function $\phi : [0,1] \to R$ with $\phi(0) = 0$, $\phi(1) = 1$ and $\phi'(t) = 0$ a.e. in $[0,1]$.*

The following version of the mean value theorem for abstract functions is a consequence of Theorem 1.5.1.

Theorem 1.5.2. *Suppose that the hypotheses of Theorem 1.5.1 hold and, in addition, assume that $\|f'_+(t)\| \leq M$ for all $t \in [a,b] - \Gamma$. Then*

$$\|f(b) - f(a)\| \leq M(b-a).$$

The proof follows from Theorem 1.5.1 since $\overline{\text{co}}(\{f'_+(t) : t \in [a,b] - \Gamma\}) \subset \{x \in E : \|x\| \leq M\}$.

Observe that if $f'_+(t) = 0$ for all $t \in J$, then f is a constant function on J. As an immediate consequence of Theorem 1.5.2, we have the following.

Corollary 1.5.1. *Assume that f is a differentiable function from the interval J into the Banach space E. Then*

$$\|f(t+h) - f(t) - f'(t)h\| \leq |h| \sup\{\|f'(s) - f'(t)\| : s \text{ is between } t \text{ and } t+h\},$$

for all $t, t+h \in J$.

Proof. Set $g(s) = f(s) - f'(t)s$ for all $s \in J$. Then the function g is differentiable and $g'(s) = f'(s) - f'(t), s \in J$. By Theorem 1.5.2,

$$\begin{aligned}\|f(t+h) - f(t) - f'(t)h\| &= \|g(t+h) - g(t)\| \\ &\leq |h| \sup\{\|g'(s)\| : s \text{ is between } t \text{ and } t+h\},\end{aligned}$$

which proves the conclusion.

For a function $f : J \times E \to E$ which is Frèchet differentiable, the mean value theorem can be expressed by the following result, which is useful in the consideration of boundary value problems.

Theorem 1.5.3. *Let $f \in C[J \times B[x_0, r], E]$ and let the Frèchet derivative $f_x(t,x)$ exist and be continuous for each $t \in J$ and $x \in B[x_0, r] = \{x \in E : \|x - x_0\| \leq r\}$. Then for $x_1, x_2 \in B[x_0, r]$ and $t \in J$,*

$$f(t, x_1) - f(t, x_2) = \int_0^1 f_x(t, sx_1 + (1-s)x_2)(x_1 - x_2)ds.$$

Proof. Define $F(s) = f(t, sx_1 + (1-s)x_2)$, for $0 \le s \le 1$. Since $B[x_0, r]$ is convex, $F(s)$ is well defined. Using the chain rule for Frèchet derivatives, we have

$$F'(s) = f_x(t, sx_1 + (1-s)x_2)(x_1 - x_2). \tag{1.5.3}$$

Noting that $F(1) = f(t, x_1)$ and $F(0) = f(t, x_2)$, we obtain the desired result by integrating (1.5.3) with respect to s from 0 to 1.

The analog of Theorem 1.5.3 in weak topology may be stated as follows:

Theorem 1.5.4. *Suppose that $f \in C_\omega[J, E]$ and $f(t)$ is weakly differentiable on J. Let $a, b \in J$ and $a < b$. Then*

$$f(b) - f(a) \in (b-a)\overline{\text{co}}(\{f'(s) : s \in [a, b]\}).$$

Proof. The weak differentiability of $f(t)$ implies that $(\phi f)(t)$, is differentiable and $(\phi f)'(t) = \phi(f'(t))$, $\phi \in E^*$. For each $\phi \in E^*$, by virtue of the mean value theorem, there exists a $t_\phi \in [a, b]$ such that

$$\phi(f(b)) - \phi(f(a)) = (b-a)(\phi f)'(t_\phi) = (b-a)\phi(f'(t_\phi)).$$

Letting $A = \{f'(t) : t \in [a, b]\}$, we have

$$\phi\left(\frac{f(b) - f(a)}{b - a}\right) \in \phi(A), \text{ for all } \phi \in E^*.$$

It only remains to show that $\dfrac{f(b) - f(a)}{b - a} \in \overline{\text{co}}A$ which is a consequence of the following lemma.

Lemma 1.5.1. *Let E be a real Banach space, $A \subset E$ and $x \in E$. If $\phi(x) \in \phi(A)$ for all $\phi \in E^*$ then $x \in \overline{\text{co}}(A)$.*

Proof. Let x_0 be any fixed element of A. Let $M = \overline{\text{co}}(A - \{x_0\})$. Clearly M is a closed, convex set and $\theta \in M$. Suppose $x - x_0 \notin M$. Then there exists a $\phi \in E^*$ such that $\phi(x - x_0) > 1$ and $\phi(z) \le 1$, $z \in M$. But, by the hypotheses of the lemma, there exists a $y \in A$ such that $\phi(x) = \phi(y)$ and thus $\phi(x - x_0) = \phi(y - x_0)$. Note that $y - x_0 \in M$. Observing that $1 < \phi(x - x_0) = \phi(y - x_0) \le 1$ is a contradiction, we conclude that $x - x_0 \in M$. Now, given $\varepsilon > 0$, there exists $\{y_i\} \in A$ and $\{\alpha_i\}$ with $\alpha_i > 0$, $\sum_{i=1}^{n} \alpha_i = 1$ such that

$$\|x - x_0 - \sum_{i=1}^{n} \alpha_i(y_i - x_0)\| < \varepsilon.$$

But $\|x - x_0 - \sum_{i=1}^{n} \alpha_i(y_i - x_0)\| = \|x - \sum_{i=1}^{n} \alpha_i y_i\| < \varepsilon$ which implies that $x \in \overline{\text{co}}(A)$. The lemma is proved.

1.6 Measures of Noncompactness

Let A be a bounded subset in a Banach space E. The diameter of A is defined by

$$\text{dia}(A) = \sup\{\|x-y\| \ : x, y \in A\}$$

Clearly $0 \leqslant dia(A) < \infty$. Kuratowski's measure of noncompactness of A is defined by

$$\alpha(A) = \inf\{d > 0 : A \text{ is covered by a finite number of sets with diameter} \leqslant d\}.$$

In particular, given $d > \alpha(A)$, there exists a finite number of sets $S_1, S_2, ..., S_n \subset A$ such that $dia(S_i) \leqslant d$ and $\bigcup_{i=1}^{n} S_i = A$. In other words, $\alpha(A)$ can be regarded as a measure of the extent to which A is not compact. Note also that $\alpha(A) \leqslant \text{dia}(A)$ and $\alpha(A) \leqslant 2d$ if $\sup_{x \in A} \|x\| \leqslant d$. The various properties of α that will be useful later are listed in the following theorem.

Theorem 1.6.1. *Let A, B be bounded subsets of E. Then*

(i) $\alpha(A) = 0$ *if and only if* $\overline{A}$ *is compact, where* $\overline{A}$ *denotes the closure of* A*;*

(ii) $\alpha(A) = \alpha(\overline{A})$*;*

(iii) $\alpha(\lambda A) = |\lambda|\alpha(A)$, $\lambda \in R$ *where* $\lambda A = \{\lambda x : x \in A\}$*;*

(iv) $\alpha(A \bigcup B) = \max(\alpha(A), \alpha(B))$;

(v) $\alpha(A) \leqslant \alpha(B)$ *if* $A \subset B$;

(vi) $\alpha(A+B) \leqslant \alpha(A) + \alpha(B)$ *where* $A + B = \{x + y : x \in A$ *and* $y \in B\}$; *in particular, if* $A = \{x_n\}, B = \{y_n\}$ *are two countable sets of points in* E*, then*

$$\alpha(\{x_n\}) - \alpha(\{y_n\}) \leqslant \alpha(\{x_n - y_n\});$$

(vii) α *is continuous with respect to the Hausdorff metric;*

(viii) $\alpha(A) = \alpha(co(A))$ *where* $co(A)$ *is the convex hull of* A*;*

(ix) if $\{A_n\}$ *is a family of nonempty bounded subsets of* E *such that* $A_{n+1} \subset A_n$ *for* n = *1,2,...,* *and* $\lim_{n\to\infty} \alpha(A_n) = 0$, *then* $\bigcap_{n=1}^{\infty} \overline{A}_n$ *is nonempty and compact.*

Proof. We shall only indicate the proofs of (iii), (vi), (viii) and (ix), since the other properties follow easily from the definition of $\alpha(A)$.

Let $0 < \varepsilon$ and let $S_1, S_2, ..., S_n \subset$ of E be such that $dia(S_i) \leqslant \alpha(A) + \varepsilon$ and $A \subset \bigcup_i S_i$. Then $\lambda A \subset \bigcup_i (\lambda S_i), \lambda \in R$. Since $dia(\lambda S_i) = |\lambda| dia(S_i)$, we have

$\alpha(\lambda A) \leqslant |\lambda|(\alpha(A)+\varepsilon)$. This proves that $\alpha(\lambda A) \leqslant |\lambda|\alpha(A)$, since ε is arbitrary. If $\lambda \neq 0$, then $\alpha(A) = \alpha\Big(\frac{1}{\lambda}(\lambda A)\Big) \leqslant \frac{1}{|\lambda|}\alpha(\lambda A)$ and consequently, $|\lambda|\alpha(A) \leqslant \alpha(\lambda A)$, proving (iii).

To prove (vi); let $\varepsilon > 0$. Then there exist sets $S_1, ..., S_n$ and $T_1, ..., T_m$ of E such that $dia(S_i) \leqslant \alpha(A) + \frac{\varepsilon}{2}$, $dia(T_j) \leqslant \alpha(B) + \frac{\varepsilon}{2}$, $A \subset \bigcup_i S_i$ and $B \subset \bigcup_j T_j$. Hence $A+B \subset \bigcup_{i,j}(S_i+T_j)$. Since $dia(S_i+T_j) \leqslant dia(S_i)+dia(T_j) \leqslant \alpha(A) + \alpha(B) + \varepsilon$, we obtain $\alpha(A+B) \leqslant \alpha(A) + \alpha(B) + \varepsilon$ which implies (vi).

To prove (viii); let $\varepsilon > \alpha(A)$ and let $S_1, S_2, ..., S_n$ be subsets of E such that $dia(S_i) \leqslant \varepsilon$ and $\bigcup_i S_i \supset A$. Also let $M \geqslant 1$ be such that $\|y\| \leqslant M$ whenever $y \in co(S_i)$. Let $\eta > 0$ and let Γ be a finite subset of $[0,1]$ such that $t \in [0,1]$ implies there is a $\gamma \in \Gamma$ with $|\gamma - t| \leqslant \frac{\eta}{(nM)}$. For each $i \in \{1, ..., n\}$ define

$$T_i = \{x \in E : inf\{\|x-y\| : y \in co(S_i)\} \leqslant \eta\}.$$

Since $dia(co(S_i)) = dia(S_i) \leqslant \varepsilon$, it is easy to check that $dia(T_i) \leqslant \varepsilon + 2\eta$. Furthermore, if $\lambda_i \in [0,1]$ and $\gamma_i \in \Gamma$ are such that $|\lambda_i - \gamma_i| \leqslant \frac{\eta}{(nM)}$, then

$$\lambda_i co(S_i) \subset \gamma_i T_i. \tag{1.6.1}$$

For if $\lambda_i y \in \lambda_i co(S_i)$, then $\lambda_i y = \gamma_i y + (\lambda_i - \gamma_i)y$ where $|(\lambda_i - \gamma_i)y| \leqslant \eta$. Now let $\{U_j\}$ be the (finite) collection of subsets of E of the form

$$U_j = \sum_{i=1}^{n} \gamma_i T_i, \quad \text{where} \quad \gamma_i \in \Gamma \quad \text{and} \quad |1 - \sum_{i=1}^{n} \gamma_i| \leqslant \eta.$$

By the properties of the diameter of a set and the fact that $dia(T_i) \leq \varepsilon + 2\eta$ we have that

$$\begin{aligned} dia(U_j) &\leqslant \sum_{i=1}^{n} dia(\gamma_i T_i) = \sum_{i=1}^{n} \gamma_i \, dia(T_i) \\ &\leqslant \Big(\sum_{i=1}^{n} \gamma_i\Big)(\varepsilon + 2\eta) \leqslant (1+\eta)(\varepsilon + 2\eta) \\ &= \varepsilon + \eta(\varepsilon + 2 + 2\eta). \end{aligned}$$

To complete the proof we need only to show that $\bigcup_j U_j \supset co(A)$. For if this is true then $\alpha(co(A)) \leqslant \varepsilon + \eta(\varepsilon + 2 + 2\eta)$ for all $\varepsilon > \alpha(A)$ and $\eta > 0$, which implies $\alpha(co(A)) \leqslant \alpha(A)$. Since the reverse inequality is evident, property (viii) will follow. So let $\omega \in co(A)$ be such that

$$\omega = \sum_{k=1}^{p} \xi_k x_k \quad \text{where} \quad x_k \in A, \xi_k \geqslant 0, \quad \text{and} \quad \sum_{k=1}^{p} \xi_k = 1.$$

Let $J_i, i = 1, ..., n$, be mutually disjoint subsets of {1,...,p} such that $x_k \in S_i$ for $k \in J_i$ (some J_i may be empty). Set $\lambda_i = \sum_{k \in J_i} \xi_k$ for $i = 1, ..., n$, where $\lambda_i = 0$ if $J_i = \phi$. Now for each $i \in \{1, ..., n\}$ choose $\gamma_i \in \Gamma$ such that $|\gamma_i - \lambda_i| \leqslant \frac{\eta}{(nM)}$. Then $\sum_i \lambda_i = 1$ and

$$|\sum_i \gamma_i - 1| = |\sum_i (\gamma_i - \lambda_i)| \leqslant \sum_i \frac{\eta}{(nM)} \leqslant \eta;$$

so $U_j = \sum_{i=1}^{n} \gamma_i T_i$ for some fixed j. Since

$$\omega = \sum_{i=1}^{n} \Big(\sum_{k \in J_i} \xi_k y_k \Big)$$

and, if $\lambda_i > 0$,

$$\sum_{k \in J_k} \xi_k y_k = \lambda_i \sum_{k \in J_i} \xi_k \lambda_i^{-1} y_k \in co(S_i),$$

it follows from (1.6.1) that $\omega \in \sum_{i=1}^{n} \lambda_i co(S_i) \subset \sum_{i=1}^{n} \gamma_i T_i = U_j$. Hence $co(A) \subset \bigcup_j U_j$.

To prove (ix); for each n choose $x_n \in A_n$ and define $A = \{x_n : n \geq 1\}$. Now let $\varepsilon > 0$ and let N be an integer such that $\alpha(A_N) \leqslant \varepsilon$. Then

$$\alpha(A) = \alpha(\{x_1, ..., x_{N-1}\} \cup \{x_N, x_{N+1}, ...\}) \leqslant \alpha(A_N) \leqslant \varepsilon,$$

and so $\alpha(A) = 0$ and $\overline{A}$ is compact. Consequently, by relabeling if necessary, we may assume that $\lim_{n \to \infty} x_n = x \in E$. It is immediate that $x \in \overline{A_n}$ for all $n \geqslant 1$ since $x_i \in A_k \subset A_n$ for all $k \geqslant n$. Thus $x \in \bigcap_{j=1}^{\infty} \overline{A_n}$ and the assertions of the lemma now follow easily.

Remark 1.6.1. *A general version of property (iii) of Theorem 1.6.1 can be stated as follows: For any bounded subset S of R (the real line) and any bounded set $A \subset E$, $\alpha(S \cdot A) = \Big(\sup_{t \in S} |t|\Big)\alpha(A)$, where $S \cdot A \equiv \{sx : s \in S, x \in A\}$.*

Also, using the definition of $\alpha(\cdot)$, it is easy to show that $\alpha(A \times B) = max(\alpha(A), \alpha(B))$ where $A \times B$ is the cartesian product of two bounded subsets A, B of Banach spaces E, F respectively, with $\|(x, y)\| = max(\|x\|, \|y\|), x \in A, y \in B$.

While considering bounded subsets of $C(I, E), I$ being any compact subset of the real line, it is convenient to use the following notation: for any set $H \subset$

$C[I,E], H(t)$ and $H(I)$ denote the sets given by $\{\phi(t) : \phi \in H\}$ and $\bigcup_{t\in I}\{\phi(t) : \phi \in H\}$ respectively. A useful property of the measure of noncompactness α is in the following result, which in some sense, also provides a generalization of the theorem of Ascoli-Arzela (Theorem 1.2.5).

Theorem 1.6.2. *If $H = \{x_k\}$, where $x_k \in C[I,E]$, is any bounded equicontinuous family of functions, then*

$$\sup_{t\in I} \alpha(H(t)) = \alpha(H)$$

The proof of Theorem 1.6.2, is an immediate consequence of

Lemma 1.6.1. *If $H \subset C(I,E)$ is a bounded, equicontinuous set, then*

(a) $\alpha(H) = \alpha(H(I))$;

(b) $\alpha(H(I)) = \sup_{t\in I} \alpha(H(t))$.

Proof. In order to prove (a), let us first show that $\alpha(H(I)) \leqslant \alpha(H)$. By the definition of the measure of noncompactness α, given $\varepsilon > 0$ we can find a finite covering $\{H_i\}_{1\leqslant i\leqslant n}$ of H such that, for any $i = 1,2,...,n$, $dia(H_i) \leqslant \alpha(H) + \varepsilon$. As H is an equicontinuous set corresponding to the given ε it is possible to consider a covering $V_j (j = 1,2,...,m)$ of the compact interval I such that

$$\|\phi(t_1) - \phi(t_2)\| < \varepsilon, \quad \text{for every} \quad \phi \in H,$$

whenever $t_1, t_2 \in V_j$. For any two indices i, j, let the set $B_{ij} \subset H(I)$ be defined by

$$B_{ij} = H_i(V_j) = \{\phi(t) : \phi \in H_i, t \in V_j\}.$$

It is clear that $\{B_{ij}\}(1 \leqslant i \leqslant n, 1 \leqslant j \leqslant m)$ is a finite covering of $H(I)$ and

$$dia(B_{ij}) = sup\{\|\phi(t_1) - \psi(t_2)\| : \phi, \psi \in H_i, t_1, t_2 \in V_j\}.$$

Furthermore,

$$\|\phi(t_1) - \psi(t_2)\| \leqslant \|\phi(t_1) - \psi(t_1)\| + \|\psi(t_1) - \psi(t_2)\|.$$

But, for $t_1, t_2 \in V_j, \|\psi(t_1) - \psi(t_2)\| \leqslant \varepsilon$ for any $\psi \in H$ and therefore

$$\|\phi(t_1) - \psi(t_2)\| \leqslant \|\phi(t_1) - \psi(t_1)\| + \varepsilon.$$

Thus we have, for $i = 1,2,...,n$,

$$\begin{aligned} dia(B_{ij}) &\leqslant sup\{\|\phi(t_1) - \psi(t_1)\| : \phi, \psi \in H_i, t_1 \in V_j\} + \varepsilon \\ &\leqslant sup\{\|\phi(t) - \psi(t)\| : \phi, \psi \in H_i, t \in I\} + \varepsilon \\ &= dia(H_i) + \varepsilon \\ &< \alpha(H) + 2\varepsilon. \end{aligned}$$

Therefore we have the inequality $\alpha(H(I)) < \alpha(H) + 2\varepsilon$ and, ε being arbitrary, this proves that $\alpha(H(I)) \leqslant \alpha(H)$.

Let us now show that $\alpha(H) \leqslant \alpha(H(I))$. Given $\varepsilon > 0$, by the equicontinuity of H, we can cover I by a finite number of neighborhoods $V(t_i)\,(i = 1, 2, ..., n)$ such that $\|\phi(t) - \phi(t_i)\| < \varepsilon$, for any $\phi \in H$, whenever $t \in V(t_i)$. On the other hand, it is also possible for the given ε to cover $H(I)$ by a finite number of sets $B_j\,(j = 1, 2, ..., m)$ such that $dia(B_{ij}) < \alpha(H(I)) + \varepsilon$.

Let Ω be the finite set of all the maps $i \to \mu(i)$ of $[1, 2, ..., n] \subset N$ into $[1, 2, ..., m] \subset N$, where N is the set of positive integers. For any $\mu \in \Omega$, let us denote by L_μ the set of all $\phi \in H$ such that for every $i \in [1, 2, ..., n]$ we have $\phi(t_i) \in B_{\mu(i)}$. It is easy to see that H is covered by the union of $L_\mu, \mu \in \Omega$. Further, let $\phi, \psi \in L_\mu$. Since for any $t \in I$, there is an index i such that $t \in V(t_i)$, we then have $\|\phi(t) - \phi(t_i)\| < \varepsilon$ and $\|\psi(t) - \psi(t_i)\| < \varepsilon$. Since $\|\phi(t_i) - \psi(t_i)\| < \alpha(H(I)) + \varepsilon$, we obtain

$$\begin{aligned}\|\phi(t) - \psi(t)\| &\leqslant \|\phi(t) - \phi(t_i)\| + \|\phi(t_i) - \psi(t_i)\| + \|\psi(t_i) - \psi(t)\| \\ &< \alpha(H(I)) + 3\varepsilon,\end{aligned}$$

and thus, $\alpha(H) \leqslant \alpha(H(I)) + 3\varepsilon$. The proof of part (a) of the lemma is complete since ε is arbitrary.

To prove part (b); we first observe that, for every $t \in I$, $H(t) \subset H(I)$ and therefore $\alpha(H(t)) \leqslant \alpha(H(I))$. This, in turn, implies that $\sup\limits_{t \in I} \alpha(H(t)) \leqslant \alpha(H(I))$.On the other hand, since H is equicontinuous, given $\varepsilon > 0$,we can divide I into n intervals $V(t_1), ..., V(t_n)$ such that on these intervals we have $\|\phi(t) - \psi(t_i)\| < \varepsilon$ for every $\phi \in H$. Also, for any $i = 1, 2, ..., n$, we can find a finite covering $\{H_j^{(i)}\}_{1 \leqslant j \leqslant m}$ of H such that $\{H_j^{(i)}(t_i)\}_{1 \leqslant j \leqslant m}$ is a finite covering of $H(t_i)$ satisfying

$$max_{1 \leqslant j \leqslant m}(dia(H_j^{(i)}(T_i))) < \alpha(H(t_i)) + \varepsilon.$$

Then, setting $B_{ij} = H_j^{(i)}(V(t_i))$, we have $\{B_{ij}\} \quad (i = 1, 2, ..., n,\ j = 1, 2, ..., m)$ is a finite covering of $H(I)$ and

$$dia(B_{ij}) = sup\{\|\phi(t) - \psi(\tau)\| : \phi, \psi \in H_j^{(i)},\ t, \tau \in V(t_i)\}.$$

Moreover,

$$\begin{aligned}\|\phi(t) - \psi(\tau)\| &\leqslant \|\phi(t) - \phi(t_i)\| + \|\phi(t_i) - \psi(t_i)\| + \|\psi(t_i) - \psi(\tau)\| \\ &< 2\varepsilon + \|\phi(t_i) - \psi(t_i)\|,\end{aligned}$$

which implies that

$$dia(B_{ij}) \leqslant 2\varepsilon + sup\{\|\phi(t_i) - \psi(t_i)\| : \phi, \psi \in H_j^{(i)}\} = 2\varepsilon + dia(H_j^{(i)}(t_i)).$$

Therefore

$$\alpha(H(I)) \leqslant \max_{\substack{1\leqslant i\leqslant n\\ 1\leqslant j\leqslant m}} (dia(B_{ij})) < 3\varepsilon + \max_{1\leqslant i\leqslant n} \alpha(H(t_i))$$
$$\leqslant 3\varepsilon + \sup\{\alpha(H(t)) : t \in I\}$$

and, ε being arbitrary, this completes the proof of part (b).

A useful generalization of Theorem 1.6.2, which is needed for the study of boundary value problems in Banach spaces, is given by

Theorem 1.6.3. *Let A be a bounded subset of $C^1[I, E]$. Let $A(t)$ and $A'(t)$ denote the sets $\{\phi(t) : \phi \in A\}$ and $\{\phi'(t) : \phi \in A\}$ respectively. Suppose that $A' = \{\phi' : \phi \in A\}$ is an equicontinuous set. Then*

$$max[\sup_{t\in I} \alpha(A(t)), \sup_{t\in I} \alpha(A'(t))] = \alpha(A).$$

For convenience, we first prove the following two lemmas.

Lemma 1.6.2. *Let $A \subseteq C^1(I, E)$ be bounded. Then*

$$\alpha(A) \geqslant \max[\sup_{t\in I} \alpha(A(t)), \sup_{t\in I} \alpha(A'(t))].$$

Proof. Let $\varepsilon > 0$ and $d = \alpha(A)$. By definition of $\alpha(A)$, there exist sets $T_1, ..., T_k \subset C^1[I, E]$ such that $A \subseteq \bigcup_{i=1}^{k} T_i$ and $dia(T_i) < d + \varepsilon$, for each i. For any $t_0 \in I$ observe that

$$A(t_0) \subseteq \bigcup_{i=1}^{k} T_i(t_0),\ A'(t_0) \subseteq \bigcup_{i=1}^{k} T_i'(t_0)$$

and

$$dia(T_i(t_0)) < d + \varepsilon,\ dia(T_i'(t_0)) < d + \varepsilon.$$

Thus $\alpha(A(t_0)) < d + \varepsilon,\quad \alpha(A'(t_0)) < d + \varepsilon$ and the proof follows since ε and $t_0 \in I$ are arbitrary.

Lemma 1.6.3. *Let $A \subset C^1(I, E)$ with A bounded and A' equicontinuous. Then*

$$\alpha(A) \leqslant \max[\sup_{t\in I} \alpha(A(t)), \sup_{t\in I} \alpha(A'(t))].$$

Proof. Since A and A' are equicontinuous, given $\varepsilon > 0$ there exists a partition of I such that if $t \in [t_i, t_{i+1}]$ and $\phi \in A$, then $\|\phi(t) - \phi(t_i)\| < \varepsilon$ and $\|\phi'(t) - \phi'(t_i)\| < \varepsilon$. Let $d = \alpha(B)$ where $B = \bigcup_{i=1}^{n} [A(t_i) \cup A'(t_i)]$. By the definition of the measure of noncompactness α, there exists sets $S_1, ..., S_k$ such that $B \subseteq \bigcup_{i=1}^{k} S_i$

and $dia(S_i) < d + \varepsilon$, for each i. Let ϕ be the set of mappings from $\{1, ..., n\}$ into $\{1, ..., k\}$. Obviously ϕ is a finite family. For each $f, g \in \phi$, let

$$T_{f,g} = \{\phi \in A : \phi(t_i) \in S_{f(i)}, \phi'(t_i) \in S_{g(i)},\ i = 1, ..., n\}.$$

Note that $A \subseteq \bigcup_{f,g\in\phi} T_{f,g}$. Also, if $\phi, \psi \in T_{f,g}$ and $t \in [t_i, t_{i+1}]$, then

$$\begin{aligned}\|\phi(t) - \psi(t)\| &\leqslant \|\phi(t_i) - \phi(t)\| + \|\psi(t_i) - \psi(t)\| + \|\psi(t_i) - \phi(t_i)\| \\ &< \varepsilon + \varepsilon + d + \varepsilon\end{aligned}$$

and

$$\|\phi'(t) - \psi'(t)\| \leqslant \|\phi'(t_i) - \phi'(t)\| + \|\psi'(t_i) - \psi'(t)\| + \|\psi'(t_i) - \phi'(t_i)\| < \varepsilon + \varepsilon + d + \varepsilon.$$

Thus, $\max[\sup_{t\in I} \|\phi(t) - \psi(t)\|, \sup_{t\in I} \|\phi'(t) - \psi'(t)\|] < d + 3\varepsilon$ and so, $dia(T_{f,g}) \leqslant d + 3\varepsilon$. This further yields

$$\alpha(A) \leqslant d + 3\varepsilon = \alpha(B) + 3\varepsilon$$

Using property (iv) of Theorem 1.6.1 and the definition of the set B, we obtain

$$\begin{aligned}\alpha(A) \leqslant \alpha(B) + 3\varepsilon &\leqslant \max_{i=1,\dots,n} [\alpha(A(t_i)), \alpha(A'(t_i))] + 3\varepsilon \\ &\leqslant max[\sup_{t\in I} \alpha(A(t)), \sup_{t\in I} \alpha(A'(t))] + 3\varepsilon.\end{aligned}$$

Since ε is arbitrary, the assertion of the lemma follows.

Proof of Theorem 1.6.3. Combining Lemmas 1.6.2 and 1.6.3 we obtain the desired conclusion of Theorem 1.6.3.

An interesting relation between $\alpha(A), \alpha(A(I))$ and $\alpha(A'(I))$ is given by

Lemma 1.6.4. *Under the assumptions of Lemma 1.6.2,* $\alpha(A) \geqslant \alpha(A(I))$ *and* $\alpha(A) \geqslant \frac{1}{2}\alpha(A'(I))$.

Proof. Let $d = \alpha(A)$ and $\varepsilon > 0$. Then there exists sets $S_1, ..., S_k \subseteq C^1[I, E]$ with $\bigcup_{i=1}^{k} S_i \supseteq A$ and $dia(S_i) < d + \varepsilon$ for each i. Since A is continuous, there exists a partition $\{t_j\}$, $j = 1, 2, ..., n$ of I such that for $t \in [t_j, t_{j+1}]$ and $\phi \in A$, we have $\|\phi(t_j) - \phi(t)\| < \varepsilon$. Let

$$T_{i,j} = \{x \in E : \|x - \phi(t_j)\| < \varepsilon \quad \text{for some} \quad \phi \in S_i\}.$$

Observe that if $x, y \in T_{ij}$, then

$$\|x - y\| \leqslant \|x - \phi(t_j)\| + \|\phi(t_j) - \psi(t_j)\| + \|\psi(t_j) - y\| < \varepsilon + (d + \epsilon) + \varepsilon,$$

since $\phi, \psi \in S_i$. Hence $dia(T_{ij}) < d + 3\varepsilon$. Moreover, if $\phi \in A$, $t \in I$, that is, $\phi \in S_i$ and $t \in [t_j, t_{j+1}]$, then $\|\phi(t) - \phi(t_j)\| < \varepsilon$ which implies that $\phi(t) \in T_{ij}$ and $A(I) \subset \bigcup T_{ij}$. Thus it follows that $\alpha(A) \geqslant \alpha(A(I))$.

Suppose that $d = \alpha(A)$, $\varepsilon > 0$, $A \subseteq \bigcup_{i=1}^{k} S_i$ with $dia(S_i) < d + \varepsilon$ for each i. Choose $\phi_i \in S_i$. The equicontinuity of the set $\{\phi_1', ..., \phi_k'\}$ implies that there exists a partition $\{t_j\}$, $j = 1, 2, ..., n$ such that if $t, s \in [t_j, t_{j+1}]$, then $\|\phi_j'(t) - \phi_j'(s)\| < \varepsilon$. Defining $V_{i,j} = \{\phi'(t) : \phi \in S_i,\, t \in [t_j, t_{j+1}]\}$, we see that

$$\begin{aligned}\|\phi'(t) - \psi'(s)\| &\leqslant \|\phi'(t) - \phi_i'(t)\| + \|\phi_i'(t) - \phi_i'(s)\| + \|\phi_i'(s) - \psi'(s)\| \\ &\leqslant 2d + 3\varepsilon,\end{aligned}$$

whenever $\phi'(t), \psi'(s) \in V_{i,j}$. Furthermore, if $\phi \in A$, $t \in S_i$ and $t \in [t_j, t_{j+1}]$ for some i, j, then $\phi^(t) \in V_{i,j}$ and the above inequality leads to

$$\alpha(A'(I)) \leqslant 2\alpha(A).$$

The proof is complete.

We now wish to consider two types of measures of noncompactness which have properties similar to those of $\alpha(A)$. For any bounded $A \subset E$, the Hausdorff measure of noncompactness of A is defined by $\gamma(A) = inf\{r > 0 : A$ is covered by a finite number of spheres of radius r$\}$.

Another kind of measure of noncompactness which is due to DeBlasi is defined by $\beta(A) = inf\{t > 0$: there exists a compact set $C \subset E$ such that $A \subset C + t\overline{S}\}$, where $\overline{S}$ denotes the closed unit ball in E. This definition is flexible enough to be adapted for the measure of weak noncompactness in the weak topology which is considered next.

Let A be a nonvoid bounded subset of E. We shall define the measure of noncompactness of A in the weak topology as follows:

$$\beta^\star(A) = \inf\{t > 0 : \quad \text{there exists a} \quad C \in K^\omega \quad \text{such that} \quad A \subset C + t\overline{S}\}$$

where K^ω denotes the family of all weakly compact subsets of E and $\overline{S}$ is the closed unit ball in the strong topology of E, i.e., $\overline{S} = \{x \in E : \|x\| \leqslant 1\}$. Though the properties of $\beta^\star$ are quite similar to those of α (*or* γ, β), we list them in the following theorem for convenience of reference.

Theorem 1.6.4. *Let A and B be bounded subsets of E and $\{x_n\}$, $\{y_n\}$ be bounded sequences in E. Then*

(i) $A \subset B$ *implies* $\beta^\star(A) \leqslant \beta^\star(B)$;

(ii) $\beta^\star(A) = \beta^\star(\overline{A}^\omega)$ *where* $\overline{A}^\omega$ *denotes the weak closure of* A;

(iii) $\beta^\star(A) = 0$ *if and only if* $\overline{A}^\omega$ *is weakly compact;*

(iv) $\beta^\star(A \bigcup B) = max\{\beta^\star(A), \beta^\star(B)\}$;

(v) $\beta^\star(A) = \beta^\star(co(A))$;

(vi) $\beta^\star(A+B) \leqslant \beta^\star(A)+\beta^\star(B)$; *in particular*, $\beta^\star(\{x_n\})-\beta^\star(\{y_n\}) \leqslant \beta^\star(\{x_n-y_n\})$;

(vii) $\beta^\star(x + A) = \beta^\star(A)$ *if* $x \in E$;

(viii) $\beta^\star(tA) = t\beta^\star(A), t > 0$;

(ix) $\beta^\star(A) \leqslant \mathrm{dia}(A)$ *and* $\beta^\star(A) \leqslant \alpha(A)$, *where* $\alpha(A)$ *and* $\beta^\star(A)$ *are the measures of noncompactness of* A.

Proof. Assertions (i) and (iv)-(ix) are immediate consequences of the definition of $\beta^\star$. The proof of the "if" part of (iii) is trivial. To prove the "only if" part of (iii); let S be the unit ball in E, $E^{\star\star}$ the bidual of E, $\overline{S}$ the closure of S in $E^{\star\star}$ with the weak topology, i.e., $\sigma(E^{\star\star}, E^\star)$. Since A is bounded, it suffices to prove that its weak closure in $E^{\star\star}$ is contained in E. Observing that $A \subset C + t\overline{S}$, one can conclude that $\overline{A} \subset C + t\overline{S}$ because, as C is weakly compact and $t\overline{S}$ is weakly closed, the right hand side member is weakly closed in $E^{\star\star}$. Therefore $\overline{A} \subset E + t\overline{S}$ for all $t > 0$ and $\overline{A} \subset E$ since E is strongly closed in $E^{\star\star}$.

To prove (ii); note that from the definition of $\beta^\star(A)$, there exist $C \in K^\omega$ and $t > 0$ such that $A \subset C + t\overline{S}$. Then $A \subset \overline{co}(C) + t\overline{S}$ and $\overline{co}(C)$ is weakly compact by Theorem 1.2.11. Since $\overline{co}(C)$ is weakly compact and $t\overline{S}$ is weakly closed, the sum $\overline{co}(C) + t\overline{S}$ is weakly closed. Thus, $\overline{A}^\omega \subset \overline{co}(C) + t\overline{S}$ implies $\beta^\star(\overline{A}^\omega) \leqslant t$ and $\beta^\star(\overline{A}^\omega) \leqslant \beta^\star(A)$. The reverse inequality is obvious and therefore the proof is complete.

Let us state the following known result due to Radstrom.

Lemma 1.6.5. *Let* L, M, N *ne nonvoid subsets of* E. *Suppose that* M *is strongly closed and convex and* N *is bounded with* $L + N \subset M + N$. *Then* $L \subset M$.

Using Lemma 1.6.5, we can prove a result which characterizes the reflexivity of the space E in terms of $\beta^\star$.

Theorem 1.6.5. *Let* E *be a Banach space. Then*

(i) $\beta^\star(\overline{S}) = 0$, *if* E *is reflexive.*

(ii) $\beta^\star(\overline{S}) = 1$, *if* E *is not reflexive.*

Proof. It is known that E is reflexive if and only if $\overline{S}$ is weakly compact. To prove (i); let E be reflexive. Then $\overline{S}$ is weakly compact and trivially $\beta^\star(\overline{S}) = 0$. To show (ii); suppose that E is not reflexive. Since $\overline{S} \subset \{0\} + 1\overline{S}$, we have $\beta^\star(\overline{S}) = 1$. For the purpose of an indirect proof, suppose $\beta^\star(\overline{S}) < 1$. From the definition of $\beta^\star$, there exist $C \in K^\omega$ and $t > 0$ such that $\beta^\star(\overline{S}) < t < 1$ and $\overline{S} \subset C + t\overline{S}$. Thus, $\overline{S} \subset \overline{co}(C) + t\overline{S}$ and $(1-t)\overline{S} + t\overline{S} \subset \overline{co}(C) + t\overline{S}$. Since $\overline{co}(C)$

is strongly closed and convex, Lemma 1.6.5 implies that $(1-t)\overline{S} \subset \overline{co}(C)$. Therefore, by Theorem 1.2.11, $\overline{co}(C)$ is weakly compact. But $(1-t)\overline{S}$ being a weakly closed subset of the weakly compact set $\overline{co}(C)$ must also be weakly compact. Since the map $x \to (1-t)^{-1}x$ is weakly continuous, $\overline{S}$ is weakly compact and hence E is reflexive. This contradiction establishes that $\beta^\star(\overline{S}) = 1$ if E is not reflexive.

Note that statements (i) and (ii) of Theorem 1.6.2 remain valid if the closed unit ball $\overline{S}$ is replaced by the unit sphere S. Two more interesting properties of $\beta^\star$ are given in the following.

Theorem 1.6.6. *(a) Let $A \subset E$ be a bounded set. Then $\beta^\star(A + \delta\overline{S}) = \beta^\star(A) + \delta\beta^\star(\overline{S})$, $\delta \geqslant 0$;*

(b) Let $\{A_n\}_{n=1}^{\infty}$ be a sequence of nonvoid weakly closed subsets of E such that $A_1 \supset A_2 \supset ... \supset A_n ...$ and $\beta^\star(A_n) \to 0$ as $n \to \infty$. Suppose that A_1 is bounded. Then $\bigcap_{n=1}^{\infty}$ is nonempty.

Proof. By property (vi) of Theorem 1.6.4, it is clear that $\beta^\star(A + \delta(\overline{S}) \leqslant \beta^\star(A) + \delta\beta^\star(\overline{S})$. For the case when E is reflexive, $\beta^\star(\overline{S}) = 0$. Moreover, the boundedness of A and $A + t\overline{S}$ implies that they are weakly relatively compact and hence, $\beta^\star(A) = \beta^\star(A + \delta\overline{S}) = 0$. On the other hand, if E is not reflexive, we have to prove that $\beta^\star(A + \delta\overline{S}) \geqslant \beta^\star(A) + \delta$. It is easy to observe that $\delta \leqslant \beta^\star(A + \delta\overline{S})$, since otherwise, for any point $x \in A$, $A - \{x\} + \delta\overline{S} \supset \delta\overline{S}$ and

$$\delta > \beta^\star(A + \delta\overline{S}) = \beta^\star(A - \{x\} + \delta\overline{S}) \geqslant \beta^\star(\delta\overline{S}) = \delta\beta^\star(\overline{S}) = \delta$$

which is a contradiction. By the definition of $\beta^\star(A + \delta\overline{S})$, there exist $C \in K^\omega$ and $t > \beta^\star(A + \delta\overline{S})$ such that $A + \delta\overline{S} \subset C + t\overline{S}$. This means that $A + \delta\overline{S} \subset \overline{co}(C) + t\overline{S} = \overline{co}(C) + \delta\overline{S} + (t-\delta)\overline{S}$, where $\overline{co}(C) + (t-\delta)\overline{S}$ is strongly closed. Applying Lemma 1.6.1, we obtain that $A \subset \overline{co}(C) + (t-\delta)\overline{S}$. Since $\overline{co}(C)$ is weakly compact, $\beta^\star(A) \leqslant t - \delta$ which implies $\beta^\star(A) + \delta \leqslant \beta^\star(A + \delta\overline{S})$. This completes the proof of (a).

To prove (b); let $x_n \in A_n$, $n = 1, 2, ...,$. Then

$$\beta^\star(\bigcup_{n=1}^{\infty}\{x_n\}) = \beta^\star(\bigcup_{n=k}^{\infty}\{x_n\}) \leqslant \beta^\star(A_k).$$

Since $\lim_{n\to\infty} \beta^\star(A_n) = 0$, we have $\beta^\star(\bigcup_{n=1}^{\infty}\{x_n\}) = 0$ and $\overline{\left(\bigcup_{n=1}^{\infty} x_n\right)}^{w}$ is weakly compact. By Theorem 1.2.12, $\{x_n\}_{n=1}$ contains a subsequence which converges weakly to some point $x \in E$. Clearly $x \in A_n$, $n = 1, 2, ...,$. since all the sets A_n are weakly closed. Thus $x \in \bigcap_{n=1}^{\infty} A_n$, proving that $\bigcap_{n=1}^{\infty} A_n$ is nonempty.

1.7 Comparison Results

An important technique in the theory of differential equations is concerned with estimating a function satisfying a differential inequality by the extremal solutions of the corresponding differential equation. The results that are given in this section are the various types of comparison theorems which we employ several times in the text.

A basic comparison result is given in the following theorem.

Theorem 1.7.1. *Let Ω be an open (t,u)-set in R^2 and $g \in C[\Omega, R]$. Suppose that $[t_0, t_0 + a)$ is the largest interval in which the maximal solution $r(t, t_0, u_0)$ of*

$$u' = g(t,u), \; u(t_0) = u_0, \tag{1.7.1}$$

exists. Let $m \in C[[t_0, t_0 + a), R]$ be such that $(t, m(t)) \in \Omega$ for $t \in [t_0, t_0 + a)$, $m(t_0) \leq u_0$ and for a fixed Dini derivative

$$Dm(t) \leq g(t, m(t)), \; t \in [t_0, t_0 + a) - S,$$

S being an at most countable subset of $[t_0, t_0 + a)$. Then

$$m(t) \leq r(t, t_0, u_0), \; t \in [t_0, t_0 + a).$$

A useful convergence result in the context of comparison technique is the following.

Lemma 1.7.1. *Let $g \in C[\Omega, R]$ where Ω is an open set in R^2 and let $(t_0, u_0) \in \Omega$. Let $[t_0, t_0 + a)$ be the largest interval of existence of the maximal solution $r(t, t_0, u_0)$ of (1.7.1). Suppose further that $[t_0, T] \subset [t_0, t_0 + a)$. Then there is an $\varepsilon_0 > 0$ such that , for $0 < \varepsilon < \varepsilon_0$, the maximal solution $r(t, \varepsilon)$ of*

$$u' = g(t,u) + \varepsilon, \; u(t_0) = u_0 + \varepsilon,$$

exists on $[t_0, T]$ and

$$\lim_{\varepsilon \to 0} r(t, \varepsilon) = r(t, t_0, u_0)$$

uniformly on $[t_0, T]$.

The following theorem is a comparison result involving the estimation of a function which is allowed to be discontinuous.

Theorem 1.7.2. *Let $g \in C[R_+ \times R_+, R]$ and let $g(t,u)$ be nondecreasing in u for each $t \in R_+$. Suppose that $m(t) \geq 0$ is a right continuous function on $[t_0, t_0 + a) \subset R_+$ with isolated discontinuities at t_k, $k = 1, 2, \ldots$, $t_k > t_0$ such that*

$$|m(t_k) - m(t_k^-)| \leq \lambda_k,$$

where $\sum_{k=1}^{\infty} \lambda_k$ is convergent and

$$Dm(t) \leq g(t, m(t)), \; t \in [t_i, t_{i+1}), \; i = 0, 1, 2, \ldots.$$

Then $m(t_0) \le u_0$ implies

$$m(t) \le r(t, t_0, u_0 + \sum_{k=1}^{\infty} \lambda_k), \; t \in [t_0, t_0 + a),$$

where $r(t, t_0, u_0)$ is the maximal solution of (1.7.1) *existing on* $[t_0, t_0 + a)$.

Proof. By Theorem 1.7.1, we have

$$m(t) \le r_0(t, t_0, u_0), \; t \in [t_0, t_1)$$

where $r_0(t, t_0, u_0)$ is the maximal solution of (1.7.1). This implies that $m(t_1^-) \le r_0(t_1, t_0, u_0)$. Also,

$$m(t_1) \le m(t_1^-) + \lambda_1 \le r_0(t_1, t_0, u_0) + \lambda_1.$$

Hence, by Theorem 1.7.1, we get

$$m(t) \le r_1(t, t_1, u_1), \; t \in [t_1, t_2),$$

where $r_1(t, t_1, u_1)$ is the maximal solution of (1.7.1) passing through (t_1, u_1) such that $u_1 = r_0(t_1, t_0, u_0) + \lambda_1$. We now define a function $\rho_0(t)$ as follows:

$$\rho_0(t) = \begin{cases} r_0(t, t_0, u_0) + \lambda_1, \; t \in [t_0, t_1], \\ r_1(t, t_1, u_1), \; t \in [t_1, t_2). \end{cases}$$

Note that $\rho_0(t)$ is well defined. By the monotonicity of $g(t, u)$, we get

$$\begin{aligned} \rho_0'(t) &= r_0'(t, t_0, u_0) = g(t, r_0(t, t_0, u_0)) \\ &\le g(t, r_0(t, t_0, u_0) + \lambda_1) = g(t, \rho_0(t)), \; t \in [t_0, t_1] \end{aligned}$$

and

$$\rho_0'(t) = r_1'(t, t_1, u_1) = g(t, r_1(t, t_1, u_1)) = g(t, \rho_0(t)), \; t \in [t_1, t_2).$$

We therefore have, by applying Theorem 1.7.1,

$$\rho_0(t) \le r_0(t, t_0, u_0 + \lambda_1), \; t \in [t_0, t_2),$$

where $r_0(t, t_0, u_0 + \lambda_1)$ is the maximal solution of (1.7.1) through $(t_0, u_0 + \lambda_1)$. Clearly, $m(t) \le \rho_0(t) \le r_0(t, t_0, u_0 + \lambda_1)$, $t \in [t_0, t_2)$. Proceeding similarly and arguing as before, we obtain finally

$$m(t) \le r(t, t_0, u_0 + \sum_{k=1}^{\infty} \lambda_k), \; t \in [t_0, t_0 + a),$$

where $r(t, t_0, u_0)$ is the maximal solution of (1.7.1).

In some situations we need to employ a vector version of the Theorem 1.7.1. For this purpose we require the following notion which is essential for the existence of the maximal solution of a vector differential equation.

Let $f \in C[R^n, R^n]$. Then the function $f(u)$ is said to be *quasimonotone nondecreasing* in u, if for each $i = 1, 2, \ldots, n$, $f_i(u)$ is nondecreasing in u_j, $j = 1, 2, \ldots, n$, $j \ne i$. We shall use vectorial inequalities with the understanding that the same inequalities hold between their respective components.

Theorem 1.7.3. *Let $g \in C[[t_0, t_0+a) \times R^n_+, R^n]$ and let $g(t,u)$ be quasimonotone nondecreasing in u for each $t \in [t_0, t_0+a)$. Let $m \in C[[t_0, t_0+a), R^n_+]$ and*

$$Dm(t) \leq g(t, m(t))$$

for $t \in [t_0, t_0+a) - S$, where S is an at most countable subset of $[t_0, t_0+a)$. Then $m(t_0) \leq u_0$ implies

$$m(t) \leq r(t, t_0, u_0), \; t \in [t_0, t_0+a)$$

where $r(t, t_0, u_0)$ is the maximal solution of the vector differential equation

$$u' = g(t,u), \; u(t_0) = u_0,$$

existing on $[t_0, t_0+a)$.

While considering boundary value problems, the following comparison result is useful.

Theorem 1.7.4. *Suppose that*

(i) $u, v, z \in C^2[[0,1], R]$, $z(t) > 0$ for $t \in [0,1]$ and $g \in C[[0,1] \times R \times R, R]$;

(ii) for $t \in (0,1)$, $u'' \geq g(t,u,u')$, $B_\mu u(\mu) \leq b_\mu$, $\mu = 0,1$, and $v'' \leq g(t,v,v')$, $B_\mu v(\mu) \geq b_\mu$, $\mu = 0,1$, where $B_\mu u(\mu) \equiv \alpha_\mu u(\mu) + (-1)^{\mu+1}\beta_\mu u'(\mu)$, $\mu = 0,1$ and $\alpha_0, \alpha_1 \geq 0$, $\beta_0, \beta_1 > 0$;

(iii) for every $\lambda > 0$ and $t \in (0,1)$,

$$\lambda z'' < g(t, v+\lambda z, v'+\lambda z') - g(t,v,v'), \; B_0 z(0) > 0, \; B_1 z(1) > 0.$$

Then $u(t) \leq v(t)$, $t \in [0,1]$.

Proof. Suppose that the claim of the theorem is false. Then there exists a minimal $\lambda > 0$ such that

$$u(t) \leq v(t) + \lambda z(t), \; t \in [0,1].$$

Clearly, for some $t_0 \in [0,1]$, we have

$$u(t_0) = v(t_0) + \lambda z(t_0).$$

If $t_0 \in (0,1)$, we will also have

$$u'(t_0) = v'(t_0) + \lambda z'(t_0),$$

and

$$u''(t_0) \leq v''(t_0) + \lambda z''(t_0).$$

Using the assumptions (ii) and (iii), we get

$$\begin{aligned}-u''(t_0) \geq & - v''(t_0) - \lambda z''(t_0) \\ > & - v''(t_0) - g(t_0, v(t_0) + \lambda z(t_0), v'(t_0) + \lambda z'(t_0)) \\ & + g(t_0, v(t_0), v'(t_0)) \\ \geq & - g(t_0, v(t_0) + \lambda z(t_0), v'(t_0) + \lambda z'(t_0)) \\ = & - g(t_0, u(t_0), u'(t_0))\end{aligned}$$

which is a contradiction.

If $t_0 = 0$, we have

$$u'(0) \leq v'(0) + \lambda z'(0).$$

From the boundary conditions this implies

$$\lambda(\alpha_0 z(0) - \beta_0 z'(0)) \leq 0.$$

Since $\lambda > 0$, this contradicts $B_0 z(0) > 0$. Similarly, for $t_0 = 1$, we can arrive at a contradiction and this completes the proof.

In order to study differential equations with delay, we need the following results.

Theorem 1.7.5. *Let $g \in C[[t_0, t_0 + a) \times R_+, R]$ and $m \in C[[t_0 - \tau, t_0 + a), R_+]$, $\tau > 0$. Let S be a countable subset of $[t_0, t_0 + a]$ and for every $t_1 \in (t_0, t_0 + a)$, $t_1 \notin S$ for which*

$$\max_{-\tau \leq s \leq 0} m(t_1 + s) = m(t_1),$$

the differential inequality

$$D_- m(t_1) \leq g(t_1, m(t_1)) \ (\text{or } D^+ m(t_1) \leq g(t_1, m(t_1)))$$

hold. Then $\max_{-\tau \leq s \leq 0} m(t_0 + s) \leq u_0$ implies

$$m(t) \leq r(t, t_0, u_0), \ t \in [t_0, t_0 + a)$$

where $r(t, t_0, u_0)$ is the maximal solution of (1.7.1) existing on $[t_0, t_0 + a)$.

Let C $= C[[-\tau, 0], R]$ and let C$^+$ denote the set of nonnegative functions belonging to C. For any $m \in C[[t_0 - \tau, t_0 + a), R_+]$, let the symbol m_t, for $t \in [t_0, t_0 + a)$ denote the element of C$^+$, defined by $m_t(s) = m(t + s)$, $-\tau \leq s \leq 0$.

Theorem 1.7.6. *Let $m \in C[[t_0 - \tau, t_0 + a), R_+]$ and satisfy*

$$D_- m(t) \leq g(t, m(t), m_t), \ t \in (t_0, t_0 + a),$$

where $g \in C[[t_0, t_0 + a) \times R_+ \times C^+, R]$ and $g(t, u, \phi)$ is nondecreasing in ϕ for each (t, u). Let $r(t, t_0, \phi_0)$ be the maximal solution of

$$u' = g(t, u, u_t), \ u_{t_0} = \phi_0 \in C^+,$$

existing for $t \in [t_0, t_0 + a)$. Then $m_{t_0} \leq \phi_0$ implies

$$m(t) \leq r(t, t_0, \phi_0), \ t \in [t_0, t_0 + a).$$

1.8 Notes and Comments

The preliminary material presented in Section 1.2 is from standard books on Functional Analysis. For example see Hille and Philips [1], Barbu [1], Dieudonne [1] and Deimling [1]. Section 1.3, which contains the results on cones and their properties, is taken from Guo and Lakshmikantham [1]. Section 1.4 dealing with the directional derivative is from Mazur [1]. See also Lumer [1], for semi-inner products. For the contents of Section 1.5 relative to mean value theorems, see Mcleod [1], Dieudonne [1], Cramer, Lakshmikantham and Mitchell [1]. In Section 1.6, we consider measure of noncompactness which is taken from Kuratowskii [1], Chandra, Lakshmikantham and Mitchell [1]. See also De Blasi [1,2]. In Section 1.7, we have provided needed comparison rsults that are taken from Lakshmikantham and Leela [1,4].

Chapter 2

Fundamental Theory

2.1 Introduction

We begin in Section 2.2 introducing the strict and nonstrict inequalities relative to the cone and prove results concerning such inequalities. Flow invariance relative to a closed set in a Banach space is also investigated. Results are provided when the interior of the cone is nonempty, and empty in which case a notion of distance set is employed. Section 2.3 considers local existence theorems under suitable noncompactness conditions and Lyapunov like functions. The existence of maximal and minimal solutions of IVP is discussed in Section 2.4, relative to a cone. The existence and uniqueness of solutions of IVP utilizing a general uniqueness condition is discussed in Section 2.5. In Section 2.6, we consider the existence results in closed sets where a boundary condition is necessary. The global existence results are discussed in Section 2.7. using an estimate in terms of norm and a Lyapunov like function, showing the necessity of different proofs to get the same result. In section 2.8, we introduce the notion of Euler solutions, and in Section 2.9, we consider the flow invariance of a closed set employing the idea of nonsmooth analysis. Finally, Section 2.10 provides some notes and comments.

2.2 Differential Inequalities in Cones

Recall that a proper subset K of E, where E is a real Banach space, is said to be a cone if $\lambda K \subset K, \lambda \geq 0, K + K \subset K, K = \overline{K}$, the closure of K and $K \cap \{-K\} = \theta$, where θ denotes the null element of the Banach space E. Let K^0 denote the interior of K and assume that K^0 is nonempty. The cone K induces the order relations in E defined by

$$x \leq y \quad \text{iff} \quad y - x \in K, \qquad x < y \quad \text{iff} \quad y - x \in K^0.$$

Let K^* be the set of all continuous linear functionals C on E such that $Cx \geq 0$ for all $x \in K$ and K_0^* be the set of all continuous linear functionals on E

such that $Cx > 0$ for all $x \in K^0$. A function f from E into E is said to be quasimonotone nondecreasing if

$$x \le y \quad \text{and} \quad Cx = Cy \quad \text{for some} \quad C \in K_0^*, \quad \text{then} \quad Cf(x) \le Cf(y).$$

To prove the basic results on differential inequalities, we need the following result of Mazur which we merely state.

Lemma 2.2.1. *Let K be a cone with nonempty interior K^0. Then*

(i) $x \in K$ is equivalent to $Cx \ge 0$ for all $C \in K^*$ and

(ii) $x \in \partial K$ implies that there exists a $C \in K_0^*$ such that $Cx = 0$, where ∂K is the boundary of K.

We are now in a position to prove the following results.

Theorem 2.2.1. *Let K be a cone with nonempty interior K^0.*

Assume that

(i) $u, v \in C^1(R_+, E), f \in C(R_+ \times E, E)$ and $f(t,x)$ is quasimonotone nondecreasing for each $t \in R_+$;

(ii) $u'(t) \le f(t, u(t)), v'(t) \ge f(t, v(t)), t \in (t_0, \infty)$,

one of the inequalities being strict. Then, $u(t_0) < v(t_0)$ implies that $u(t) < v(t), t \ge t_0$.

Proof. Suppose that the assertion of the theorem is false. Then, there exists a $t_1 > t_0$ such that

$$v(t_1) - u(t_1) \in \partial K$$

and

$$v(t) - u(t) \in K^0, \quad t \in [t_0, t_1).$$

By Lemma 2.2.1, there exists a $C \in K_0^*$ such that $C(v(t_1) - u(t_1)) = 0$. Setting $m(t) = C(v(t) - u(t))$, we find that

$$m(t_1) = 0 \qquad \text{and} \qquad m(t) > 0 \qquad \text{for} \qquad t_0 \le t < t_1.$$

Consequently, we arrive at $m'(t_1) \le 0$ and at $t = t_1$, we have

$$u(t_1) \le v(t_1) \quad \text{and} \quad C(u(t_1)) = C(v(t_1)).$$

Now, using the quasimonotonocity of f and (ii), it follows that

$$\begin{aligned} m'(t_1) &= C(v'(t_1) - u'(t_1)) \\ &> C(f(t_1, v(t_1)) - f(t_1, u(t_1))) \ge 0, \end{aligned}$$

since one of the inequalities in (ii) is assumed to be strict. This is a contradiction which proves the stated result.

To prove the corresponding result when both inequalities are nonstrict, we need an extra assumption of one-sided uniqueness condition. This is the next result.

Theorem 2.2.2. *Let the assumptions of Theorem 2.2.1 hold with both inequalities in assumption (ii) being nonstrict. Suppose further that*

$$f(t,u) - f(t,v) \leq L(u-v), \quad L > 0,$$

whenever $u \geq v$, L *being a constant. Then* $u(t_0) \leq v(t_0)$ *implies* $u(t) \leq v(t), \quad t_0 \leq t < \infty$.

Proof. We shall reduce to Theorem 2.2.1 so that we can arrive at the stated result. We set

$$v_\epsilon(t) = v(t) + \epsilon \exp(2L(t - t_0)),$$

with $\epsilon = \frac{1}{n} y_0$, where $y_0 \in K^0$ and $\|y_0\| = 1$, so that we have

$$v_\epsilon(t_0) > v(t_0) \qquad \text{and} \qquad v_\epsilon(t) \geq v(t).$$

Now, using the one-sided Lipschitz condition, we get

$$\begin{aligned} f(t, v_\epsilon(t)) - f(t, v(t)) &\leq L(v_\epsilon(t) - v(t)) \\ &= L\epsilon \exp(2L(t - t_0)), \end{aligned}$$

which yields,

$$\begin{aligned} v_\epsilon'(t) &= v'(t) + 2\epsilon L \exp(2L(t - t_0)) \\ &\geq f(t, v(t)) + 2\epsilon L \exp(2L(t - t_0)) \\ &\geq f(t, v_\epsilon(t)) + \big[\exp(2L(t - t_0))\big](2\epsilon L - \epsilon L) \\ &> f(t, v_\epsilon(t)). \end{aligned}$$

Now, we apply Theorem 2.2.1 to $u(t), v_\epsilon(t)$, to get

$$u(t) < v_\epsilon(t), \qquad t_0 \leq t < \infty$$

which, as $\epsilon \to 0$, yields the desired inequality $u(t) \leq v(t), \quad t_0 \leq t < \infty$. The proof is complete.

As we have seen, the assumption that K^0 is non-empty is crucial for validity of Theorems 2.2.1 and 2.2.2. Since there are cones whose interior is empty such as cones consisting of nonnegative functions in L^P-spaces, it is necessary to prove differential inequalities results corresponding to Theorems 2.2.1, 2.2.2, in cones with empty interior. Since the proofs of such results are closely related to results on flow-invariance, we shall begin discussing flow-invariance of closed sets or cones.

Let $F \subset E$ be a closed set and let $f \in C(R_+ \times E, E)$. Consider the differential equation

$$x' = f(t, x), \qquad x(t_0) = x_0 \in F. \tag{2.2.1}$$

The set F is said to be *flow-invariant* with respect to f if every solution $x(t)$ of (2.2.1) on (t_0, ∞) is such that $x(t) \in F$ for $t_0 \leq t < \infty$. A set $A \subset E$ is called a *distance set* if to each $x \in E$ there corresponds a point $y \in A$ such that $d(x, A) = \|x - y\|$. A function $g \in C(R_+ \times R_+, R_+)$ is said to be a *uniqueness function* if the following holds: if $m \in C(R_+, R_+)$ is such that $m(t_0) \leq 0$ and $D^+m(t) \leq g(t, m(t))$, whenever $m(t) > 0$, then $m(t) \leq 0$ for $t_0 \leq t < \infty$.

We are now in a position to prove the following result on flow invariance of F.

Theorem 2.2.3. *Let $F \subset E$ be closed and a distance set. Suppose further that*

(i) $\lim_{h \to 0} \frac{1}{h} d(x + hf(t, x), F) = 0, \qquad t \in R_+, \quad x \in \partial F;$

(ii) $\|f(t, x) - f(t, y)\| \leq g(t, \|x - y\|), \quad x \in E - F, \quad y \in \partial F$, *where g is a uniqueness function. Then F is flow invariant with respect to f.*

Proof. Let $x(t)$ be a solution of (2.2.1). Assume that $x(t) \in F$ for $t_0 \leq t < t_0 + a$, where $t_0 + a < \infty$ is maximal, that is, $x(t)$ leaves the set F at $t = t_0 + a$ for the first time. Let $x(t_1) \notin F, \quad t_1 \in (t_0 + a, \infty)$ and let $y_0 \in \partial F$ be such that $d(x(t_1), F) = \|x(t_1) - y_0\|$. Set, for $t \in [t_0, \infty)$,

$$m(t) = d(x(t), F),$$

and

$$v(t) = \|x(t) - y_0\|.$$

For $h > 0$ sufficiently small, we have, letting $x = x(t_1)$,

$$\begin{aligned} m(t_1, h) &\leq \|x(t_1 + h) - y_0 - hf(t_1, y_0)\| + d(y_0 + hf(t_1, y_0), F) \\ &\leq \|x + hf(t_1, x) - y_0 - hf(t_1, y_0)\| + \|x(t_1 + h) \\ &\quad - x - hf(t_1, x)\| + d(y_0 + hf(t_1, y_0), F) \\ &\leq \|x - y_0\| + hg(t_1, \|x - y_0\|) + o(h). \end{aligned}$$

Since, $m(t_1) = v(t_1) > 0$, we obtain

$$D^+m(t_1) \leq g(t_1, m(t_1)).$$

This implies, in view of the facts g is a uniqueness function and $m(t_0) = 0$, that $m(t) \leq 0, \quad t_0 \leq t < \infty$. This contradicts $d(x(t_1), F) = m(t_1) > 0$. The proof is complete. □

Remark 2.2.1. Observe that Theorem 2.2.3 is true when $F = K$, K a cone. Although K is not assumed to have nonempty interior, Theorem 2.2.3 requires that K must be distance set. This is, however, a weaker assumption because the cones in L^P-spaces are distance sets whose interior is empty. Note also that every closed convex set in a reflexive Banach space is a distance set.

The requirement in Theorem 2.2.3 that F be a distance set can be dispensed with if the boundary condition (i) holds locally uniformly, that is, for every $(t, \overline{x}) \in (t_0, \infty) \times \partial F$ there exists a $\delta = \delta(t, \overline{x}) > 0$ such that

$$\lim_{h \to 0} \frac{1}{h} d(x + hf(t, x), F) = 0$$

uniformly for $\|x - \overline{x}\| < \delta$ and $x \in \partial F$. In this case, it is possible to choose a minimal sequence $\{x_n\}$ in F such that $\|x(t_1) - x_n\|$ approaches $d(x(t_1), F)$ as $n \to \infty$. This observation leads to the following.

Corollary 2.2.1. *Let the assumptions of Theorem 2.2.3 hold except that F is not a distance set and the boundary condition (i) holds locally uniformly. Then F is flow invariant relative to f.*

We shall next consider an interesting result which shows the equivalence of the quasimonotonicity property and the boundary condition.

If $f \in C(D, E)$, D being a subset of E, then

(a) the quasimonotonic nondecreasing property of f relative to the cone K means that whenever $y, z \in D, y \leq z$ and $\phi(z - y) = 0$ for some $\phi \in K_0^*$, we have $\phi(f(z) - f(y)) \geq 0$;

(b) the corresponding boundary condition takes the form

$$\lim_{h \to 0^+} \frac{1}{h} d(z - y + h(f(z) - f(y)), K) = 0, \quad y, z \in D, \quad z - y \in K.$$

Setting $x = z - y, \quad g(x) = f(x + y) - f(y)$ and noting that $x \in D \cap K$, (a) and (b) can be expressed as

(P_1) $x \in K \cap D, \quad \phi x = 0$ for some $\phi \in K_0^*$ implies $\phi(g(x)) \geq 0$;

(P_2) $\displaystyle\lim_{h \to 0^+} \frac{1}{h} d(x + hg(x), K) = 0, x \in K \cap D,$

respectively. The following result shows that (P_1) and (P_2) are equivalent, which implies that the quasimonotonicity property and the boundary condition are equivalent.

Theorem 2.2.4. *Let $g \in C(D, E)$, D being a subset of E, then properties (P_1) and (P_2) are equivalent.*

Proof. We shall first prove that (P_2) implies (P_1). Let $x \in K$ and $\phi x = 0$ for some $\phi \in K_0^*$. Because of (P_2), there exists, for each $h > 0$, a $z_h \in K$ such that

$$\|x + hg(x) - z_h\| = o(h)$$

and hence

$$\phi(x + hg(x) - z_h) = o(h).$$

Since

$$\begin{aligned}\phi(g(x)) = \frac{1}{h}\phi(x + hg(x)) &= \frac{1}{h}\phi(x + hg(x) - z_h) + \frac{1}{h}\phi(z_h) \\ &= o(1) + \frac{1}{h}\phi(z_h),\end{aligned}$$

and $\phi(z_h) \geq 0$, we obtain the desired inequality $\phi(g(x)) \geq 0$ by taking the limit as $h \to 0^+$.

To show (P_1) implies (P_2); let us suppose that (P_2) does not hold. This means that

$$d(x + h_n g(x), K) \geq h_n \delta \tag{2.2.2}$$

where $\delta > 0$ and $h_n \to 0$ as $n \to \infty$. Let A be the set

$$A = \{y \in E : \|y - (x + hg(x))\| < h\delta \quad \text{for some} \quad h > 0\}.$$

We shall show that A is open, convex and that $A \cap K = \theta$. It is clear that A is open. To prove that A is convex; let $x_1, x_2 \in A$ and $y = \alpha x_1 + \beta x_2$ where $\alpha, \beta > 0$ with $\alpha + \beta = 1$. From the inequalities

$$\|x_i - x - h_i g(x)\| < h_i \delta, \qquad i = 1, 2,$$

we have

$$\|y - x - (\alpha h_1 + \beta h_2) g(x)\| < (\alpha h_1 + \beta h_2)\delta.$$

Hence $y \in A$ and the convexity of the set A is proved. To show that $A \cap K = \theta$, assume that $y \in A \cap K$, $\|y - x - hg(x)\| < h\delta$, where $h > 0$. We can also write $y = x + hz$ where $\|z - g(x)\| < \delta$. Fixing an h_n such that $0 < h_n < h$, consider the point $y_n = x + h_n z$. The point $y_n \in K$ because it is on the line connecting x and y and K is convex. It is also in A, since $\|y_n - x - h_n g(x)\| < h_n \delta$. But this contradicts (2.2.2).

We have thus far shown that the set A is open, convex and $A \cap K = \theta$. By Theorem 1.2.11, there is a hyperplane separating K and A, that is, there exists a $\phi \in K^*$ such that

$$\phi(K) \geq \alpha, \quad \phi(A) < \alpha.$$

Since x can be approximated by points from A, we have $\phi x = \alpha$, and since $\lambda x \in K$ for $\lambda \geq 0$, it follows that

$$\lambda\alpha = \phi(\lambda x) \geq \alpha \qquad \text{for all} \qquad \lambda \geq 0.$$

This implies that $\alpha = 0$ and $\phi \in K_0^*$. The point $x + g(x)$ being in A, we get

$$\phi(g(x)) = \phi(x + g(x)) < 0,$$

which contradicts (P_1). The proof is therefore complete. □

We can now prove a differential inequality result without requiring that K^0 be nonempty.

Theorem 2.2.5. *Let K be a cone in E. Assume that*

(i) $u, v \in C^1(R_+, E), f \in C(R_+ \times E, E)$ and $f(t, x)$ is quasimonotone nondecreasing in x relative to K, for each $t \in R_+$;

(ii) $u'(t) - f(t, u(t)) \leq v'(t) - f(t, v(t)), \quad t \in (t_0, \infty)$;

(iii) $\|f(t, x) - f(t, y)\|\| \leq g(t, |x - y|), \quad x \in E - K, \quad y \in \partial K$, *where* g *is a uniqueness function;*

(iv) K *is a distance set.*

Then $u(t_0) \leq v(t_0)$ *implies that* $u(t) \leq v(t), \quad t \geq t_0$.

Proof. The main idea is to reduce the theorem to Theorem 2.2.3. We note that $m(t) = v(t) - u(t)$ satisfies

$$m'(t) = H(t, m(t)), \qquad t \in (t_0, \infty), \quad m(t_0) \geq 0,$$

where

$$H(t, w) = f(t, u(t) + w) - f(t, u(t)) + q(t),$$

$$q(t) = v'(t) - f(t, v(t)) - u'(t) + f(t, u(t)) \geq 0.$$

It is enough to show that the cone K is flow invariant with respect to H, since flow invariance of K implies that $m(t) \in K, \quad t \geq t_0$, which is equivalent to the conclusion of the theorem.

It is easy to see that $H(t, w)$ satisfies (iii). Let $w \in K$ and $\phi w = 0$ for some $\phi \in K_0^*$. Then

$$\phi(H(t, w)) = \phi(q(t)) + \phi(f(t, u(t) + w) - f(t, u(t))) \geq 0$$

since $\phi(q(t)) \geq 0$ and f is quasimonotone nondecreasing. This shows that H satisfies (P_1). By Theorem 2.2.4 (P_1) is equivalent to (P_2) and hence H satisfies (P_2). Now, applying Theorem 2.2.3, we get $m(t) \in K, \quad t \geq t_0$, and the proof is complete.

2.3 Local Existence

Consider the IVP for differential system

$$x' = f(t, x), \quad x(t_0) = x_0, t_0 \in R_+, \tag{2.3.1}$$

where $f \in C(R_0, E)$ and $R_0 = \{(t, x) \in [t_0, t_0 + a] \times B[x_0, b]\}$, $B[x_0, b] = \{x \in E : \|x - x_0\| \leq b\}$. We also need to consider the scalar IVP

$$u' = g(t, u), \quad u(t_0) = u_0 \geq 0, \tag{2.3.2}$$

where $g \in C([t_0, t_0 + a] \times [0, 2b], R)$.

We shall begin by proving the following local existence result parallel to Peano's theorem. To prove such a local existence result, just continuity of f is not sufficient and we need to impose some kind of compactness condition in terms of measure of noncompactness α.

Suppose that f maps bounded sets into relatively compact sets. In this case, we say that f is compact and we have $\alpha(f(t, A)) = 0$. Suppose that f admits a splitting $f = f_1 + f_2$ where f_1 is compact and f_2 is Lipschitzian.

Then we have

$$\alpha(f(t, A)) \leq \alpha(f_1(t, A) + f_2(t, A)) \leq \alpha(f_2(t, A)) \leq L\alpha(A),$$

where L is the Lipschitzian constant for f_2.

A local existence theorem employing a general compactness type condition is the following.

Theorem 2.3.1. *Assume that*

(i) $f \in C(R_0, E)$, $\quad \|f(t, x)\| \leq M$ *on* R_0 *and* $\alpha = \min(a, \dfrac{b}{M+1})$,

(ii) (a) $\alpha(f(t, A)) \leq g(t, \alpha(A))$, *for every bounded set* $A \subset B[x_0, b]$,

(b) f *is uniformly continuous on* R_0;

(iii) $g \in C([t_0, t_0 + a] \times [0, 2b], R)$, $g(t, 0) \equiv 0$ *and* $u(t) \equiv 0$ *is the unique solution of* (2.3.1).

Then the Cauchy problem (2.3.1) *has a solution on* $[t_0, t_0 + \alpha]$.

Before we proceed to prove this theorem, we have to consider the following results concerning approximate solutions.

Theorem 2.3.2. *Assume that* $f \in C(R_0, E)$, $\|f(t, x)\| \leq M$ *on* R_0 *and* $\alpha = \min(a, \dfrac{b}{M+1})$. *Let* $\{\epsilon_n\}$ *be a sequence such that* $0 \leq \epsilon_n \leq 1$, $\lim_{n \to \infty} \epsilon_n = 0$. *Then for each positive integer n, the Cauchy problem (2.3.1) has an* ϵ_n*-approximate solution* $x_n(t)$ *on* $[t_0, t_0 + \alpha]$ *satisfying (i)* $x_n(t)$ *is continuously differentiable on* $[t_0, t_0 + \alpha]$ *and (ii)* $\|x_n'(t) - f(t, x_n(t))\| \leq \epsilon_n$ *for* $t_0 \leq t \leq t_0 + \alpha$.

To prove this theorem, we first need the result which shows that a continuous function $F(x)$ can be approximated by locally Lipschitzian functions.

Lemma 2.3.1. *Let* $F \in C(\Omega, E)$, *where* $\Omega \subset E$ *is open. Then, for each* $\epsilon > 0$, *there exists a locally Lipschitzian function* $F_\epsilon(x) : \Omega \to E$ *such that* $\|F(x) - F_\epsilon(x)\| < \epsilon$ *on* Ω.

Proof. Let $N(x, \epsilon) = \{y \in \Omega : \|F(x) - F(y)\| < \frac{\epsilon}{2}\}$. Then $\Omega = \bigcup_{x \in \Omega} N(x, \epsilon)$. Since any metric space is paracompact, there exists a locally finite refinement $\{Q_\lambda : \lambda \in \Lambda\}$ of $\{N(x, \epsilon) : x \in \Omega\}$, where each Q_λ is nonempty and open, that is, an open cover of Ω has a neighborhood $N(x)$ with $N(x) \bigcap Q_\lambda$ nonempty for only finitely many $\lambda \in \Lambda$, and for each $\lambda \in \Lambda$ there exists a $x \in \Omega$ with $x \in Q_\lambda \bigcap N(x, \epsilon)$. Define the functions $\mu_\lambda : E \to R_+$ and $\phi_\lambda : E \to [0, 1]$ by

$$\mu_\lambda(\boldsymbol{x}) = \begin{cases} 0 & \text{if} \quad x \notin Q_\lambda, \\ d(x, \partial Q_\lambda) & \text{if} \quad x \in Q_\lambda, \end{cases}$$

where $d(x, A) = \inf_{y \in A} \|x - y\|$ and ∂Q_λ is the boundary of Q_λ;

$$\phi_\lambda(x) = \mu_\lambda(x) \Big[\sum_{\beta \in \Lambda} \mu_\beta(x) \Big]^{-1}.$$

Clearly $\mu_\lambda(x)$ is Lipschitzian with the constant 1 and, because of the fact that $\{Q_\lambda\}$ is locally finite, each $\phi_\lambda(x)$ is well defined and locally Lipschitzian. Let $\{x_\lambda\}$ be a set of points such that $x_\lambda \in Q_\lambda$ for each λ. Define $F_\epsilon : \Omega \to E$ by

$$F_\epsilon(x) = \sum_{\lambda \in \Lambda} \phi_\lambda(x) F(x_\lambda).$$

Then F_ϵ is locally Lipschitzian. Moreover, for $x \in \Omega$,

$$\|F(x) - F_\epsilon(x)\| = \| \sum_{\lambda \in \Lambda} \phi_\lambda(x) [F(x) - F(x_\lambda)] \| \leq \sum_{\lambda \in \Lambda} \phi_\lambda(x) \|F(x) - F(x_\lambda)\|.$$

Suppose that $\phi_\lambda(x) = 0$. Then $x \in Q_\lambda \subset N(x_0, \epsilon)$ for some x_0 and $x_\lambda \in Q_\lambda$.
Consequently, $\|F(x) - F(x_\lambda)\| < \epsilon$ and hence,

$$\|F(x) - F_\epsilon(x)\| \leq \epsilon \sum_{\lambda \in \Lambda} \phi_\lambda(x) = \epsilon.$$

Proof of Theorem 2.3.2. By Dugundji's extension theorem (Theorem 1.2.8), $f(t, x)$ has a continuous extension $\tilde{f} : R \times E \to E$ such that $\|\tilde{f}(t, x)\| \leq M$ on $R \times E$. By Lemma 2.3.1, there exist, for every $0 < \epsilon_n < 1$, a function $\tilde{f}_{\epsilon_n} : R \times E \to E$ which is locally lipschitzian in x satisfying

$$\|\tilde{f}_{\epsilon_n}(t, x) - \tilde{f}(t, x)\| \leq \epsilon_n.$$

In particular, we have $\|\tilde{f}_{\epsilon_n}(t, x) - f(t, x)\| \leq \epsilon_n$ and $\|\tilde{f}_{\epsilon_n}(t, x)\| \leq M + 1$ on R_0. Let $x_n(t)$ be the unique solution of

$$x' = \tilde{f}_{\epsilon_n}(t, x), \qquad x(t_0) = x_0,$$

which exists on $[t_0, t_0 + \alpha]$, $\alpha = \min(a, \dfrac{b}{M+1})$. Hence, we have

$$\|x_n'(t) - f(t, x_n(t))\| \leq \|\tilde{f}_{\epsilon_n}(t, x_n(t)) - f(t, x_n(t))\| \leq \epsilon_n,$$

for $t_0 \leq t \leq t_0 + \alpha$. The theorem is proved.
Proof of Theorem 2.3.1. By Theorem 2.3.2, we have approximate solution $\{x_n(t)\}$ on $[t_0, t_0 + \alpha]$ such that

$$x_n'(t) = f(t, x_n(t)) + y_n(t), \qquad x_n(t_0) = x_0$$

and $\|y_n(t)\| \leq \epsilon_n$, where $\epsilon_n \to 0$ as $n \to \infty$. Since $\{x_n(t)\}$ is equicontinuous and uniformally bounded, it is enough to show, to appeal to Ascoli-Arzela's theorem (Theorem 1.2.5), that the set $\{x_n(t)\} = \{x_n(t) : n \geq 1\}$ is relatively compact, that is, $\alpha(\{x_n(t)\}) \equiv 0$ on $[t_0, t_0+\alpha]$. Then, using standard arguments, it is easy to show that limit of the uniform convergent subsequence of $\{x_n(t)\}$ is a solution of (2.3.1) on $[t_0, t_0+\alpha]$.

Let $m(t) = \alpha(B_k(t))$ where $B_k(t) = \{x_n(t) : n \geq k\}$ and note that $m(t_0) = 0$. Also $m(t)$ is continuous, because, by property (vi) of Theorem 1.6.1, we have

$$|m(t) - m(s)| \leq \alpha(\{x_n(t) - x_n(s) : n \geq k\}) \leq 2(M+1)|t-s|.$$

We shall show that $D\_m(t) \leq g(t, m(t))$ on $[t_0, t_0+\alpha]$, where

$$D\_m(t) = \liminf_{h \to 0^+} \frac{1}{h}[m(t) - m(t-h)].$$

By (vi) of Theorem 1.6.1, we obtain

$$\frac{1}{h}[m(t) - m(t-h)] \leq \alpha\Big\{\frac{x_n(t) - x_n(t-h) : n \geq k}{h}\Big\}.$$

This implies, because of the mean value theorem 1.5.1 and property (viii) of Theorem 1.6.1, that

$$D\_m(t) \leq \liminf_{h \to 0^+} \alpha(B'_k(J_h)),$$

where $J_h = [t-h, t]$, $B'_k(J_h) = \bigcup_{t \in J_h} B'_k(t)$ and $B'_k(t) = \{x'_n(t) : n \geq k\}$. Also,

$$\alpha(B'_k(J_h)) \leq \alpha(f(J_h, B_k(J_h)) + \{y_n(J_h)\}) \leq \alpha(f(J_h, B_k(J_h)) + 2\epsilon_k.$$

The equicontinuity of $\{x_n(t)\}$ and the uniform continuity of f imply that

$$f(J_h, B_k(J_h)) \to f(t, B_k(t)),$$

with respect to the Hausdorff metric. We therefore obtain, in view of the assumption (ii),

$$D\_m(t) \leq g(t, m(t)) + 2\epsilon_k, \qquad t \in [t_0, t_0+\alpha].$$

By Theorem 1.7.1 we then have

$$\alpha(\{x_n(t) : n \geq 1\}) = m(t) \leq r_k(t, t_0, 0), \qquad t \in [t_0, t_0+\alpha],$$

where $r_k(t, t_0, 0)$ is the maximal solution of $u' = g(t, u) + 2\epsilon_k$, $u(t_0) = 0$. By Lemma 1.7.1 and (iii) we have $\lim_{k \to \infty} r_k(t, t_0, 0) = r(t, t_0, 0)$, where $r(t, t_0, 0)$ is the maximal solution of (2.3.2). Thus $\alpha(\{x_n(t)\}) \equiv 0$ on $[t_0, t_0+\alpha]$, proving the theorem.

We can also employ a Lyapunov-like function in this context. Let $\Omega = \{A : A \subset B[x_0, b]\}$ and let $V : [t_0, t_0+a] \times \Omega \to R_+$ be such that V is continuous in t, $V(t, A) \equiv 0$ if and only if $\bar{A}$ is compact and for $t \in [t_0, t_0+a]$, $A, B \in \Omega$,

$$|V(t, A) - V(t, B)| \leq L|\alpha(A) - \alpha(B)|.$$

Define

$$D\_V(t, A) = \liminf_{h \to 0^+} \frac{1}{h}[V(t, A) - V(t-h, A_h(f))]$$

where $A_h(f) = \{x - hf(t, x) : x \in A\}$.

Theorem 2.3.3. . *Assume that the hypotheses (i) and (iii) of Theorem 2.3.1 are valid. Suppose, instead of (ii)(a), that a Lyapunov-like function V, as defined above, exists and satisfies*

$$D_-V(t,A) \leq g(t, V(t,A)) \tag{2.3.3}$$

on $[t_0, t_0+a] \times \Omega$. Then the conclusion of Theorem 2.3.1 remains true.

Proof. Let $x_n(t)$ be the sequence of approximate solutions given by Theorem 2.3.2. As observed in the proof of the Theorem 2.3.1, it is enough to prove $\alpha(x_n(t)) \equiv 0$. In this case we set

$$m(t) = V(t, x_n(t)),$$

and see that $m(t_0) = 0$. In view of the properties of V, it is clear that $m(t)$ is continuous. For small $h > 0$, we have

$$m(t) - m(t-h) \leq V(t, \{x_n(t)\}) - V(t-h, \{x_n(t) - hf(t, x_n(t))\})$$
$$+L|\alpha(\{x_n(t) - hf(t, x_n(t))\}) - \alpha(\{x_n(t-h)\})|.$$

This, together with (vi) of Theorem 1.6.1 and (2.3.3), yields

$$D_-m(t) \leq g(t, m(t)) + L \liminf_{h \to 0^+} \frac{1}{h} \alpha[(\{x_n(t) - hf(t, x_n(t)) - x_n(t-h)\}]. \tag{2.3.4}$$

Now,

$$\|x_n(t) - hf(t, x_n(t)) - x_n(t-h)\| \leq \int_{t-h}^{t} \|x_n'(s) - f(t, x_n(t)\| ds$$
$$\leq \int_{t-h}^{t} \|x_n'(s) - f(s, x_n(s))\| ds + \int_{t-h}^{t} \|f(s, x_n(s)) - f(t, x_n(t)\| ds.$$

By uniform continuity of f, given $\epsilon > 0$ there exists a $\delta = \delta(\epsilon) > 0$ such that $\|f(t_1, x) - f(t_2, y)\| \leq \frac{\epsilon}{4}$ if $\|t_1 - t_2\| < \delta$ and $\|x - y\| < \delta$. Since $\|x_n(t) - x_n(s)\| \leq (M+1)|t-s|$, we have $\|x_n(t) - x_n(s)\| < \delta$ if $|t-s| < \frac{\delta}{M+1}$. Thus, if h is sufficiently small,

$$\int_{t-h}^{t} \|f(s, x_n(s)) - f(t, x_n(t))\| ds < \frac{\epsilon h}{4}.$$

Also, $\int_{t-h}^{t} \|x_n'(s) - f(s, x_n(s))\| ds \leq \epsilon_n h < \frac{\epsilon h}{4}$, if n is sufficiently large. Consequently, we get

$$\frac{L}{h} \alpha(\{x_n(t) - hf(t, x_n(t) - x_n(t-h)\}) \leq \frac{2L}{h}(\frac{\epsilon h}{4} + \frac{\epsilon h}{4}) = L\epsilon.$$

Since ϵ is arbitrary, the relation (2.3.4) gives

$$D_-m(t) \leq g(t, m(t))$$

on $[t_0, t_0 + \alpha]$, which implies, by Theorem 1.7.1, that

$$V(t, \{x_n(t)\}) = m(t) \leq r(t, t_0, 0), \qquad t \in [t_0, t_0 + \alpha], \tag{2.3.5}$$

where $r(t, t_0, 0)$ is the maximal solution of (2.3.2). Since $r(t, t_0, 0) \equiv 0$ by assumption (iii) and $V(t, A) \equiv 0$ iff $\bar{A}$ is compact, the relation (2.3.5) proves the Theorem.

We note the proof of Theorem 2.3.3 is somewhat different from the proof of Theorem 2.3.1. If $V(t, A) = \alpha(A)$, then the condition (2.3.3) reduces to

$$D_-V(t, A) = \liminf_{h \to 0^+} \frac{1}{h}[\alpha(A) - \alpha(A_h(f))] \leq g(t, \alpha(A)),$$

where, as before, $A_h(f) = \{x - hf(t, x) : x \in A\}$. It is easy to see that this condition is weaker than the condition assumed in Theorem 2.3.1, namely,

$$\alpha(f(t, A)) \leq g(t, \alpha(A)).$$

The uniform continuity assumption on f in Theorem 2.3.1, can be dispensed with if condition (ii)(a) is improved to

(ii*) $\qquad \alpha(f(I, A)) \leq g(\alpha(A)),$

where $I = [t_0, t_0 + a]$ and A is any bounded subset of $B[x_0, b]$.

Theorem 2.3.4. *Theorem 2.3.1 remains valid if the condition (ii) is replaced by (ii*).*

Proof. We follow the proof of Theorem 2.3.1.

By (ii*), we obtain

$$\alpha(f(J_h, B_k(J_h))) \leq \alpha(f(I \times B_k(J_h))) \leq g(\alpha(B_k(J_h))).$$

Noting $g(\alpha(B_k(J_h))) \to g(\alpha(B_k(t)))$ as $h \to 0^+$, we get

$$D_-m(t) \leq g(m(t)) + 2\epsilon_k, \qquad t \in [t_0, t_0 + \alpha].$$

The rest of the proof proceeds as before and thus, the theorem is proved.

Remark 2.3.1 Theorem 2.3.1 is true if condition (ii) is replaced by

$$\lim_{h \to 0^+} \alpha(f(J_h, A)) \leq g(t, \alpha(A))$$

for every $A \subset B[x_0, b]$.

2.4 Existence of Extremal Solutions

Having local existence results given in Sec.2.3 and strict differential inequalities result Theorem 2.2.1, one can prove the existence of extremal solutions relative to a cone K. We shall utilize a special case of Theorem 2.2.3 which we give below for convenience and employ Theorem 2.2.1 to obtain existence of extremal solutions. We consider only the existence of maximal solution relative to a cone K and the proof for the existence of minimal solutions can be given by following similar arguments. However, utilizing the notion of maximal solution, we shall prove a comparison result relative to a cone $K \subset E$, where E is a real Banach space. We shall also prove similar comparison results without using compactness type assumption but using a different set of conditions which are interesting in themselves.

Utilizing Theorem 2.2.1, we prove the existence of the maximal solution of

$$x' = f(t, x), \qquad x(t_0) = x_0, \tag{2.4.1}$$

relative to the cone K.

Theorem 2.4.1. *Let K be a cone with nonempty interior K^0. Suppose that*

(i) $f \in C(R_0, E)$ and $f(t, x)$ is quasimonotone nondecreasing in x for each $t \in [t_0, t_0 + a]$, where $R_0 = \{(t, x) : t \in [t_0, t_0 + a], \|x - x_0\| \leq b\}$;

(ii) f is uniformly continuous on R_0 (and hence we may assume that a, b are such that $\|f(t, x)\| \leq M$ on R_0);

(iii) $g \in C([t_0, t_0 + a] \times R_+, R)$ with $g(t, 0) \equiv 0$ and the only solution of the scalar differential equation

$$u' = g(t, u), \qquad u(t_0) = 0 \tag{2.4.2}$$

is the trivial solution;

(iv) $\alpha(A) - \alpha(\{x - hf(t, x) : x \in A\}) \leq hg(t, \alpha(A))$, for $h > 0$, $t \in [t_0, t_0 + a]$, where A is a bounded subset of $B[x_0, b]$ and α is the measure of non-compactness.

Then there exists a maximal solution of (2.4.1) relative to K on $[t_0, t_0 + \eta]$ where $\eta = \min(a, \dfrac{b}{M+1})$.

Proof. Let $y_0 \in K^0$ with $\|y_0\| = 1$. Consider the system

$$x' = f(t, x) + \frac{1}{n} y_0, \qquad x(t_0) = x_0 + \frac{1}{n} y_0, \tag{2.4.3}$$

for each positive integer n. We have

$$\|f(t, x) + \frac{1}{n} y_0\| \leq \|f(t, x)\| + \frac{1}{n} \leq M + 1.$$

Applying Theorem 2.3.3 with $V(t, A) = \alpha(A)$, we deduce that there exists a solution $x_n(t)$ of (2.4.3) for each n and a solution $x(t)$ of (2.4.1) on $[t_0, t_0+\eta]$. The equicontinuity of the family $\{x_n(t)\}$ follows easily. Noting that $\alpha(\{x_n(t_0)\}) = \alpha(\{x_0 + \frac{1}{n}y_0\}) = 0$, because of the fact $x_0 + \frac{1}{n}y_0$ converges to x_0, we can conclude, as in Theorem 2.3.3, that the set $\{x_n(t)\}$ is relatively compact for each $t \in [t_0, t_0 + \eta]$. We can then apply Ascoli's theorem to obtain a subsequence of $\{x_n(t)\}$ which converges uniformly to a continuous function $r(t)$ on $[t_0, t_0 + \eta]$. Using standard arguments it is easily seen that $r(t)$ is a solution of (2.4.1) on $[t_0, t_0 + \eta]$. Let $x(t)$ be any solution of (2.4.1) on $[t_0, t_0 + \eta]$.

Then

$$x'(t) - f(t, x(t)) = 0 < \frac{1}{n}y_0 = x_n'(t) - f(t, x_n(t))$$

and

$$x(t_0) = x_0 < x_0 + \frac{1}{n}y_0 = x_n(t_0).$$

Applying Theorem 2.2.1, we get

$$x(t) \leq x_n(t), \qquad t \in [t_0, t_0 + \eta]$$

and hence

$$x(t) \leq \lim_{n \to \infty} x_n(t) \equiv r(t), \qquad t \in [t_0, t_0 + \eta].$$

This shows that $r(t)$ is the desired maximal solution. The proof is complete.

Having established the existence of the maximal solution of (2.4.1) relative to K, it is now easy to prove a comparison result. Clearly, such a result generalizes well known comparison theorems in finite dimensional spaces.

Theorem 2.4.2. *Suppose that the assumptions of Theorem 2.4.1 are satisfied. Let $m \in C^1([t_0, t_0 + \eta], E)$ and*

$$m'(t) \leq f(t, m(t)), \qquad t \in [t_0, t_0 + \eta].$$

If $m(t_0) \leq x_0$, then $m(t) \leq r(t)$, $t \in [t_0, t_0 + \eta]$, where $r(t)$ is the maximal solution of (2.4.1).

Proof. Let $x_n(t)$ be a solution of (2.4.3) as in Theorem 2.4.1. Note that

$$x(t_0) = x_0 < x_0 + \frac{1}{n}y_0 = x_n(t_0)$$

and

$$m'(t) - f(t, m(t)) \leq 0 < \frac{1}{n}y_0 = x_n'(t) - f(t, x_n(t)),$$

for $t \in [t_0, t_0 + \eta]$. By Theorem 2.2.1,

$$m(t) < x_n(t), \text{ for each } n \text{ and } t \in [t_0, t_0 + \eta].$$

Hence, $m(t) \leq \lim_{\eta \to \infty} x_n(t) \equiv r(t), \qquad t \in [t_0, t_0 + \eta].$

Corollary 2.4.1. *Let the hypotheses of Theorem 2.4.1 hold and let* $f(t,0) \equiv 0$. *Then the maximal solution* $r(t)$ *of* (2.4.1) *such that* $r(t_0) = x_0 \in K$ *remains in* K *for* $t \in [t_0, t_0 + \eta]$.

The proof follows by choosing $m(t) \equiv 0$ in Theorem 2.4.2.

If the compactness type condition in Theorem 2.4.1 were strengthened, one could prove Theorems 2.4.1 and 2.4.2 without assuming the uniform continuity of f. One has to utilize the argument of Theorem 2.3.4. We give below such a result without proof.

Theorem 2.4.3. *Suppose that (i)* $f \in C(R_0, E)$, $\|f(t,x)\| \leq M$ *on* R_0 *and* $\eta = \min(a, \frac{b}{M+1})$; *(ii)* $\alpha(f(I \times A)) \leq g(\alpha(A))$ *where* $I = [t_0, t_0 + a]$ *and* A *is a bounded subset of* $B[x_0, b]$; *(iii)* $g \in C([0, 2b], R)$, $g(0) = 0$ *and* $u(t) \equiv 0$ *is the only solution of* (2.4.2); *(iv)* $f(t,x)$ *is quasimonotone nondecreasing in* x *for each* $t \in [t_0, t_0 + a]$. *Then there exists a maximal solution on* $[t_0, t_0 + \eta]$ *for the problem* (2.4.1) *relative to the cone* K, *provided the interior of* K *is non-empty.*

If, further, $m \in C^1([t_0, t_0+\eta], E)$ *and* $m'(t) \leq f(t, m(t))$, $t \in [t_0, t_0+\eta]$ *with* $m(t_0) \leq x_0$, *then* $m(t) \leq r(t)$ *on* $[t_0, t_0 + \eta]$, *where* $r(t)$ *is the maximal solution of* (2.4.1) *on* $[t_0, t_0 + \eta]$.

One could prove Theorems (2.4.1) and (2.4.2) under a different set of conditions without directly imposing compactness type conditions. Such results are useful in discussing later the method of quasilinearization. We proceed by proving a useful convergence result.

Theorem 2.4.4. *Assume that*

(i) $f \in C([t_0, t_0 + a] \times E, E)$ *and there exist constants* M *and* $b > 0$ *such that* $\|f(t,x)\| \leq M$ *on* R_0 *and* $Ma < b$, *where* $R_0 = \{(t,x) : t_0 \leq t \leq t_0 + a, \|x - x_0\| \leq b\}$;

(ii) $f_n \in C([t_0, t_0 + a] \times E, E)$ *for* $n \geq 1$ *and the sequence* $\{f_n\}$ *converges uniformly to* f *on* R_0;

(iii) for each $n \geq 1$, $x_n(t)$ *is a solution of* $x' = f_n(t,x)$, $x_n(t_0) = y_n$, *existing on* $[t_0, t_0 + a]$ *such that* $\lim_{n\to\infty} x_n(t) = x(t)$ *whenever* $\lim_{n\to\infty} y_n = x_0$.

Then the sequence $\{x_n(t)\}$ *converges uniformly to* $x(t)$ *and* $x(t)$ *is a solution of* (2.4.1).

Proof. By the uniform convergence of f_n to f, for $\epsilon = 1$ there exists $N > 0$ such that $\|f_n(t,x) - f(t,x)\| \leq 1$ for all $(t,x) \in R_0$ and $n \geq N$. Thus, $\|f_n(t,x)\| \leq 1+M$ for all $(t,x) \in R_0$ and $n \geq N$. For $\epsilon > 0$, let $\delta = \frac{\epsilon}{1+M}$. If $t, s \in [t_0, t_0+a]$ with $|t - s| < \delta$, then

$$\|x_n(t) - x_n(s)\| \leq |\int_s^t \|f_n(\tau, x_n(\tau)\| d\tau| \leq |t - s|(1 + M) < \epsilon.$$

Therefore the family $\{x_n(t)\}_{n=N}^{\infty}$ is equicontinuous. Using this fact and the pointwise convergence of $\{x_n(t)\}$ to $x(t)$, it is easily seen that $x(t)$ is continuous on $[t_0, t_0 + a]$. Thus, $x(t)$ is uniformly continuous on $[t_0, t_0 + a]$. The standard arguments show that $x(t)$ is a solution of (2.4.1).

Let us now consider, for each $n \geq 1$, the initial value problem (2.4.3), namely,

$$x_n' = f(t, x_n) + \frac{1}{n} y_0, \qquad x_n(t_0) = x_0 + \frac{1}{n} y_0,$$

where $y_0 \in K^0$ with $\|y_0\| = 1$.

For convenience we list below some needed hypotheses;

(H_1) $f(t,x)$ is quasimonotone nondecreasing in x for each $t \in [t_0, t_0 + a]$;

(H_2) for each $n \geq 1$, a solution of (2.4.3) exists on $[t_0, t_0 + a]$;

(H_3) a solution of (2.4.1) exists on $[t_0, t_0 + a]$;

(H_4) K is a regular cone in E, that is, every monotone bounded sequence has a limit;

(H_5) there exists a subsequence $\{x_{n_k}(t)\}$ of $\{x_n(t)\}$ which converges to a function $r(t)$.

The following theorem gives two sets of conditions guaranteeing the existence of the maximal solution of (2.4.1) and also a comparison result.

Theorem 2.4.5. *Suppose that K is a cone with nonempty interior and that assumptions (i) of theorem 2.4.4 hold. Assume further that either*

(a) (H_1), (H_2), (H_3), and (H_4) hold

or

(b) (H_1), (H_2), and (H_5) hold.

Then there exists a maximal solution of (2.4.1) on $[t_0, t_0 + a]$. Moreover, if either (a) or (b) holds, $y \in C([t_0, t_0 + a], E)$ and

$$y'(t) \leq f(t, y(t)), \qquad y(t_0) \leq x_0$$

then $y(t) \leq r(t, t_0, x_0)$ on $[t_0, t_0 + a]$, where $r(t, t_0, x_0)$ is the maximal solution of (2.4.1).

Proof.

(I) Let

$$x_n'(t) = f(t, x_n(t)) + \frac{1}{n} y_0$$

and

$$x_{n+1}'(t) = f(t, x_{n+1}(t)) + \frac{1}{n+1} y_0.$$

Then

$$x'_n(t) > f(t, x_n(t)) + \frac{1}{n+1} y_0$$

and

$$x'_{n+1}(t) \leq f(t, x_{n+1}(t)) + \frac{1}{n+1} y_0$$

with $x_n(t_0) > x_{n+1}(t_0)$. By Theorem 2.2.1 and the fact that $f(t, x) + \frac{1}{n} y_0$ is quasimonotone nondecreasing in x, $x_n(t) > x_{n+1}(t)$ for all $t \in [t_0, t_0 + a]$. Suppose that $x(t)$ is a solution of (2.4.1) on $[t_0, t_0 + a]$. Then

$$x'_n(t) = f(t, x_n(t)) + \frac{1}{n} y_0 > f(t, x_n(t))$$

and

$$x_n(t_0) = x_0 + \frac{1}{n} y_0 > x_0 = x(t_0).$$

As before we are lead to $x_n(t) > x(t)$ on $[t_0, t_0 + a]$ for $n \geq 1$. Hence $x_n(t)$ is a decreasing seqence that is bounded below and, by the regularity of K, $x_n(t)$ converges. Clearly $f(t, x_n) + \frac{1}{n} y_0$ converges uniformly to $f(t, x)$ on $[t_0, t_0 + a]$. By Theorem 2.4.4, $x_n(t)$ converges uniformly to a solution $r(t)$ of (2.4.1). But

$$r(t) = \lim_{n \to \infty} x_n(t) \geq \lim_{n \to \infty} x(t) = x(t)$$

which implies that $r(t, t_0, x_0)$ is the maximal solution of (2.4.1) on $[t_0, t_0 + a]$.

(II) The subsequence $\{x_{n_k}(t)\}$ converges to a solution of (2.4.1) by Theorem 2.4.4. That this solution is maximal follows directly as in the proof of part (I).

(III) Consider a solution $x_n(t) = x_n(t, t_0, x_0 + \frac{1}{n} y_0)$ of (2.4.3). Then $x'_n(t) > f(t, x_n(t))$ and $y'(t) \leq f(t, y(t))$ with $x_n(t_0) = x_0 + \frac{1}{n} y_0 > x_0 \geq y(t_0)$.

By Theorem 2.2.1, $y(t) < x_n(t)$. Thus taking the limit either of the sequence used in (I) or the subsequence used in (II), we obtain the result.

2.5 Existence and Uniqueness

In this section, we shall prove an existence and uniqueness theorem under an assumption more general than the Lipschitz condition.

Theorem 2.5.1. *Assume that*

(i) $f \in C[R_0, E]$ *where* $R_0 = [t_0, t_0 + a] \times B[x_0, b]$ *and* $\|f(t, x)\| \leq M_0$ *on* R_0*;*

(ii) $g \in C[[t_0, t_0 + a] \times [0, 2b], R_+]$, $0 \le g(t, u) \le M_1$ *on* $[t_0, t_0 + a] \times [0, 2b]$, $g(t, 0) \equiv 0$, $g(t, u)$ *is nondecreasing in* u *for each* t *and* $u \equiv 0$ *is the unique solution of the scalar differential equation*

$$u' = g(t, u), \qquad u(t_0) = 0 \tag{2.5.1}$$

on $[t_0, t_0 + a]$;

(iii) $\|f(t, x) - f(t, y)\| \le g(t, \|x - y\|)$ *on* R_0.

Then the successive approximation defined by

$$x_{n+1}(t) = x_0 + \int_{t_0}^{t} f(s, x_n(s))\mathrm{d}s, \qquad n = 0, 1, 2, \ldots, \tag{2.5.2}$$

exists on $[t_0, t_0 + \alpha]$, *where* $\alpha = \min(a, \frac{b}{M})$, $M = \max(M_0, M_1)$, *as continuous functions and converge uniformly to the unique solution* $x(t)$ *of the Cauchy problem* (2.3.1) *on* $[t_0, t_0 + \alpha]$.

Proof. It is easy to see, by induction, that the successive approximations (2.5.2) are defined and continuous on $[t_0, t_0 + \alpha]$ and

$$\|x_{n+1}(t) - x_0\| \le b, \qquad n = 0, 1, 2, \ldots.$$

We shall now define the successive approximations for problem (2.5.1) as follows:

$$\begin{cases} u_0(t) = M(t - t_0) \\ u_{n+1}(t) = \int_{t_0}^{t} g(s, u_n(s))\mathrm{d}s, \qquad t_0 \le t \le t_0 + \alpha. \end{cases} \tag{2.5.3}$$

An easy induction proves that the successive approximations (2.5.3) are well defined and satisfy

$$0 \le u_{n+1}(t) \le u_n(t), \qquad t_0 \le t \le t_0 + \alpha.$$

Since $|u_n'(t)| \le M_1$, we conclude by Ascoli-Arzela theorem and monotonicity of the sequence $\{u_n(t)\}$ that $\lim_{n\to\infty} u_n(t) = u(t)$ uniformly on $[t_0, t_0 + \alpha]$. It is also clear that $u(t)$ satisfies (2.5.1). Hence, by (ii), $u(t) \equiv 0$ on $[t_0, t_0 + \alpha]$. Now,

$$\|x_1(t) - x_0\| \le \int_{t_0}^{t} \|f(s, x_0)\|\mathrm{d}s \le M(t - t_0) \equiv u_0(t).$$

Assume that $\|x_k(t) - x_{k-1}(t)\| \le u_{k-1}(t)$ for a given k. Since

$$\|x_{k+}(t) - x_k(t)\| \le \int_{t_0}^{t} \|f(s, x_k(s)) - f(s, x_{k-1}(s))\|\mathrm{d}s,$$

using the nondecreasing character of $g(t, u)$ in u and the assumption (iii), we get

$$\|x_{k+1}(t) - x_k(t)\| \leq \int_{t_0}^{t} g(s, u_{k-1}(s))ds \equiv u_k(t).$$

Thus, by induction, the inequality

$$\|x_{n+1}(t) - x_n(t)\| \leq u_n(t), \qquad t_0 \leq t \leq t_0 + \alpha,$$

is true for all n. Also

$$\|x'_{n+1}(t) - x'_n(t)\| \leq \|f(t, x_n(t)) - f(t, x_{n-1}(t))\|$$
$$\leq g(t, \|x_n(t) - x_{n-1}(t)\|) \leq g(t, u_{n-1}(t)).$$

Let $n \leq m$. Then one can easily obtain

$$\|x'_n(t) - x'_m(t)\| \leq g(t, u_{n-1}(t)) + g(t, u_{m-1}(t)) + g(t, \|x_n(t) - x_m(t)\|).$$

Since $u_{n+1}(t) \leq u_n(t)$ for all n, it follows that

$$D^+\|x_n(t) - x_m(t)\| \leq g(t, \|x_n(t) - x_m(t)\|) + 2g(t, u_{n-1}(t)),$$

where D^+ is the Dini derivative. An application of Theorem 1.7.1 gives

$$\|x_n(t) - x_m(t)\| \leq r_n(t), \qquad t_0 \leq t \leq t_0 + \alpha,$$

where $r_n(t)$ is the maximal solution of

$$\nu' = g(t, \nu) + 2g(t, u_{n-1}(t)), \quad \nu_n(t_0) = 0,$$

for each n. Since, as $n \to \infty$, $2g(t, u_{n-1}(t)) \to 0$ uniformly on $[t_0, t_0 + \alpha]$, it follows by Lemma 1.7.1 that $r_n(t) \to 0$ uniformly on $[t_0, t_0 + \alpha]$. This implies that $x_n(t)$ converges uniformly to $x(t)$ and it is now easy to show that $x(t)$ is a solution of (2.3.1) by standard argument.

To show this solution is unique; let $y(t)$ be another solution of (2.3.1) existing on $[t_0, t_0 + \alpha]$. Define $m(t) = |x(t) - y(t)|$ and note that $m(t_0) = 0$.

Then

$$D^+m(t) \leq |x'(t) - y'(t)| = |f(t, x(t)) - f(t, y(t))| \leq g(t, m(t)),$$

using the assumption (iii). Again applying Theorem 1.7.1, we have

$$m(t) \leq r(t) \qquad t_0 \leq t \leq t_0 + \alpha,$$

where $r(t)$ is the maximal solution of (2.5.1). But, by assumption (ii), $r(t) \equiv 0$ and this proves that $x(t) \equiv y(t)$. Hence the limit of successive approximations is the unique solution of (2.3.1). The proof is complete.

2.6 Existence in Closed Sets

In the preceding sections we have considered the existence of solutions of (2.3.1) when the initial value x_0 is an interior point of the domain of f. Sometimes, one is interested in solutions when x_0 is on the boundary of the domain of f, and this necessitates the consideration of existence of solutions in closed sets.

Let F be a closed subset of a Banach space E and let $f \in C([t_0, t_0+\alpha] \times F, E)$. In order that the problem

$$x' = f(t, x), \qquad x(t_0) = x_0 \in F, \tag{2.6.1}$$

has a solution $x(t) \in F$ on some interval $[t_0, t_0 + \alpha]$, f must clearly satisfy

$$\lim_{h \to 0^+} \frac{1}{h} d(x + hf(t, x), F) = 0, \tag{2.6.2}$$

for $x \in F$ and $t \in [t_0, t_0 + \alpha]$ where $d(x, F)$ is the distance function. This is because, in that case, $x(t) = x_0 + (t - t_0)f(t_0, x_0) + o(t - t_0)$ as $t \to t_0$. We note that (2.6.2) is obviously satisfied at every interior point of F and hence (2.6.2) is a boundary condition. Having seen that a condition of the type (2.6.2) is essential to discuss the existence problem (2.6.1), we proceed to consider the same. However, we first observe that the approximate solutions constructed in Theorem 2.3.2 are of no use in this case, since all of them may lie in the exterior of F. It turns out that the polygonal approximate solutions are more appropriate for our discussion, since using the boundary condition (2.6.2) makes it possible to construct polygonal approximate solutions in such a way that at least the corners of the polygons are in F. Happily, this is good enough to solve the Cauchy problem (2.6.1). Accordingly, we shall first extend Theorem 2.3.2.

Let $F \subset E$ be a locally closed set, that is, for each $x_0 \in F$ there exists a $b > 0$ such that $F_0 = F \cap B[x_0, b]$ is closed in E, where $B[x_0, b] = \{x \in E : \|x - x_0\| \le b\}$. We list the following assumptions:

(A_1) $f \in C([t_0, t_0 + \alpha] \times F, E)$, $x_0 \in F$, F is locally closed, $\|f(t, x)\| \le M - 1$, $M \ge 1$ on $[t_0, t_0 + \alpha] \times F_0$ and $\alpha = \min(a, \frac{b}{M})$;

(A_2) $\liminf_{h \to 0^+} \frac{1}{h} d(x + hf(t, x), F) = 0, \qquad t \in [t_0, t_0 + a], \quad x \in F;$

(A_3) $g \in C([t_0, t_0 + \alpha] \times R_+, R)$, $g(t, 0) \equiv 0$ and $u \equiv 0$ is the unique solution of (2.5.1) on $[t_0, t_0 + a]$.

Theorem 2.6.1. *Let (A_1) and (A_2) hold. Let $\{\epsilon_n\}$ be a sequence of numbers such that $\epsilon_n \in (0, 1)$ and $\lim_{n \to \infty} \epsilon_n = 0$. Then, for each positive integer n, the Cauchy problem (2.6.1) has an ϵ_n-approximate solution x_n on $[t_0, t_0 + \alpha]$ into $B[x_0, b]$ such that the following properties are satisfied: there is a sequence $\{t_i^n\}_{i=0}^{\infty}$ in $[t_0, t_0 + \alpha]$ such that*

(i) $t_0^n = t_0$, $t_i^n - t_{i-1}^n \le \epsilon_n$, $i = 1, 2, \ldots$, *and* $\lim_{i \to \infty} t_i^n = t_0 + \alpha$;

(ii) $x_n(t_0^n) = x_0$ *and* $\|x_n(t) - x_n(s)\| \le M|t-s|, \quad t,s \in [t_0, t_0+\alpha]$;

(iii) $x_n(t_{i-1}^n) \in F_0$ *and* $x_n(t)$ *is linear on* $[t_{i-1}^n, t_i^n]$ *for each* i;

(iv) *if* $t \in (t_{i-1}^n, t_i^n)$, *then* $\|x_n'(t) - f(t_{i-1}^n, x_n(t_{i-1}^n))\| \le \epsilon_n$;

(v) *if* $(t,y) \in [t_{i-1}^n, t_i^n] \times F_0$, *with* $|y - x_n(t_{i-1}^n)| \le M(t_i^n - t_{i-1}^n)$, *then* $\|f(t,y) - f(t_{i-1}^n, x_n(t_{i-1}^n))\| \le \epsilon_n$.

Proof. We shall proceed to construct $x_n(t)$ and $\{t_i^n\}$ by induction on i and as before, for convenience, we shall write $x(t)$, $\{t_i\}$ and ϵ in place of $x_n(t)$, $\{t_i^n\}$ and ϵ_n respectively, dropping the subscript or superscript n. We assume that $x(t)$ is defined on $[t_0, t_{i-1}]$, $t_{i-1} \le t_0 + \alpha$ and that properties (i)–(v) hold on $[t_0, t_{i-1}]$. Choose, $\delta_i \in [0, \epsilon]$ such that δ_i is the largest number satisfying:

(1) $t_{i-1} + \delta_i \le t_0 + \alpha$;

(2) if $t \in [t_{i-1}, t_{i-1} + \delta_i]$ and $x \in F_0$ with $\|x - x(t_{i-1})\| \le M\delta_i$, then $\|f(t,x) - f(t_{i-1}, x(t_{i-1}))\| \le \epsilon$;

(3) $d(x(t_{i-1}) + \delta_i f(t_{i-1}, x(t_{i-1})), F) \le \frac{\epsilon}{2}\delta_i$.

Since $\delta_i > 0$, we let $t_i = t_{i-1} + \delta_i$ and choose, $x(t_i) \in F$ such that

$$\|x(t_{i-1}) + (t_i - t_{i-1}) f(t_{i-1}, x(t_{i-1})) - x(t_i))\| \le \epsilon(t_i - t_{i-1}),$$

which is possible in view of (3). We then define

$$x(t) = \frac{x(t_i) - x(t_{i-1})}{t_i - t_{i-1}}(t - t_{i-1}) + x(t_{i-1})$$

for $t \in [t_{i-1}, t_i]$.

If $t, s \in [t_{i-1}, t_i]$, then

$$\begin{aligned} \|x(t) - x(s)\| &\le \frac{\|x(t_i) - x(t_{i-1})\|}{t_i - t_{i-1}}|t-s| \\ &\le |t-s|\left[\|f(t_{i-1}, x(t_{i-1}))\| + \epsilon\right] \\ &\le M|t-s|, \end{aligned}$$

which shows that $x(t)$ satisfies (ii) on $[t_0, t_i]$. Also,

$$\|x(t_i) - x_0\| \le M(t_i - t_0) \le b.$$

This implies that $x(t_i) \in F_0$ and that (iii) holds. If $t \in (t_{i-1}, t_i)$, then $x'(t)$ exists and hence

$$\|f(t_{i-1}, x(t_{i-1})) - x'(t)\| \le \|f(t_{i-1}, x(t_{i-1})) - \frac{x(t_i) - x(t_{i-1})}{t_i - t_{i-1}}\| \le \epsilon.$$

Hence (iv) holds. The fact that (v) holds is clear from (2).

To complete the proof, we need to show that $\lim_{i\to\infty} t_i = t_0 + \alpha$. Assume, for contradiction, that $t_i < t_0 + \alpha$ for all $i = 0, 1, 2, \ldots$, and let $\lim_{i\to\infty} t_i = \tau < t_0 + \alpha$. Since $|x(t_i) - x(t_j)| \leq M|t_i - t_j|$ and F_0 is closed, $\lim_{i\to\infty} x(t_i) = z \in F_0$ exists. The continuity of f implies that there exists an $\eta > 0$ such that

$$\|f(t,y) - f(\tau,z)\| \leq \frac{\epsilon}{3}$$

whenever $|t - \tau| \leq 2\eta$ and $\|y - z\| \leq 2\eta M$.

Using the boundary condition (A_2), we see that there exists a $\gamma > 0$ such that

$$d(z + \gamma f(\tau, z), F) \leq \gamma\frac{\epsilon}{3} \tag{2.6.3}$$

and $\gamma < \min\{\epsilon, \eta, t_0 + \alpha - \tau\}$. As $\lim_{i\to\infty} t_i = \tau$, we can choose N sufficiently large so that for all $i \geq N$,

$$\tau - t_i < \gamma \quad \text{and} \quad \|z - x(t_i)\| \leq \gamma M.$$

As $\gamma < \eta$, if $|t - t_i| < \gamma, \|y - x(t_i)\| \leq \gamma M$, it follows that $|t - \tau| \leq 2\eta$, $\|y - z\| \leq 2\eta M$ and hence

$$\|f(t,y) - f(t_i, x(t_i))\| \leq \|f(t,y) - f(\tau,z)\| + \|f(\tau,z) - f(t_i,x(t_i))\| \leq \frac{2\epsilon}{3}.$$

That is to say, (1) and (2) are valid with δ_i replaced by γ for each $i \geq N$ and $\gamma > \delta_i$. Since δ_i is the largest number chosen for which (1), (2) and (3) hold, we must have

$$d(x(t_i) + \gamma f(t_i, x(t_i)), F) > \gamma\frac{\epsilon}{2}, \quad \text{for all} \quad i \geq N.$$

But, the continuity of the distance function $x \to d(x, F)$ yields

$$d(z + \gamma f(\tau, z), F) = \lim_{i\to\infty} d(x(t_i) + \gamma f(t_i, x(t_i)), F) \geq \gamma\frac{\epsilon}{2}$$

and this contradicts (2.6.3). Therefore $\lim_{i\to\infty} t_i = t_0 + \alpha$ and $x(t_0 + \alpha) = \lim_{i\to\infty} x(t_i)$. This completes the proof of the Theorem.

We shall next show that if these constructed approximate solutions converge, then their limit is a solution of (2.6.1).

Lemma 2.6.1. *Suppose that the assumptions of Theorem 2.6.1 hold and that $\lim_{n\to\infty} x_n(t) = x(t)$ for each $t \in [t_0, t_0 + \alpha]$. Then $x(t)$ is a solution for the problem (2.6.1) on $[t_0, t_0 + \alpha]$.*

Proof. Since the sequence $\{x_n(t)\}$ is equicontinuous on $[t_0, t_0 + \alpha]$, by (ii) of Theorem 2.6.1, it follows that $\{x_n(t)\}$ converges uniformly to $x(t)$ on $[t_0, t_0 + \alpha]$ and that $x(t)$ is continuous. Obviously $x(t_0) = x_0$. Also, if $t \in [t_0, t_0 + \alpha]$ and for each n, i_n is such that $t \in (t^n_{i_n}, t^n_{i_n+1})$, then

$$\lim_{n\to\infty} \|x(t) - x_n(t^n_{i_n})\| \leq \lim_{n\to\infty} \|x(t) - x_n(t)\| + |t - t^n_{i_n}|M = 0.$$

Since, $x_n(t^n_{i_n}) \in F_0$ and F_0 is closed, we have $x(t) \in F_0$. The continuity of f yields that for every $\epsilon > 0$, there exists a $\delta = \delta(\epsilon) > 0$ such that

$$\|f(s, x) - f(t, x(t))\| < \epsilon$$

if $|s - t| < \delta$, $x \in F_0$ and $\|x - x(t)\| < \delta$. Also,

$$\|x_n'(t) - f(t^n_{i_n}, x_n(t^n_{i_n}))\| \leq \epsilon_n$$

for $t \in (t^n_{i_n}, t^n_{i_n+1})$ and $t^n_{i_n+1} - t^n_{i_n} \leq \epsilon_n$. Hence, choosing n large enough such that $\epsilon_n \leq \delta$, we obtain, for $t \in [t_0, t_0 + \alpha]$,

$$\|x(t) - x_0 - \int_{t_0}^{t} f(s, x(s))\mathrm{d}s\| \leq \max_{t_0 \leq t \leq t_0+\alpha} \|x(t) - x_n(t)\| + (\epsilon_n + \epsilon)\alpha,$$

which proves that $x(t)$ is a solution of (2.6.1).

We now need to look for additional conditions on f which would guarantee that the approximate solutions $\{x_n(t)\}$ converge. We begin with a simple, but illustrative, existence result, employing such an assumption.

Theorem 2.6.2. *Assume that the conditions (A_1), (A_2) and (A_3) hold. Suppose further that for $t \in [t_0, t_0 + a]$, $\quad x, y \in F_0$,*

$$\|f(t, x) - f(t, y)\| \leq g(t, |x - y|) \tag{2.6.4}$$

and that $g(t, u)$ is nondecreasing in u for each t. Then the problem (2.6.1) has a unique solution on $[t_0, t_0 + \alpha]$.

Proof. Let n, m be positive integers and let

$$m(t) = \|x_n(t) - x_m(t)\|, \qquad t \in [t_0, t_0 + \alpha].$$

If $t \in (t^n_i, t^n_{i+1}) \cap (t^m_j, t^m_{j+1})$, then

$$\begin{aligned} D^+m(t) &\leqslant \|x_n'(t) - x_m'(t)\| \\ &\leqslant \|f(t, x_n(t^n_i)) - f(t, x_m(t^m_j))\| + \|f(t, x_n(t^n_i)) - f(t^n_i, x_n(t^n_i))\| \\ &\quad + \|f(t, x_m(t^m_j)) - f(t^m_j, x_m(t^m_j))\| + \|x_n'(t) - f(t^n_i, x_n(t^n_i))\| \\ &\qquad + \|x_m'(t) - f(t^m_j, x_m(t^m_j))\|. \end{aligned}$$

By (iv) of Theorem 2.6.1 and the assumption (2.6.4), we get

$$D^+m(t) \leqslant g(t, |x_n(t^n_i) - x_m(t^m_j)|) + 2(\epsilon_n + \epsilon_m).$$

Now, using (i) and (ii) of Theorem 2.6.1, we see that

$$\begin{aligned} \|x_n(t^n_i) - x_m(t^m_j)\| &\leqslant \|x_n(t^n_i) - x_n(t)\| + \|x_n(t) - x_m(t)\| + \|x_m(t^m_j) - x_m(t)\| \\ &\leqslant M(\epsilon_n + \epsilon_m) + \|x_n(t) - x_m(t)\|. \end{aligned}$$

We thus have, in view of monotony of $g(t, u)$ in u,

$$D^+m(t) \leqslant g(t, m(t) + \beta_{m,n}) + \eta_{m,n},$$

where $\beta_{m,n} = M(\epsilon_n + \epsilon_m)$ and $\eta_{m,n} = 2(\epsilon_n + \epsilon_m)$. Setting $\nu(t) = m(t) + \beta_{m,n}$, we see that

$$D^+\nu(t) \leqslant g(t, \nu(t)) + \eta_{m,n}.$$

This inequality holds for all but a countable number of $t \in [t_0, t_0 + \alpha]$. Also $\nu(t_0) = \beta_{m,n}$. Hence, by Theorem 1.7.1, we get

$$m(t) \leqslant \nu(t) \leqslant r_{m,n}(t), \qquad t \in [t_0, t_0 + \alpha],$$

where $r_{m,n}(t)$ is the maximal solution of

$$u' = g(t, u), \qquad u(t_0) = \beta_{m,n}.$$

Since $\beta_{m,n}, \eta_{m,n} \to 0$ as $n, m \to \infty$, we see by Lemma 2.6.1, that $\lim\limits_{n,m\to\infty} r_{m,n}(t) = r(t)$ uniformly on $[t_0, t_0 + \alpha]$, where $r(t)$ is the maximal solution of (2.5.1). By (A_3), it follows that $m(t) \equiv 0$ on $[t_0, t_0 + \alpha]$, as $n, m \to \infty$. Thus, the sequence $\{x_n(t)\}$ is uniformly Cauchy on $[t_0, t_0 + \alpha]$ and the existence of a solution for (2.6.1) follows by Lemma 2.6.1. The proof of uniqueness of solutions is standard. Hence the proof is complete.

2.7 Global Existence

Contrary to the case in R^n, a solution $x(t)$ of an abstract Cauchy problem need not have the property that either $x(t)$ exists on $[t_0, t_0 + a]$ or on $[t_0, T)$ with $T < t_0 + a$ and $|x(t)| \to \infty$ as $t \to T$ for a counterexample. We shall therefore have to impose conditions besides continuity in order to rule out such a behavior. In this section, we shall give different sets of assumptions which guarantee the global existence of solutions of the problem

$$x' = f(t, x), \qquad x(t_0) = x_0, \tag{2.7.1}$$

where we assume that $f \in C(R_+ \times E, E)$, E being the Banach space.

Theorem 2.7.1. *Assume that*

$$\|f(t, x)\| \leqslant g(t, \|x\|), \qquad (t, x) \in R_+ \times E, \tag{2.7.2}$$

where $g \in C(R_+ \times R_+, R_+)$, $g(t, u)$ is nondecreasing in u for each $t \in R_+$, and the maximal solution $r(t, t_0, u_0)$ of the scalar differential equation

$$u' = g(t, u), \qquad u(t_0) = u_0 \geq 0, \tag{2.7.3}$$

exists on $[t_0, \infty)$. Suppose that f is smooth enough to assure local existence of solutions of (2.7.1) for any $(t_0, x_0) \in R_+ \times E$. Then the largest interval of existence of any solution $x(t, t_0, x_0)$ of (2.7.1) such that $\|x_0\| \leqslant u_0$ is $[t_0, \infty)$. If in addition $r(t, t_0, u_0)$ is bounded on $[t_0, \infty)$ then $\lim\limits_{t\to\infty} x(t, t_0, x_0) = y \in E$.

Proof. Let $x(t, t_0, x_0)$ be any solution of (2.7.1) with $\|x_0\| \leqslant u_0$, which exists on $[t_0, \beta)$ for $t_0 < \beta < \infty$ and such that the value of β cannot be increased. Define $m(t) = \|x(t, t_0, x_0)\|$ for $t_0 \leqslant t < \beta$. Then, using the assumption (2.7.2), we obtain

$$D^+m(t) \leqslant \|x'(t, t_0, x_0)\| = \|f(t, x(t, t_0, x_0))\| \leqslant g(t, m(t)), t_0 \leqslant t < \beta,$$

and $m(t_0) \leqslant u_0$. Hence by Theorem 1.7.1, we have

$$\|x(t, t_0, x_0)\| \leqslant r(t, t_0, u_0), \qquad t_0 \leqslant t < \beta, \tag{2.7.4}$$

where $r(t, t_0, u_0)$ is the maximal solution of (2.7.3). For any t_1, t_2 such that $t_0 \leqslant t_1 < t_2 < \beta$,

$$\begin{aligned}\|x(t_1, t_0, x_0) - x(t_2, t_0, x_0)\| &= \|\int_{t_1}^{t_2} f(s, x(s, t_0, x_0))ds\| \\ &\leqslant \int_{t_1}^{t_2} g(s, \|x(s, t_0, x_0)\|)ds \leqslant \int_{t_1}^{t_2} g(s, r(s, t_0, u_0))ds \\ &\qquad\qquad = r(t_2, t_0, u_0) - r(t_1, t_0, u_0).\end{aligned} \tag{2.7.5}$$

Here we have used the monotony of $g(t, u)$ and the relation (2.7.4). Since $\lim_{t \to \beta^-} r(t, t_0, u_0)$ exists and is finite, taking limits as $t_1, t_2 \to \beta^-$ and using Cauchy criterion for convergence, it follows that $\lim_{t \to \beta^-} x(t, t_0, x_0)$ exists.

We define $x(\beta, t_0, x_0) = \lim_{t \to \beta^-} x(t, t_0, x_0)$ and consider the initial value problem

$$x' = f(t, x), \qquad x(\beta) = x(\beta, t_0, x_0).$$

By assumed local existence, we see that $x(t, t_0, x_0)$ can be continued beyond β , contradicting our assumption. Hence every solution of (2.7.1) such that $\|x_0\| \leqslant u_0$ exists on $[t_0, \infty)$.

Since $r(t, t_0, u_0)$ is bounded and nondecreasing on $[t_0, \infty)$, it follows that $\lim_{t \to \infty} r(t, t_0, u_0)$ exists and is finite. This, together with the inequalities (2.7.5) and (2.7.4) which now hold with $\beta = \infty$, yields the last part of the theorem. The proof is complete.

Remark 2.7.1 Clearly, f is bounded on bounded sets, if f satisfies condition (2.7.2), which need not be true if we relax (2.7.2) to

$$(f(t, x), x) \leqslant g(t, \|x\|)\|x\|. \tag{2.7.6}$$

This is also the case when one relaxes (2.7.2) to a more general condition by means of Lyapunov like function. If we need only global existence we could also remove the restriction of monotony on $g(t, u)$. This is the motivation for the next result.

Theorem 2.7.2. *Assume that*

(i) $f \in C(R_+ \times E, E)$, *f is bounded on bounded sets and for any* $(t_0, x_0) \in R_+ \times E$ *there exists a local solution for the problem* (2.7.1)

(ii) $V \in C(R_+ \times E, R_+)$, *$V$ is locally Lipschitzian in x,* $V(t,x) \to \infty$ *as* $\|x\| \to \infty$ *uniformly for* $[0,T]$ *for every* $T > 0$ *and for* $(t,x) \in R_+ \times E$

$$D^+V(t,x) \equiv \lim_{h \to 0} \frac{1}{h}[V(t+h, x+hf(t,x)) - V(t,x)] \leqslant g(t, V(t,x)),$$

where $g \in C(R_+ \times R_+, R)$;

(iii) the maximal solution $r(t) = r(t, t_0, u_0)$ *of* (2.7.3) *exists on* $[t_0, \infty)$ *and is positive if* $u_0 > 0$.

Then for every $x_0 \in E$ *such that* $V(t_0, x_0) \leqslant u_0$, *the problem* (2.7.1) *has a solution* $x(t)$ *on* $[t_0, \infty)$ *which satisfies the estimate*

$$V(t, x(t)) \leqslant r(t), \qquad t \geq t_0. \tag{2.7.7}$$

Proof. Let S denote the set of all functions x defined on $I_x = [t_0, c_x)$ with values in E such that $x(t)$ is a solution of (2.7.1) on I_x and $V(t, x(t)) \leqslant r(t), \quad t \in I_x$. We define a partial order $\leqslant$ on S as follows: the relation $x \leqslant y$ implies that $I_x \subseteq I_y$ and $y(t) \equiv x(t)$ on I_x. We shall first show that S is nonempty. By (i), there exists a solution $x(t)$ of (2.7.1) defined on $I_x = [t_0, c_x)$. Setting $m(t) = V(t, x(t))$ for $t \in I_x$ and using assumption (ii), it is easy to obtain the differential inequality

$$D^+m(t) \leqslant g(t, m(t)), \qquad t \in I_x.$$

Now, by Theorem 1.7.1, it follows that

$$V(t, x(t)) \leqslant r(t), \qquad t \in I_x, \tag{2.7.8}$$

where $r(t)$ is the maximal solution of (2.7.3). This shows that $x \in S$ and so S is nonempty.

If $(x_\beta)_\beta$ is a chain $(S, \leqslant)$, then there is a uniquely defined map y on $I_y = [t_0, \sup_\beta c_{x_\beta})$ that coincides with x_β on I_{x_β}. Clearly $y \in S$ and hence y is an upper bound of $(x_\beta)_\beta$ in $(S, \leqslant)$. Then Zorn's lemma assures the existence of a maximal element z in $(S, \leqslant)$. The proof of the theorem is complete if we show that $c_z = \infty$. Suppose that it is not true, so that $c_z < \infty$. Since, $r(t)$ is assumed to exist on $[t_0, \infty)$, $r(t)$ is bounded on I_z. Since $V(t,x) \to \infty$ as $\|x\| \to \infty$ uniformly in t on $[t_0, c_z]$, the relation $V(t, z(t)) \leqslant r(t)$ on I_z implies that $\|z(t)\|$ bounded on I_z. By (i), this shows that there is an $M > 0$ such that

$$\|f(t, z(t))\| \leq M, \qquad t \in I_z.$$

We then have, for all $t_1, t_2 \in I_z, \quad t_1 \leqslant t_2$,

$$\|z(t_2) - z(t_1)\| \leqslant \int_{t_1}^{t_2} \|f(s, z(s))\| \mathrm{d}s \leqslant M(t_2 - t_1),$$

which shows that z is Lipschitzian on I_z and consequently has a continuous extension z_0 on $[t_0, c_z]$. By continuity, we get

$$z_0(c_z) = x_0 + \int_{t_0}^{c_z} f(s, z_0(s)) \mathrm{d}s.$$

This implies that $z_0(t)$ is a solution of (2.7.1) on $[t_0, c_z]$ and, clearly, $V(t, z_0(t)) \leqslant r(t), \quad t \in [t_0, c_z]$. Consider the problem

$$x' = f(t, x), \qquad x(t_0) = z_0(c_z).$$

By the assumed local existence there exists a solution $x_0(t)$ on $[c_z, c_z + \delta), \quad \delta > 0$.

Define

$$z_1(t) = \begin{cases} z_0(t) & \text{for} \quad t_0 \leqslant t \leqslant c_z \\ x_0(t) & \text{for} \quad c_z \leqslant t \leqslant c_z + \delta. \end{cases}$$

Clearly $z_1(t)$ is a solution of (2.7.1) on $[t_0, c_z + \delta)$ and, by repeating the arguments that were used to obtain (2.7.8), we get

$$V(t, z_1(t)) \leqslant r(t), \quad t \in [t_0, c_z + \delta).$$

This contradicts the maximality of z and hence $c_z = \infty$. The proof is complete.

The global existence of solutions of (2.7.1) can also be proved under dissipative type conditions. We shall give a result of this nature in closed sets.

Theorem 2.7.3. *Let $F \subset E$ be closed. Suppose that*

(i) $f \in C(R_+ \times F, E)$;

(ii) $\liminf_{h \to 0^+} \frac{1}{h} d(x + hf(t, x), F) = 0$ *for* $(t, x) \in R_+ \times F$;

(iii) $[x - y, f(t, x) - f(t, y)]_- \leqslant \lambda(t)\|x - y\|$, *for* $t \in R_+, \quad x, y \in F$, *where* $\lambda \in C[R_+, R]$. *Suppose further that one of the following conditions hold:*

(a) for each $T \in [t_0, \infty)$ *and* $L > 0$, *there exists a* $M = M(T, L) > 0$ *such that* $\|f(t, x)\| \leq M$ *for* $(t, x) \in [t_0, T] \times F$ *with* $\|x\| \leqslant L$;

(b) for each $T \in [t_0, \infty), \quad L > 0$ *and* $\epsilon > 0$, *there exists a* $\delta = \delta(T, \epsilon, L) > 0$ *such that* $\|f(t, x) - f(s, x)\| \leqslant \epsilon$ *whenever* $(t, x), (s, x) \in [t_0, T] \times F$ *with* $|t - s| \leqslant \delta$ *and* $\|x\| \leqslant L$.

Then, for each $x_0 \in F$, *there is a unique solution* $x(t, t_0, x_0)$ *of* (2.7.1) *on* $[t_0, \infty)$.

Proof. One can conclude local existence and uniqueness of solutions of (2.7.1) on the basis of Theorem 2.7.1. Assume that the solution $x(t, t_0, x_0)$ of (2.7.1) exists on $[t_0, T), T < \infty$ and is noncontinuable. First, let

$$m(t) = \|x(t, t_0, x_0) - x_0\|, \qquad \text{for} \quad t \in [t_0, T).$$

Then, for $t \in [t_0, T)$,

$$\begin{aligned}m'_+(t) =&[x(t, t_0, x_0) - x_0, f(t, x(t, t_0, x_0))]_+ \\ &\leqslant [x(t, t_0, x_0) - x_0, f(t, x(t, t_0, x_0)) - f(t, x_0)]_+ + \|f(t, x_0)\| \\ &\leqslant \lambda(t)m(t)|f(t, x_0)|.\end{aligned}$$

Since $m(t_0) = 0$, this differential inequality implies that there is a $L > 0$ such that $\|x(t, t_0, x_0)\| \leqslant L, \quad t \in [t_0, T)$.

If (a) holds, then we have

$$\|x(t, t_0, x_0) - x(s, t_0, x_0)\| \leqslant M|t - s|, \quad \text{for} \quad t, s \in [t_0, T)$$

This shows, since F is closed, that $\lim_{t \to T^-} x(t, t_0, x_0)$ exists and belongs to F. This contradicts the noncontinuability of $x(t, t_0, x_0)$. If, on the other hand, (b) holds, let ϵ and h be positive numbers with $h > \delta$. Setting

$$m(t) = \|x(t + h, t_0, x_0) - x(t, t_0, x_0)\|, \qquad t \in [t_0, T - h),$$

we obtain

$$\begin{aligned}m'_+(t) =&[x(t + h, t_0, x_0) - x(t, t_0, x_0), f(t + h, x(t + h, t_0, x_0)) - f(t, x(t, t_0, x_0))]_+ \\ &\leqslant \lambda(t)m(t) + \|f(t + h, x(t + h, t_0, x_0)) - f(t, x(t, t_0, x_0))\| \\ &\leqslant \lambda(t)m(t) + \epsilon.\end{aligned}$$

Again, this differential inequality implies that $\lim_{t \to T^-} x(t, t_0, x_0)$ exists and is in F, contradicting the fact that $x(t, t_0, x_0)$ is noncontinuable. The proof is therefore complete.

2.8 Existence of Euler Solutions

We consider the IVP.

$$x'(t) = f(t, x), \qquad x(t_0) = x_0, \tag{2.8.1}$$

where $f : E \to E$, E being the Banach space . Let

$$\pi = \{t_0, t_1, \ldots, t_N = T\} \tag{2.8.2}$$

be a partition of $[t_0, T]$ and consider the interval $[t_0, t_1]$. Note that the right side of the differential equation

$$x'(t) = f(t_0, x_0), \qquad x(t_0) = x_0$$

is a constant on $[t_0, t_1]$ and hence, the IVP (2.8.1) clearly has a unique solution $x(t) = x(t, t_0, x_0)$ on $[t_0, t_1]$. We define the node $x_1 = x(t_1)$ and iterate next by considering on $[t_1, t_2]$, the IVP

$$x'(t) = f(t_1, x_1), \qquad x(t_1) = x_1.$$

The next node is $x_2 = x(t_2) = x(t_2, t_1, x_1)$ and proceed this way till the entire arc $x_\pi = x_\pi(t)$ has been defined on $[t_0, T]$. We employ the notation x_π to emphasize the role played by the particular partition π in finding the arc x_π which is the Euler polygonal arc corresponding to the partition π. The diameter μ_π of the partition π is given by

$$\mu_\pi = \max[t_i - t_{i-1} : 1 \leq i \leq N]. \tag{2.8.3}$$

Definition 2.8.1. By an Euler solution of (2.8.1), we mean any arc $x = x(t)$ which is the uniform limit of Euler polygonal arcs x_π, corresponding to some sequence π_j such that $\pi_j \to 0$ i.e. as the diameter $\mu_{\pi_j} \to 0$ as $j \to \infty$.

Clearly, the corresponding number N_j of the position points in π_j and the nodes also go to ∞. Also, the Euler arc satisfies the initial condition $x(t_0) = x_0$. We can now prove the following result on the existence of Euler solution for IVP (2.8.1).

Theorem 2.8.1. *Assume that*

(i) $\|f(t,x)\| \leq g(t, \|x(t)\|)$, *where* $g \in C([t_0, T] \times \mathbb{R}_+, \mathbb{R}_+)$, $g(t,u)$ *is nondecreasing in* (t,u)*;*

(ii) *the maximal solution* $r(t) = r(t, t_0, u_0)$ *of the scalar differential equation*

$$u' = g(t,u), \qquad u(t_0) = u_0, \tag{2.8.4}$$

exists on $[t_0, T]$*. Then*

(a) *there exists at least one Euler solution* $x(t) = x(t, t_0, x_0)$ *of the IVP* (2.8.1) *which satisfies the Lipschitz condition;*

(b) *any Euler solution* $x(t)$ *of* (2.8.1) *satisfies the relation*

$$\|x(t) - x_0\| \leq r(t, t_0, u_0) - u_0, \qquad t \in [t_0, T], \tag{2.8.5}$$

where $u_0 = \|x_0\|$.

Proof. Let π be the partition of $[t_0, T]$ defined by (2.8.2) and let $x_\pi = x_\pi(t)$ denote the corresponding arc with nodes of x_π represented by $x_0, x_1, \ldots, x_N$. Let us set $x_\pi = x_i(t)$ on $t_i \leq t \leq t_{i+1}$, $i = 0, 1, \ldots, N-1$ and observe that $x_i(t_i) = x_i$, $i = 0, 1, 2, \ldots, N$. On the interval (t_i, t_{i+1}), we have

$$\|x_\pi'(t)\| = \|f(t_i, x_i)\| \leq g(t_i, \|x_i\|). \tag{2.8.6}$$

On $[t_0, t_1]$, we obtain

$$\begin{aligned}
\|x_i(t) - x_0\| &= \|x_0 + \int_{t_0}^{t} f(t_0, x_0)\mathrm{d}s - x_0\| \leq \int_{t_0}^{t} \|f(t_0, x_0)\|\mathrm{d}s \\
&\leq \int_{t_0}^{t} g(t_0, \|x_0\|)\mathrm{d}s \leq \int_{t_0}^{t} g(s, r(s))\mathrm{d}s \\
&= r(t, t_0, \|x_0\|) - \|x_0\| \\
&\leq r(T, t_0, \|x_0\|) - \|x_0\| \equiv M, \textit{ say}.
\end{aligned}$$

Here we have employed the properties of the norm and the integral, monotone character of $g(t, u)$ in u, and the fact that $r(t, t_0, u_0) \geq 0$ is nondecreasing in t. Similarly, we get, on $[t_1, t_2]$,

$$\begin{aligned}
\|x_2(t) - x_0\| &= \|x_1 + \int_{t_1}^{t} f(t_1, x_1)\mathrm{d}s - x_0\| \\
&= \|x_0 + \int_{t_0}^{t_1} f(t_0, x_0)\mathrm{d}s + \int_{t_1}^{t} f(t_1, x_1)\mathrm{d}s - x_0\| \\
&\leq \int_{t_0}^{t_1} \|f(t_0, x_0)\|\mathrm{d}s + \int_{t_1}^{t} \|f(t_1, x_1)\|\mathrm{d}s \\
&\leq \int_{t_0}^{t_1} g(s, r(s))\mathrm{d}s + \int_{t_1}^{t} g(s, r(s))\mathrm{d}s \\
&= \int_{t_0}^{t} g(s, r(s))\mathrm{d}s \leq r(T, t_0, \|x_0\|) - \|x_0\| = M.
\end{aligned}$$

Proceeding in this way, we obtain on $[t_i, t_{i+1}]$,

$$\|x_i(t) - x_0\| \leqslant r(T, t_0, \|x_0\|) - \|x_0\| = M.$$

Hence, it follows that

$$\|x_\pi(t) - x_0\| \leqslant M, \qquad on [t_0, T].$$

Also, from (2.8.6), we have

$$\|x'_\pi(t)\| \leqslant g(T, r(T)) = r'(T, t_0, \|x_0\|) \equiv K, (say).$$

Consequently, using similar arguments, we can find for $t_0 \leqslant s \leqslant t \leqslant T$,

$$\begin{aligned}
\|x_\pi(t) - x_\pi(s)\| &\leqslant \int_{t_0}^{t} \|f(\tau, x_\pi)\|\mathrm{d}\tau - \int_{t_0}^{s} \|f(\tau, x_\pi)\|\mathrm{d}\tau \\
&\leqslant \int_{t_0}^{t} g(\tau, r(\tau))\mathrm{d}\tau - \int_{t_0}^{s} g(\tau, r(\tau))\mathrm{d}\tau \\
&= \int_{s}^{t} g(\tau, r(\tau))\mathrm{d}\tau = r(t) - r(s) \\
&= r'(\sigma)|t - s| \leqslant K(t - s)
\end{aligned}$$

for some $\sigma, s \leqslant \sigma \leqslant t$, proving $x_\pi(t)$ satisfies Lipschitz condition with constant K on $[t_0, T]$.

Now, let π_j be a sequence of partitions of $[t_0, T]$ such that $\pi_j \to 0$, i.e., $\mu_{\pi_j} \to 0$ and therefore $N_j \to \infty$. Then, the corresponding polygonal arcs x_π on $[t_0, T]$ all satisfy

$$x_{\pi_j}(t_0) = x_0, \quad \|x_{\pi_j}(t) - x_0\| \leqslant M \quad \text{and} \quad \|x'_{\pi_j}(t)\| \leqslant K.$$

Hence the family $\{x_{\pi_j}\}$ is equicontinuous and uniformly bounded, and as a consequence, Ascoli-Arzela Theorem guarantees the existence of a subsequence which converges uniformly to a continuous function $x(t)$ on $[t_0, T]$ and that is also absolutely continuous on $[t_0, T]$. Thus, by definition, $x(t)$ is an Euler solution of the IVP (2.8.1) on $[t_0, T]$ and the claim of the theorem follows. The inequality (2.8.5) in part (b) is inherited by $x(t)$ from the sequence of polygonal arcs generating it when we identify T with t. Hence the proof is complete.

If $f(t, x)$ in (2.8.1) is assumed continuous, then one can show that $x(t)$ actually satisfies (2.8.1).

Theorem 2.8.2. *Under the assumptions of Theorem 2.8.1, if we also suppose that f is continuous, then $x(t)$ is a solution of IVP* (2.8.1).

Proof. Let x_{π_j} denote a sequence of polygonal arcs for IVP (2.8.1) converging uniformly to an Euler solution $x(t)$ on $[t_0, T]$. Clearly, the arcs $x_{\pi_j}(t)$ all lie in $\overline{B}(x_0, M) = \{x \in E : |x - x_0| \leqslant M\}$ and satisfy Lipshitz condition with some constant K. Since a continuous function is uniformly continuous on compact sets, for any given $\epsilon > 0$, one can find a $\delta > 0$ such that

$$\|x - x^*\| < \delta, \qquad |t - t^*| < \delta \qquad \text{implies} \quad \|f(t, x) - f(t^*, x^*)\| < \epsilon$$

for t, $t^* \in [t_0, T]$, $x, x^* \in \{x_{\pi_j}\}$. Let j be sufficiently large so that the particular diameter μ_{π_j} satisfies $\mu_{\pi_j} < \delta$ and $K\mu_{\pi_j} < \delta$ for any t which is not one of the infinitely many points at which x_{π_j} has a node, we have $x'_{\pi_j}(t) = f(\tilde{t}, x_{\pi_j}(\tilde{t}))$ for some $\tilde{t}$ within $\mu_{\pi_j} < \delta$ of t. Since $|x_{\pi_j}(t) - x_{\pi_j}(\tilde{t})| \leqslant K\mu_{\pi_j} < \delta$, we get

$$\|x'_{\pi_j}(t) - f(t, x_{\pi_j}(t))\| = \|f(\tilde{t}, x_{\pi_j}(\tilde{t})) - f(t, x_{\pi_j}(t))\| < \epsilon$$

It follows for any $t \in [t_0, T]$, we obtain

$$\begin{aligned}
&\|x_{\pi_j}(t) - x_{\pi_j}(t_0) - \int_{t_0}^t f(s, x_{\pi_j}(s)\mathrm{d}s\| \\
&= \|x_{\pi_j}(t_0) + \int_{t_0}^t x'_{\pi_j}(s)\mathrm{d}s - x_{\pi_j}(t_0) - \int_{t_0}^t f(s, x_{\pi_j}(s)\mathrm{d}s\| \\
&= \|\int_{t_0}^t x'_{\pi_j}(s)\mathrm{d}s - \int_{t_0}^t f(s, x_{\pi_j}(s)\mathrm{d}s\| \\
&\leqslant \int_{t_0}^t \|x'_{\pi_j}(s) - f(s, x_{\pi_j}(s)\|\mathrm{d}s \\
&\leqslant \epsilon(t - t_0) < \epsilon(T - t_0).
\end{aligned}$$

Letting $j \to \infty$, we have from this inequality,

$$\|x(t) - x_0 - \int_{t_0}^{t} f(s, x(s))\| < \epsilon(T - t_0).$$

Since $\epsilon > 0$ is arbitrary, it follows that

$$x(t) = x_0 + \int_{t_0}^{t} f(s, x(s))\mathrm{d}s, \qquad t \in [t_0, T],$$

which implies that $x(t)$ is continuously differentiable and therefore,

$$x'(t) = f(t, x(t)), \qquad x(t_0) = x_0, \qquad t \in [t_0, T].$$

The proof is complete.
Remark 2.8.1 One can extend the notion of Euler solution of (2.8.1) from the interval $[t_0, T]$ to $[t_0, \infty)$, if we define f and g on $[t_0, \infty)$ instead of $[t_0, T]$ and assume that maximal solution $r(t)$ exists on $[t_0, \infty)$. Then we can show that an Euler solution exists on every compact interval $[t_0, T]$, $t_0 < T < \infty$.

2.9 Flow Invariance

Consider the IVP

$$x'(t) = f(t, x(t)), \quad x(t_0) = x_0 \in F, \tag{2.9.1}$$

where F is a closed set in E and $f : E \to E$.
Definition. The set F is said to be flow invariant with respect to f if every solution $x(t)$ of (2.9.1) on $[t_0, \infty]$ is such that $x(t) \in F$ for $t_0 \leqslant t \leqslant \infty$.
A set B is called a distance set if for each $x \in E$, there corresponds a point $y \in B$ such that $d(x, B) = \|x - y\|$.
A function $g \in C(\mathbb{R}_+^2, \mathbb{R}_+)$ is said to be a uniqueness function if the following holds:
If $m \in C[\mathbb{R}_+, \mathbb{R}_+]$ is such that $m(t_0) \leqslant 0$ and $\mathrm{D}^+m(t) \leqslant g(t, m(t))$ whenever $m(t) > 0$, then $m(t) \leqslant 0$ for $t_0 \leqslant t < \infty$.

We shall first prove a result on flow invariance for set F.

Theorem 2.9.1. *Let $F \subset E$ be closed and distance set. Suppose further that, for each t,*

(i) $\lim_{h \to 0} \frac{1}{h}\mathrm{d}(x + hf(t, x), F) = 0, \qquad t \geq t_0$ *and* $x \in \partial F$;

(ii) $\|f(t, x) - f(t, y)\| \leqslant g(t, \|x(t) - y(t)\|)$, $x, y \in \Omega$, $x \in E - F$, $y \in \partial F$ *and* $t \geq t_0$, *g being the uniqueness function.*

Then F is flow invariant with respect to f.

Proof. Let $x(t)$ be a solution of (2.9.1) for $t \geq t_0$. Assume that $x(t) \in F$ for $t_0 \leqslant t < t_0 + a < \infty$, $[t_0, t_0 + a)$ is the maximal interval of existence i.e. $x(t)$ leaves the set F at $t = t_0 + a$ for the first time. Let $x(t_1) \notin F$, $t_1 \in (t_0 + a, \infty)$ and let $y_0 \in \partial F$ be such that

$$\mathrm{d}(x(t_1), F) = \|x(t_1) - y_0\|.$$

Set for $t \in [t_0, \infty), m(t) = \mathrm{d}(x(t), F)$ and $v(t) = \|x(t) - y_0\|$. For $h > 0$ sufficiently small, we have, letting $x = x(t_1)$.

$$\begin{aligned} m(t_1 + h) &\leqslant \|x(t_1 + h) - y_0 - hf(t_1, y_0)\| + d(y_0 + hf(t_1, y_0), F) \\ &\leqslant \|x + hf(t_1, x) - y_0 - hf(t_1, y_0)\| + d(y_0 + hf(t_1, y_0), F) \\ &\qquad + \|x(t_1 + h) - x - hf(t_1, x)\| \\ &\leqslant \|x - y_0\| + hg(t_1, \|x - y_0\|) + o(h). \end{aligned}$$

Setting $m(t_1) \equiv v(t_1) > 0$, we obtain

$$\mathrm{D}^+ m(t_1) \leqslant g(t_1, m(t_1)).$$

This implies, in view of the fact that g is a uniqueness function and $m(t_0) = 0$, that $m(t) \leqslant 0$, $t_0 \leqslant t < \infty$. This contradicts $\mathrm{d}(x(t_1, F) = m(t_1) > 0$. The proof is complete.

Next we consider weak flow invariance of f. The system (F, f), where $F \subset E$ is closed, is said to be weakly flow invariant, provided that for all $x_0 \in F$, there exists an Euler solution $x(t)$ of (2.9.1) on $[t_0, \infty)$ such that $x_0 = x(t_0)$ and $x(t) \in F, t \geq t_0$. In order to prove weak invariance, we have to employ the notion of proximal normal. Let $F \subset E$ be a closed set. Assume that for any $x \in E$ such that x and F are disjoint and for any $s \in F$, there exists a $z \in E$ such that $x = s + z$. Then $x - s$ is called the Hukuhara difference. Suppose now that for any $x \in E$, there is an element $s \in F$ whose distance to x is minimal, i.e.,

$$\|x - s\| = \inf_{s_0 \in F} \|x - s_0\|,$$

then s is called a projection of x onto F. The set of all such elements is denoted by $\mathrm{Proj}_F(x)$. The element $x - s$ will be called the proximal normal direction to F at s. Any nonnegative multiple $\xi = t(x - s), t \geq 0$, is called the proximal normal to F at s. The set of all ξ obtained in this way is said to be the proximal normal cone to F at s and it is denoted by $N_F^p(s)$.

We can now prove the following result which offers sufficient conditions for the weak invariance of the system (F, f) in terms of proximal normal.

Theorem 2.9.2. *Let f satisfy the assumptions of Theorem(2.9.1). Let A be an open set containing $x(t)$ for all $t \in [t_0, T]$. Suppose that for any $(t, z) \in (t_0, T) \times A$, the proximal aiming condition is satisfied i.e., there exists $s \in \mathrm{Proj}_A(z)$ such that*

$$\langle f(t, z), z - s \rangle \leqslant 0,$$

where $\langle .,. \rangle$ is the inner product. Then we have

$$d(x(t), F) \leqslant d(x(t_0), F), \qquad t \in [t_0, T].$$

Proof. Let x_π be one polygonal arc in the sequence converging uniformly to x, as per definition of the Euler solution. As usual, denote its node at t_i by x_i, $i = 0, 1, \ldots, N$ and $x_0 = x(t_0)$. We may suppose that $x_\pi(t)$ lies in set A for all $t \in [t_0, T]$. Accordingly, there exists for each i, a point $s_i \in \mathrm{Proj}_A(x_i)$ such that

$$\langle f(t_i, x_i), x_i - s_i \rangle \leqslant 0.$$

Letting K be a priori bound on $\|x_\pi^{'}\|$, we calculate

$$\begin{aligned} d^2(x_1, F) &\leqslant \|x_1 - s_0\|^2, \qquad (\text{since} \quad x_0 \in F) \\ &= \|x_1 - x_0\|^2 + \|x_0 - s_0\|^2 + 2\langle x_1 - x_0, x_0 - s_0 \rangle \\ &\leqslant K^2(t_1 - t_0)^2 + d^2(x_0, F) + 2\int_{t_0}^{t_1} \langle x_\pi^{'}(s), x_0 - s_0 \rangle \mathrm{d}s \\ &= K^2(t_1 - t_0)^2 + d^2(x_0, F) + 2\int_{t_0}^{t_1} \langle f(t_0, x_0), x_0 - s_0 \rangle \mathrm{d}s \\ &\leqslant K^2(t_1 - t_0)^2 + d^2(x_0, F), \end{aligned}$$

since the inner product in the integral term is $\leqslant 0$. The same estimates apply at any node x_i and hence

$$d^2(x_i, F) \leqslant d^2(x_{i-1}, F) + K^2(t_i - t_{i-1})^2,$$

Repeating this recursively, we get

$$\begin{aligned} d^2(x_i, F) &\leqslant d^2(x_{i-1}, F) + K^2 \sum_{l=1}^{i} (t_l - t_{1-1})^2 \\ &\leqslant d^2(x_0, F) + K^2 \mu_\pi \sum_{l=1}^{i} (t_l - t_{l-1}) \\ &\leqslant d^2(x_0, F) + K^2 \mu_\pi (T - t_0). \end{aligned}$$

Consider now the sequence x_{π_j} of polygonal arcs converging to x. Since the above estimate holds at every node and since $\mu_{\pi_j} \to 0$, same K applying to each x_{π_j}, we can deduce that in the limit,

$$\mathrm{d}(x(t), F) \leqslant \mathrm{d}(x(t_0), F), \qquad t \in [t_0, T]$$

as claimed. The proof is complete.

2.10 Notes and Comments

Most of the results of this chapter are devoted to local and global existence of solutions, existence of extremal solutions, existence in closed sets and existence of Euler solutions. We also deal with the theory of differential inequalities in cones as well as flow invariance employing non-smooth analysis. Thus we have assembled several results dealing with these topics from the following works: Gudunov [1], Nagumo [1], Murakami [1], Martin [1], Deimling [1], Szufla [1.2.3], Ladas and Lakshmikantham [1], Lakshmikantham and Leela [1], Guo and Lakshmikantham [1]. See also Volkman [1,2,3,4].

Chapter 3

Theoretical Approximation Methods

3.1 Introduction

This chapter is devoted to the monotone flows and their convergence. In Section 3.2, we begin the method of lower and upper solutions. We first describe the existence of a solution of the IVP in the conical sector generated by the lower and upper solutions. Here, we employ for simplicity, the cone R^n_+, i.e. we use componentwise inequalities. We then consider the same result in the general setup of an arbitrary cone. Section 3.3 deals with simple monotone iterative technique and generalizations of ideas involved in fixed point theorems. In Section 3.4, we consider generalized monotone iterative technique, which includes several special cases of interest.

In section 3.5, the method of quasilinearization is discussed, where the obtained monotone sequences converge quadratically to the unique solution. The extension of the method of quasilinearization is considered in Section 3.6, which contains several intersting special cases. In Section 3.7, fixed point theorems have been developed, isolating the ideas of the method of quasilinearization, while Section 3.8 deals with the application of the fixed point theorems of Section 3.7 to ordinary and partial differential equations. Finally, Section 3.9 provides notes and comments.

3.2 Method of Upper and Lower Solutions

Consider the IVP

$$u' = f(t, u), \ u(0) = u_0, \tag{3.2.1}$$

where $f \in C(J \times E, E)$, $J = [0,\ T]$ and E a real Banach space. Let $v, w \in C^1(J,\ E)$ be such that

$$v' \leqslant f(t,v), \;\; w' \geqslant f(t,w) \;\; \text{on} \;\; J. \tag{3.2.2}$$

Then v, w are said to be lower and upper solutions of (3.2.1) respectively relative to a cone K in E defined in a natural way. In the special case when $E = R^n$ and $K = R^n_+$, one can prove the following result.

Theorem 3.2.1. *Let $v, w \in C^1(J, R^n)$ be lower and upper solutions of (3.2.1) such that $v(t) \leqslant w(t)$ on J and let $f \in C(\Omega, R^n)$, where*

$$\Omega = [(t,u) \in \; J \times R^n : v(t) \leqslant u \leqslant w(t), t \in \; J].$$

If f is quasimonotone nondecreasing in u, then there exists a solution $u(t)$ of (3.2.1) such that $v(t) \leqslant u(t) \leqslant w(t)$ on J, provided $v(0) \leqslant u(0) \leqslant w(0)$.

In fact, the conclusion of Theorem 3.2.1 is true without demanding f to be quasimonotone nondecreasing, which is restrictive. However, in this case, we need to strengthen the notion of upper and lower solutions of (3.2.1) as follows:

For each i,

$$\left.\begin{array}{lll} v_i' \leqslant f_i(t,\sigma) & \text{for all } \sigma & \text{such that } \; v(t) \leqslant \sigma \leqslant w(t) \\ & & \quad \text{and } \; v_i(t) = \sigma_i; \\ w_i' \geqslant f_i(t,\sigma) & \text{for all } \sigma & \text{such that } \; v(t) \leqslant \sigma \leqslant w(t) \\ & & \quad \text{and } \; w_i(t) = \sigma_i; \end{array}\right\} \tag{3.2.3}$$

Theorem 3.2.2. *Let $v, w \in C^1(J, R^n)$ with $v(t) \leqslant w(t)$ on J, satisfying (3.2.2)and (3.2.3). Let $f \in C(\Omega, R^n)$. Then there exists a solution u of (3.2.1) such that $v(t) \leqslant u(t) \leqslant w(t)$ on J provided $v(0) \leqslant u(0) \leqslant w(0)$.*

Since the assumptions of Theorem 3.2.1 imply the assumptions of Theorem 3.2.2 it is enough to prove Theorem 3.2.2.

Proof. Consider $P : J \times R^n \to R^n$ defined by

$$P_i(t,u) = \max[v_i(t), \; \min[u_i, w_i(t)], \;\; \text{for each } i.$$

Then $f(t, P(t,u))$ defines a continuous extention of f to $J \times R^n$, which is also bounded since f is bounded on Ω. Therefore, $u' = f(t, P(t,u))$ has a solution u on J with $u(0) = u_0$. Let us show that $v(t) \leqslant u(t) \leqslant w(t)$ and therefore a solution of (3.2.1). For $\epsilon > 0$ and $e = (1, ..., 1)$, consider $w_\epsilon(t) = w(t) + \epsilon(1+t)e$ and $v_\epsilon(t) = v(t) - \epsilon(1+t)e$. We have $v_\epsilon(0) < u_0 < w_\epsilon(0)$. Suppose that $t_1 \in J$ is such that $v_\epsilon(t) < u(t) < w_\epsilon(t)$ in $[0, t_1)$ but $u_j(t_1) = w_{\epsilon j}(t_1)$. Then we have $v(t_1) \leqslant P(t_1, u(t_1)) \leqslant w(t_1)$ and $P_j(t_1, u(t_1)) = w_j(t_1)$, hence

$$w_j'(t_1) \geqslant f_j(t_1, P(t_1, u(t_1)) = u_j'(t_1),$$

which implies $u_j'(t_1) < w_{\epsilon j}'(t_1)$, contradicting $u_j(t) < w_{\epsilon j}(t)$ for $t < t_1$. Therefore $v_\epsilon(t) < u(t) < w_\epsilon(t)$ in J. Now $\epsilon \to 0$ yields $v(t) \leqslant u(t) \leqslant w(t)$, and the proof is complete. □

Next we use the idea of differential inequalities to give a different proof of Theorem 3.2.2 which is interesting in itself. For this proof we need to have f satisfy a Lipschitz condition of the form

$$\|f_i(t,x) - f_i(t,y)\| \leqslant L_i \left(\sum_{i=1}^{n} \|x_i - y_i\| \right).$$

Proof. We first assume that v, w in (3.2.3) satisfy strict inequalities and also $v(0) < u(0) < w(0)$ and prove the conclusion for strict inequalities. If the conclusion is false there exist a $t_1 > 0$ and $i, 1 \leqslant i \leqslant n$ such that $v(t_1) \leqslant u(t_1) \leqslant w(t_1)$ and either $v_i(t_1) = u_i(t_1)$ or $u_i(t_1) = w_i(t_1)$. Then

$$\begin{aligned} f_i(t_1, u(t_1)) &= u_i'(t_1) \leqslant v_i'(t_1) < f_i(t_1, \sigma_1, ..., v_i(t_1), ... \sigma_n) \\ &= f_i(t_1, u(t_1)) \end{aligned}$$

or

$$\begin{aligned} f_i(t_1, u(t_1)) &= u_i'(t_1) \geqslant w_i'(t_1) > f_i(t_1, \sigma_1, ..., w_i(t_1), ... \sigma_n) \\ &= f_i(t_1, u(t_1)), \end{aligned}$$

which leads to a contradiction. In order to prove the result for nonstrict inequalities we consider

$$\widetilde{w}_i(t) = w_i(t) + \epsilon e^{(n+1)L_i t}, \widetilde{v}_i(t) = v_i(t) - \epsilon e^{(n+1)L_i t},$$

where $\epsilon > 0$ is sufficiently small. Further let

$$P_i(t,u) = \max\{v_i(t), \min(u_i, w_i(t))\} \text{ for each } i.$$

Then it is clear that if $\widetilde{\sigma}$ is such that $\widetilde{v}(t) \leqslant \widetilde{\sigma} \leqslant \widetilde{w}_i(t)$ and $\widetilde{\sigma}_i = \widetilde{w}_i(t)$, it follows that $\sigma = P(t, \widetilde{\sigma})$ satisfies $v(t) \leqslant \sigma \leqslant w(t)$ and $\sigma_i = w_i(t)$. Hence using the Lipschitz condition it follows that

$$\begin{aligned} \widetilde{w}_i'(t) &= w_i'(t) + \epsilon(n+1)L_i e^{(n+1)L_i t} \geqslant f_i(t,\sigma) + \epsilon(n+1)L_i e^{(n+1)L_i t} \\ &\geqslant f_i(t,\widetilde{\sigma}) + \epsilon e^{(n+1)L_i t} > f_i(t,\widetilde{\sigma}) \end{aligned}$$

for all $\widetilde{\sigma}$ such that $\widetilde{v}(t) \leqslant \widetilde{\sigma} \leqslant \widetilde{w}(t)$ and $\widetilde{\sigma}_i = \widetilde{w}_i(t)$. Here we have used the fact that $|\widetilde{\sigma}_j - P_j(t,\sigma)| \leqslant \epsilon e^{(n+1)L_i t}$ for each j. Similarly we get $\widetilde{v}_i < f_i(t,\widetilde{\sigma})$ for all $\widetilde{\sigma}$ such that $\widetilde{v}(t) \leqslant \widetilde{\sigma} \leqslant \widetilde{w}_i(t)$ and $\widetilde{\sigma}_i = \widetilde{v}_i(t)$. Since $\widetilde{v}(0) < u_0 < \widetilde{w}(0)$ we can conclude from previous argument that $\widetilde{v}(t) < u(t) < \widetilde{w}(t), t \geqslant 0$. As ϵ is arbitrary, the result follows letting $\epsilon \to 0$.□

We observe that the proofs of Theorems 3.2.1 and 3.2.2 depend crucially on the modification of f, that is $f(t, P(t,u))$ where $P(t,u) = \max[v(t), \min(u, w(t))]$. Clearly this modification makes sense only when $K = R^n_+$.

If K is an arbitrary cone, the inequalities (3.2.3) need not be changed. On the other hand, the inequalities can be formulated in terms of linear functionals from K^*, namely for $\phi \in K^*$,

$$\left.\begin{array}{lll} \phi(v' - f(t,\sigma)) \leqslant 0 & \text{for all } \sigma & \text{such that } v(t) \leqslant \sigma \leqslant w(t) \\ & & \text{and } \phi(v(t) - \sigma) = 0, \\ \phi(w' - f(t,\sigma)) \geqslant 0 & \text{for all } \sigma & \text{such that } v(t) \leqslant \sigma \leqslant w(t) \\ & & \text{and } \phi(w(t) - \sigma) = 0. \end{array}\right\} \tag{3.2.4}$$

This version of inequalities (3.2.3) allows us to consider cones K other than the standard cone.The question is whether Theorems 3.2.1 and 3.2.2 hold even when K is an arbitrary cone.

Consider the example in R^3. Let $K = [u \in R^n : (u_1^2 + u_2^2)^{1/2} \leqslant u_3]$. Take $v = (0,0,0), w = (2,0,2), f_1 = f_3 = 0$ and

$$f_2 = \begin{cases} u_1 & \text{if } u_1 \in [0,1] \\ 2 - u_1 & \text{if } u_1 \in [1,2] \\ 0 & \text{otherwise.} \end{cases}$$

The solution of (3.2.1) with $u_0 = (1,0,1)$ is $u(t) = (1,t,1)$ which does not remain in the sector $[v,w]$. Note also that f is Lipschitzian. To verify the inequalities (3.2.3), we see that they now have the form

$$\left.\begin{array}{lllll} 0 \leqslant \phi(f(z)) & \text{for} & 0 \leqslant z \leqslant w, \phi \in K^* & \text{and} & \phi(z) = 0, \\ 0 \geqslant \phi(f(z)) & \text{for} & 0 \leqslant z \leqslant w, \phi \in K^* & \text{and} & \phi(z) = \phi(w). \end{array}\right\} \tag{3.2.5}$$

Since $K \bigcap (w - K) = [\lambda w : 0 \leqslant \lambda \leqslant 1]$, all $z's$ with $0 \leqslant z \leqslant w$ have the form $z = \lambda w$. For $z = 0$ and $z = w$, (3.2.5) is true if we take into account that $f(0) = f(w) = 0$. Let us choose now $z = \lambda w, 0 < \lambda < 1$. For such a z, we have either $\phi(z) = 0$ or $\phi(z) = \phi(w)$. In both cases this implies $\phi(w) = 0$ and (3.2.5) follows from

$$\phi(f(\lambda w)) = 0 \text{ for } 0 < \lambda < 1, \phi \in K^*, \phi(w) = 0, \tag{3.2.6}$$

which has been proved already.

Let $\phi = (\phi_1, \phi_2, \phi_3) \in K^*$ and $\phi(w) = 0$. Since $w = (2,0,2)$, we see that

$$\phi_1 + \phi_2 = 0 \tag{3.2.7}$$

For $u = (0,0,1) \in K, \phi_3 = \phi(u) \geqslant 0$ and for $\overline{u} = (-\phi_1, -\phi_2, \sqrt{\phi_1^2 + \phi_2^2}) \in K$, we have

$$-\phi_1^2 - \phi_2^2 + \phi_3\sqrt{\phi_1^2 + \phi_2^2} = \phi(\overline{u}) \geqslant 0.$$

With $\phi_3 \geqslant 0$, we obtain $\phi_3 \geqslant \sqrt{\phi_1^2 + \phi_2^2}$ and from (3.2.7) we get $\phi_2 = 0$. Hence (3.2.6) holds since $f_1(\lambda w) = f_3(\lambda w) = 0.\square$

Thus, we see that Theorem 3.2.2 may not be valid even in R^3, if K is arbitrary. We therefore prove the corresponding result under suitable conditions.

Theorem 3.2.3. *Assume that the cone K is normal. Suppose further that*

(A_1) *for any bounded set B in E, $\alpha(f(J\times B)) \leqslant L\alpha(B)$, for some $L>0$ where α is the measure of noncompactness;*

(A_2) $\|f(t,u_1)-f(t,u_2)\| \leqslant L\|u_1-u_2\|$, *$t\in J$ and $u_1,u_2\in[v_0,w_0]$, where $[v_0,w_0]=\{u: v_0\leqslant u\leqslant w_0\}$;*

(A_3) *v_0, $w_0\in C^1(J,E)$ with $v_0(t)\leqslant w_0(t)$ on J such that there is an $M>0$ satisfying*

$$\begin{aligned}\phi(v_0'-f(t,\sigma)+M(v_0-\sigma)) &\leqslant 0\\ \phi(w_0'-f(t,\sigma)+M(w_0-\sigma)) &\geqslant 0\end{aligned}$$

for all $\sigma\in[v_0,w_0]$ and $\phi\in K^*$;
are satisfied. Then there exists a unique solution $u(t)$ of (3.2.1) on J such that $v_0(t)\leqslant u(t)\leqslant w_0(t)$, $t\in J$ provided $v_0(0)\leqslant u_0\leqslant w_0(0)$.

Proof. For any $\eta\in C(J,E)$ such that $v_0\leqslant\eta\leqslant w_0$ on J, define the mapping A by $A\eta=u$, where $u=u(t)$ is the unique solution of the linear IVP

$$u'=-Mu+\sigma(t), u(0)=u_0,$$

where $\sigma(t)=f(t,\eta(t))+M\eta(t)$.

We shall first show that A maps the sector $[v_0,w_0]$ into itself. For any $\phi\in K^*$, set

$$p(t)=\phi(u(t)-v_0(t))$$

so that $p(0)\geqslant 0$. Then for all $\sigma\in[v_0,w_0]$,we have

$$p'(t)\geqslant\phi(f(t,\eta)-M(u-\eta)-f(t,\sigma)+M(v_0-\sigma))$$

in view of (A_3). Choosing $\sigma=\eta$, we have $p'\geqslant -Mp$ which implies $p(t)\geqslant p(0)e^{-Mt}\geqslant 0$ on J.This proves $v_0(t)\leqslant u(t)$ on J. A similar argument shows that $u(t)\leqslant w_0(t)$ on J. Hence, $u=A\eta\in[v_0,w_0]$. Since η is arbitrary, the proof is complete. □

We can therefore define the sequence $u_n=Au_{n-1}$, with $u_0=v_0$ or w_0 satisfying $u_n\in[v_0,w_0]$ on J. Since the cone K is assumed to be normal, it follows from $u_n\in[v_0,w_0]$ that the sequence $\{u_n\}$ is uniformly bounded. This implies equicontinuity of the sequence $\{u_n\}$ by using standard arguments and the fact that f maps bounded sets into bounded sets which is a consequence of (A_1). We now set $B(t)=\{u_n(t)\}_{n=0}^{\infty}$ so that $B'(t)=\{u_n'(t)\}_{n=0}^{\infty}$ and $m(t)=\alpha(B(t))$. We can show that $\{u_n(t)\}$ is relatively compact for each $t\in J$ by following the proof of Lemma 3.3.2 which is given in the next section. Hence, we conclude by Ascoli-Arzela theorem that there exist uniformly convergent subsequences of $\{u_n\}$. Suppose that $u_n(t)-u_{n-1}(t)\to 0$ as $n\to\infty$. Then it is clear from the definition of $\{u_n(t)\}$ that the limit of any subsequence is the unique solution of (3.2.1) on J. It then follows that a selection of a subsequence is unnecessary and that the sequence $\{u_n(t)\}$ converges uniformly to the unique solution $u(t)$ on J, such that $u(t) \in [v_0, w_0], t \in J$. Thus, it is sufficient to prove that $m(t) = 0$ on J, where $m(t) = \limsup_{n\to\infty} \|u_n(t) - u_{n-1}(t)\|$. To this end, we note that $\|f(t,u)\| \leqslant N$ for $t \in J$, since f maps bounded sets into bounded sets and $u \in [v_0, w_0]$. Then, for $t_1, t_2 \in J$,

$$\begin{aligned} \|u_n(t_1) - u_{n-1}(t_1)\| &= \|u_n(t_2) - u_{n-1}(t_2)\| + 2N|t_1 - t_2| \\ &\leqslant m(t_2) + 2N|t_1 - t_2| + \epsilon \end{aligned}$$

for large n, given $\epsilon \in 0$. Hence,

$$m(t_1) \leqslant m(t_2) + 2N|t_1 - t_2| + \epsilon$$

Since t_1, t_2 can be interchanged and $\epsilon \in 0$ is arbitrary, we obtain

$$|m(t_1) - m(t_2)| \leqslant 2N|t_1 - t_2|$$

which proves that $m(t)$ is continuous on J. Now, we have

$$\begin{aligned} \|u_{n+1}(t) - u_n(t)\| \leqslant &\int_0^t [\ \|f(s, u_n(s)) - f(s, u_{n+1}(s))\| \\ &+N\|u_n(s) - u_{n-1}(s)\| + M\|u_{n+1}(s) - u_n(s)\|\ ]ds \\ \leqslant &\int_0^t [\ (L+N)\|u_n(s) - u_{n-1}(s)\| + M\|u_{n+1}(s) - u_n(s)\|\ ]ds; \end{aligned}$$

For a fixed $s \in [0, T]$, there is a sequence of integers $n_1 < n_2 < n_3 < \ldots$ such that

$$\|u_{n+1}(t) - u_n(t)\| \to m(t) \quad \text{as} \quad n = n_k \to \infty$$

and $m^*(s) = \lim_{n=n_k\to\infty} |u_n(s) - u_{n-1}(s)|$ exists uniformly on J. It therefore follows, because of the fact $m^*(s) \leqslant m(s)$, that

$$m(t) \leqslant (L + M + N) \int_0^t m(s)ds \quad \text{on} \quad J,$$

which yields that

$$m(t) \leqslant m(t_0)\ exp((L + M + N)t), \quad t \in J.$$

Since $m(t_0) = 0$, we have $m(t) \equiv 0$ on J. This completes the proof of Theorem 3.2.3.

3.3 Monotone Iterative Technique

Let us consider the IVP

$$u' = f(t, u), \quad u(0) = u_0, \tag{3.3.1}$$

where $f \in C(J \times E, E)$. Let us list the following assumptions for convenience.

(A1) For any bounded set B in E

$$\alpha(f(J \times B)) \leqslant L\alpha(B), \quad \text{for some} \quad L > 0,$$

where α denotes the Kuratowski's measure of noncompactness;

(A2) $v_0, w_0 \in C^1(J, E)$ with $v_0(t) \leqslant w_0(t)$ on J such that

$$v_0' \leqslant f(t, v_0),$$

$$w_0' \geqslant f(t, w_0) \quad \text{on} \quad J;$$

(A3) $f(t, u) - f(t, v) \geqslant -M(u - v)$ whenever $u \geqslant v$ and $u, v \in [v_0, w_0]$ for some $M > 0$, where $[v_0, w_0] = [u \in C(J, E) : v_0 \leqslant u \leqslant w_0]$. We note that when (A3) holds, f is quasimonotone.

In order to develop the monotone iterative technique, we need to consider the linear IVP

$$u' + Mu = \sigma(t), \; u(0) = u_0, \tag{3.3.2}$$

where $\sigma(t) = f(t, \eta(t)) + M\eta(t)$ and $\eta \in C(J, E)$ such that $v_0(t) \leqslant \eta(t) \leqslant w_0(t)$ on J. Clearly the linear IVP (3.3.2) possesses a unique solution on J.

For each $\eta \in (J, E)$ such that $v_0(t) \leqslant \eta(t) \leqslant w_0(t)$ on J, we define the mapping A by $A\eta = u$, where u is the unique solution of (3.3.2) corresponding to η. The following result concerning mapping A holds.

Lemma 3.3.1. *Suppose that assumptions (A1), (A2), and (A3) hold; then*

(i) $v_0 \leqslant Av_0$ *and* $w_0 \leqslant Aw_0$*;*

(ii) A is monotone on $[v_0, w_0]$*, that is, if* $\eta_1, \eta_2 \in [v_0, w_0]$ *with* $\eta \leqslant \eta_2$ *then* $A\eta \leqslant A\eta_2$.

Proof. (i) Suppose that $Av_0 = v_1$. Set $p(t) = \phi[v_1(t) - v_0(t)]$ so that $p(0) \geqslant 0$, where $\phi \in K^*$. Then

$$p' \geqslant \phi(f(t, v_0) - M(v_1 - v_0) - f(t, v_0)) = -Mp$$

in view of (A2). As a result we have $p(t) \geqslant p(0)e^{-Mt} \geqslant 0$ on J. Since $\phi \in K^*$ is arbitrary, this implies $v_1 \geqslant v_0$ on J proving $v_0 \leqslant Av_0$. Similarly we can show that $w_1 \geqslant Aw_0$.

To prove (ii), let $\eta_1, \eta_2 \in C(J, E)$ such that $\eta_1 \leqslant \eta_2$ on J and suppose that $A\eta_1 = u_1, A\eta_2 = u_2$. We set $p(t) = \phi[u_2(t) - u_1(t)]$ so that $p\ (0) \geqslant 0$, where, as before, $\phi \in K^*$. It then follows by using (A3) that

$$\begin{aligned} p\ ' &= \phi(f(t,\eta_2) - M(u_2 - \eta_2) - f(t,\eta_1) - M(u_1 - \eta_1)) \\ &\geqslant \phi(-M(\eta_2 - \eta_1) - M(u_2 - \eta_2) + M(u_1 - \eta_1)) = -Mp. \end{aligned}$$

Consequently $p(t) \geqslant p(0)e^{-Mt} \geqslant 0$ on J and this proves $A\eta_1 \leqslant A\eta_2$. The proof of the lemma is complete. □

In view of Lemma 3.3.1, we can define the sequences $\{v_n\}, \{w_n\}$ as follows:

$$v_n = Av_{n-1} \text{ and } w_n = Aw_{n-1}.$$

It is easy to see that $\{v_n\}, \{w_n\}$ are monotone sequences such that $v_n \leqslant w_n$ and $v_n, w_n \in [v_0, w_0]$. We shall now show that there exist subsequences of $\{v_n\}, \{w_n\}$ which converge uniformly on J.

Lemma 3.3.2. *Under the assumptions of Lemma 3.3.1,if the cone K is normal, then the sequences $\{v_n\}, \{w_n\}$ are uniformly bounded, equicontinuous and relatively compact on J.*

Proof. Since the cone K is assumed to be normal, it follows from $v_n, w_n \in [v_0, w_0]$ that $\{v_n\}, \{w_n\}$ are uniformly bounded. This implies the equicontinuity of the sequences by using standard estimates and the fact that f maps bounded sets into bounded sets which is a consequence of (A1). Now we set $B(t) = \{v_n(t)\}_{n=0}^{\infty}$ so that $B'(t) = \{v'_n(t)\}_{n=0}^{\infty}$ and $m(t) = \alpha(B(t))$. Using the standard arguments, we get

$$D^- m(t) \leqslant \alpha\left[\left\{\frac{v_n(t) - v_n(t-h)}{h}\right\}_{n=0}^{\infty}\right] \leqslant \alpha(\overline{\text{conv}}(v'_n(t))_{n=0}^{\infty})).$$

Thus we have

$$D^- m(t) \leqslant \lim_{h \to 0^+} \alpha(\cup_{J_h}\ B'(t)) \text{ where } J_h = [t-h, t] \subset J.$$

Evidently

$$\begin{aligned} \alpha\left(\cup_{J_h}\ B'(s)\right) &\leqslant \alpha\left(\cup_{J_h}\{f(s, v_{n-1}(s)\}_{n=1}^{\infty}\right) + 2M\alpha\left(\cup_{J_h}\{v_n(s)\}_{n=0}^{\infty}\right) \\ &\leqslant \alpha\left(\{(f(J, \cup_{J_h} B(s))\}\right) + 2M\alpha\left(\cup_{J_h} B(s)\right) \\ &\leqslant (L + 2M)\alpha\left(\cup_{J_h} B(s)\right). \end{aligned}$$

The equicontinuity of $v_n(t)$ now yields

$$D^- m(t) \leqslant (L + 2M)m(t), \ t \in J.$$

Since $m(0) = \alpha(\{v_n(0)\}_{n=0}^{\infty}) = \alpha(u_0, v_0(0)) = 0$, it is immediate that $m(t) \equiv 0$ on J , which implies the relative compactness of the sequence $\{v_n(t)\}$ for each $t \in J$. Similarly $\{w_n(t)\}$ is relatively compact for each $t \in J$. The proof of the lemma is complete. □

We now apply Ascoli's theorem to the sequences $\{v_n\}$, $\{w_n\}$ to obtain subsequences $\{v_{n_k}\}, \{w_{n_k}\}$ which converge uniformly on J . Since the sequences $\{v_n\}, \{w_n\}$ are monotone, this then shows that the full sequences converge uniformly and monotonically to continuous functions, that is, $\lim_{n\to\infty} v_n(t) = \rho(t)$ and $\lim_{n\to\infty} w_n(t) = r(t)$ on J . It then follows easily from (3.3.2) that $\rho(t)$ and $r(t)$ are solutions of IVP (3.3.1) on J.

Finally we show that $\rho(t), r(t)$ are minimal and maximal solutions of (3.3.1). To this end, let $u(t)$ be any solution of (3.3.1) on J such that $u \in [v_0, w_0]$. Assume that $v_n \leqslant u \leqslant w_n$ on J . Set $p\ (t) = \phi(u(t) - v_{n+1}(t))$, so that $p\ (0) = 0$, where, as before, $\phi \in K^*$. Then by (A3) and the assumption $v_n \leqslant u$, we have

$$\begin{aligned} p\ ' &= \phi(f(t,u) - f((t,v_n) + M(v_{n+1} - v_n)) \\ &\geqslant \phi(-M(u - v_n) + M(v_{n+1} - v_n)) = -Mp \end{aligned}$$

This implies $v_{n+1} \leqslant u$ on J . Similarly, we can show $u \leqslant w_{n+1}$ on J . Since $u \in [v_0, w_0]$, we have, by induction, $v_n \leqslant u \leqslant w_n$ on J for all n. Thus we obtain, taking the the limit as $n \to \infty, \rho(t) \leqslant u(t) \leqslant r(t)$ on J proving $\rho(t), r(t)$ are minimal and maximal solutions of (3.3.1) on J . We have therefore proved the following result which corresponds to Theorem 3.2.1. □

Theorem 3.3.1. *Let the cone K be normal and assumptions (A1), (A2), and (A3) hold. Then there exist monotone sequences $\{v_n\}$, $\{w_n\}$ which converge uniformly and monotonically to the minimal and maximal solutions $\rho(t), r(t)$, respectively, of the IVP on $[v_0, w_0]$. That is, if u is any solution of (3.3.1) in $[v_0, w_0]$, then*

$$v_0 \leqslant v_1 \leqslant ... \leqslant v_n \leqslant \rho \leqslant u \leqslant r \leqslant w_n \leqslant ... \leqslant w_1 \leqslant w_0 \quad \text{on } J$$

The ideas imbedded in Theorem 3.3.1 can be developed for monotone operators.

An operator $A : E \to E$ is said to be an increasing operator if $x_1 \leqslant x_2$ implies $Ax_1 \leqslant Ax_2$, $x_1, x_2 \in E$. The operator A is said to be completely continuous if it is continuous and compact. Note that compactness of A means that the set $A(S)$ is relatively compact for any bounded set $S \subset E$. Also, A is called a $k-$set-contraction $k \geq 0$, if it is continuous, bounded and $\alpha(A(S)) \leqslant k\alpha(S)$ for any bounded set $S \subset E$, where $\alpha(S)$ denotes the measure of noncompactness of S. A $k-$set-contraction is called a strict-set-contraction if $k < 1$. The operator A is said to be condensing if it is continuous, bounded and $\alpha(A(S)) < \alpha(S)$, with $\alpha(S) > 0$.

It is evident that if operator A is completely continuous, then A is a strict-set-contraction, and, if A is a strict-set-contraction, then A is a condensing mapping.

Theorem 3.3.2. *Let $u_0, v_0 \in E, u_0 < v_0$ and $A : [u_0, v_0] \to E$ be an increasing operator such that*

$$u_0 \leqslant Au_0, \quad Av_0 \leqslant v_0. \tag{3.3.3}$$

Suppose that one of the following two conditions is satisfied:

(H1) K is normal and A is condensing;

(H2) K is regular and A is semicontinuous, i.e., $x_n \to x$ strongly implies $Ax_n \to Ax$ weakly.

Then, A has a maximal fixed point x^* and a minimal fixed point x_* in $[u_0, v_0]$; moreover

$$x^* = \lim_{n\to\infty} v_n, \; x_* = \lim_{n\to\infty} u_n \tag{3.3.4}$$

where $v_n = Av_{n-1}$ and $u_n = Au_{n-1}$ $(n = 1, 2, 3...)$, and

$$u_0 \leqslant u_1 \leqslant ... \leqslant u_n \leqslant ... \leqslant v_n \leqslant ... \leqslant v_1 \leqslant v_0. \tag{3.3.5}$$

Proof. Since A is increasing, it follows from (3.3.3) that (3.3.5) holds. Now, we prove that $\{u_n\}$ converges to some $x_* \in E$ and $Ax_* = x_*$. When (H1) is satisfied, the set $S = \{u_0, u_1, u_2, ...\}$ is bounded and $S = A(S) \cup \{u_0\}$, hence $\alpha(S) = \alpha(A(S))$. Noticing that A is condensing, we have $\alpha(S) = 0$, i.e., S is a relatively compact set. Hence there exists a subsequence $\{u_{n_k}\} \subset \{u_n\}$ such that $u_{n_k} \to x_*$. Clearly, $u_n \leqslant x_* \leqslant v_n$ $(n = 1, 2, 3, ...)$. When $m > n_k$, we have $\theta \leqslant x_* - u_m \leqslant x_* - u_{n_k}$, and hence, by virtue of normality of K, $\|x_* - u_m\| \leqslant N \|x_* - u_{n_k}\|$, where N denotes the normal constant of K. Thus, $u_m \to x_*(m \to \infty)$. Taking limit as $n \to \infty$ on both sides of the equality $u_n = Au_{n-1}$, we get $x_* = Ax_*$ since A is continuous.

When (H2) is satisfied, $\{u_n\}$ converges to some $x_* \in E$ in view of regularity of K. Since A is semicontinuous, $u_n = Au_{n-1}$ converges to Ax_* weakly and hence $Ax_* = x_*$.

Similarly, we can prove that $\{v_n\}$ converges to some $x^* \in E$ and $Ax^* = x^*$.

Finally, we prove that x^* and x_* are the maximal and minimal fixed points of A in $[u_0, v_0]$, respectively. Let $\bar{x} \in [u_0, v_0]$ and $A\bar{x} = \bar{x}$. Since A is increasing , it follows from $u_0 \leqslant \bar{x} \leqslant v_0$, that $Au_0 \leqslant A\bar{x} \leqslant Av_0$, i.e., $u_1 \leqslant \bar{x} \leqslant v_1$. Using the same argument, we get $u_2 \leqslant \bar{x} \leqslant v_2$, and in general, $u_n \leqslant \bar{x} \leqslant v_n$ $(n = 1, 2, 3...)$. Now, taking limit as $n \to \infty$, we obtain $x_* \leqslant \bar{x} \leqslant x^*$, and our theorem is proved. □

Corollary 3.3.1. *Let the conditions of Theorem 3.3.2 be satisfied. Suppose that A has only one fixed point $\bar{x}$ in $[u_0, v_0]$. Then, for any $x_0 \in [u_0, v_0]$, the successive iterates*

$$x_n = Ax_{n-1} \quad (n = 1, 2, 3...) \tag{3.3.6}$$

converges to $\bar{x}$, i.e. $\|x_n - \bar{x}\| \to 0$ $(n \to \infty)$.

Proof. Since $u_0 \leqslant x_0 \leqslant v_0$, and A is increasing, we have

$$u_0 \leqslant \bar{x} \leqslant v_0 \quad (n = 1, 2, 3...) \tag{3.3.7}$$

By hypotheses we must have $x_* = x^* = \bar{x}$. It follows therefore from 3.3.7, 3.3.4 and normality of K, that $x_n \to \bar{x}$. □

Theorem 3.3.3. *Let $u_0, v_0 \in E$, $u_0 < v_0$ and $A : [u_0, v_0] \to E$ be an increasing operator such that* (3.3.3) *holds. Suppose that K is strongly minihedral. Then A has a maximal fixed point x^* and a minimal fixed point x_* in $[u_0, v_0]$.*

Proof. Let $D = \{x \in E \mid u_0 \leqslant x \leqslant v_0,\ Ax \geqslant x\}$. Evidently, $u_0 \in D$ and v_0 is an upper bound of D . Consequently, by the strong minihedrality of K, $x^* = \sup\ D$ exists. Now we prove that x^* is the maximal fixed point of A in $[u_0, v_0]$. In fact, $u_0 \leqslant x \leqslant x^* \leqslant v_0$ for any $x \in D$ and hence $u_0 \leqslant Au_0 \leqslant Ax \leqslant Ax^* \leq Av_0 \leqslant v_0$. Therefore, observing $Ax \geqslant x$, we obtain $x \leqslant Ax^*$ for any $x \in D$. It follows from the definition of least upper bound that $x^* \leqslant Ax^*$. On the other hand, from $x^* \leqslant Ax^*$ we know that $Ax^* \leqslant A(Ax^*)$, which implies that $Ax^* \in D$, and therefore $Ax^* \leqslant x^*$. Thus, $Ax^* = x^*$. If $\bar{x}$ is any fixed point of A in $[u_0, v_0]$, then $\bar{x} \in D$, and $\bar{x} \leqslant x^*$. in $[u_0, v_0]$, then $\bar{x} \in D$, and $\bar{x} \leqslant x^*$. This proves the maximality of x^*.

Similarly, we can prove that $x_* = \inf\ D_1$ is the minimal fixed point of A in $[u_0, v_0]$, where $D_1 = \{x \in E \mid u_0 \leqslant x \leqslant v_0,\ Ax \leqslant x\}$. □

Notice that in Theorem 3.3.3 we do not assume A to be continuous, and therefore, we cannot assert that the limits (3.3.4) exist.

Theorem 3.3.4. *Let $u_0, v_0 \in E$, $u_0 < v_0$ and $A : [u_0, v_0] \to E$ be an increasing operator such that* (3.3.3) *holds. Suppose that $A([u_0, v_0])$ is a relatively compact set of E. Then A has at least one fixed point in $[u_0, v_0]$.*

Proof. Let $F = \{x \in A([u_0, v_0]) \mid Ax \geqslant x\}$. We have $A(Au_0) \geqslant Au_0$, which implies $Au_0 \in F$, and therefore F is not empty. Since E is partially ordered by K, F is a partially ordered set. Now, suppose that G is a completely ordered subset of F .Since $A([u_0, v_0])$ is relatively compact by hypothesis and $G \subset F \subset A([u_0, v_0])$, G is relatively compact and so separable, i.e., there exists a denumerable set $V = \{y_1, y_2, ...\} \subset G$, which is dense in G. Since G is completely ordered, $z_n = \sup\ \{y_1, y_2, ...y_n\}$, $(n = 1, 2, 3...)$ exist and $z_n \in G$ (in fact, z_n equals one of y_i, $i = 1, 2, ..., n$). It follows from the relative compactness of G that there exists a subsequence $\{z_{n_i}\} \subset \{z_n\}$ such that $\{z_{n_i}\} \to z^* \in E$. Since

$$z_1 \leqslant z_2 \leqslant\ ... \leqslant z_n \leqslant\ ... ,$$

we have

$$y_n \leqslant z_n \leqslant z^*. \quad (n = 1, 2, 3...) \tag{3.3.8}$$

and $z^* \in \bar{G} \subset \bar{F} \subset \overline{A([u_0, v_0])} \subset [u_0, v_0]$. From (3.3.8) we know $z \leqslant z^*$ for any $z \in G$, and so $z \leqslant Az \leqslant Az^*$ for any $z \in G$. Thus, Az^* is an upper bound of G. On the other hand, from $z_n \leqslant Az^*$ $(n = 1, 2, 3...)$ we know $z^* \leqslant Az^*$ and therefore $Az^* \leqslant A(Az^*)$, which shows that $Az^* \in F$. Thus, Az^* is an upper bound of G in F. It follows therefore from Zorn's Lemma that F contains a maximal element x^*. Since $Ax^* \geqslant x^*$ and so $A(Ax^*) \geqslant Ax^*$, we have $Ax^* \in F$.

By the maximality of x^*, we must have $Ax^* = x^*$, □

Corollary 3.3.2. *Let $u_0, v_0 \in E$, $u_0 < v_0$, and $A : [u_0, v_0] \to E$ be an increasing operator such* (3.3.3) *holds. Suppose that K is normal and A is compact. Then A has at least one fixed point in $[u_0, v_0]$.*

Proof. Since K is normal, $[u_0, v_0]$ is bounded. Thus, the relative compactness of the set $A([u_0, v_0])$ follows from the compactness of operator A. □

Theorem 3.3.5. *Let $u_0, v_0 \in E$, $u_0 < v_0$ and $A = [u_0, v_0] \to E$ be an increasing operator such that* (3.3.3) *holds. Suppose that K is minihedral and $A([u_0, v_0])$ is a relatively compact set of E. Then A has a maximal fixed point x^* and a minimal fixed point x^* in $[u_0, v_0]$.*

Proof. Let $F = \{x \in A([u_0, v_0]) \mid Ax \geqslant x\}$. By Zorn's Lemma we have already proved in Theorem 3.3.4 that F contains a maximal element x^* and $Ax^* = x^*$. Now, we prove that x^* is the maximal fixed point of A in $[u_0, v_0]$. In fact, suppose $\bar{x}$ is any fixed point of A in $[u_0, v_0]$, then by the minihedrality of $K, v = \sup\{\bar{x}, x^*\}$ exists. From $v \geqslant \bar{x}$ and $v \geqslant x^*$ we get $Av \geqslant A\bar{x} = \bar{x}$ and $Av \geqslant Ax^* = x^*$. Hence $v \leqslant Ax$ and $Av \leqslant A(Av)$. It follows therefore $Av \in F$. By the maximality of x^*, we have $Av = x^*$ and so $x^* \geqslant \bar{x}$. Hence, x^* is the maximal fixed point of A in $[u_0, v_0]$.

In the same way we can prove that $F_1 = \{x \in A([u_0, v_0]) \mid Ax \leqslant x\}$ contains a minimal element x_* which satisfies $Ax_* = x_*$ and is the minimal fixed point of A in $[u_0, v_0]$. □

Corollary 3.3.3. *Let $u_0, v_0 \in E$, $u_0 < v_0$, and $A : [u_0, v_0] \to E$ be an increasing operator such that* (3.3.3) *holds. Suppose that K is normal and minihedral and A is compact. Then, A has a maximal fixed point x^* and a minimal fixed point x_* in $[u_0, v_0]$.*

3.4 Generalized Monotone Iterative Technique

We shall devote this section to proving general results relative to monotone iterative technique which contain, as special cases, several important results. We need the following definition which characterizes lower and upper solutions of various types. We consider the IVP

$$x' = f(t, x) + g(t, x), \quad x(0) = x_0 \tag{3.4.1}$$

where $J = [0, T]$, f, $g \in C(J \times E, E)$. We first define various possible notions of lower and upper solutions relative to (3.4.1)

Definition 3.4.1. Let $v, w \in C^1(J, E)$. Then v, w are said to be

(i) natural lower and upper solutions of (3.4.1) if

$$\left.\begin{array}{lll} v' & \leqslant & f(t,v) + g(t,v), \quad v(0) \leqslant x_0, \\ w' & \geqslant & f(t,w) + g(t,w), \quad w(0) \geqslant x_0; \end{array}\right\} \tag{3.4.2}$$

(ii) coupled lower and upper solutions of type I if

$$\left.\begin{array}{lll} v' & \leqslant & f(t,v) + g(t,w), \quad v(0) \leqslant x_0, \\ w' & \geqslant & f(t,w) + g(t,v), \quad w(0) \geqslant x_0; \end{array}\right\} \tag{3.4.3}$$

(iii) coupled lower and upper solutions of type II if

$$\left.\begin{array}{lll} v' & \leqslant & f(t,w) + g(t,v), \quad v(0) \leqslant x_0, \\ w' & \geqslant & f(t,v) + g(t,w), \quad w(0) \geqslant x_0; \end{array}\right\} \tag{3.4.4}$$

(iv) coupled lower and upper solutions of type III if

$$\left.\begin{array}{lll} v' & \leqslant & f(t,w) + g(t,w), \quad v(0) \leqslant x_0, \\ w' & \geqslant & f(t,v) + g(t,v), \quad w(0) \geqslant x_0. \end{array}\right\} \tag{3.4.5}$$

We note that whenever $v(t) \leqslant w(t)$, $t \in J$, if $f(t,x)$ is nondecreasing in x for each t and $g(t,y)$ is nonincreasing in y for each t, then the inequalities satisfied by lower and upper solutions defined in (3.4.2) and (3.4.5) reduce to (3.4.4). Hence it is enough to investigate the cases (3.4.3) and (3.4.4).

We first prove the following result.

Theorem 3.4.1. *Assume that*

(A1) v_0, $w_0 \in C(J,E)$ *are coupled lower and upper solutions of type I relative to IVP* (3.4.1) *with* $v_0(t) \leqslant w_0(t)$, $t \in J$ *and* $v_0(0) \leqslant x_0 \leqslant w_0(0)$;

(A2) f, $g \in C(J \times E, E)$, f, g *compact,* $f(t,x)$ *is nondecreasing in* x *and* $g(t,y)$ *is nonincreasing in* y *for each* $t \in J$. *Then there exist monotone sequence* $\{v_n(t)\}$, $\{w_n(t)\}$ *such that* $\{v_n(t)\} \to \rho(t), \{w_n(t)\} \to r(t)$ *as* $n \to \infty$ *and* (ρ, r) *are coupled minimal and maximal solutions of IVP* (3.4.1) *respectively, that is, they satisfy*

$$\begin{array}{ll} \rho' & = f(t,\rho) + g(t,r), \quad \rho(0) = x_0, \\ r' & = f(t,r) + g(t,\rho), \quad r(0) = x_0. \end{array}$$

Proof. For integer $n \geqslant 0$, consider the simple linear differential equations

$$v'_{n+1} = f(t,v_n) + g(t,w_n), \quad v_{n+1}(0) = x_0, \tag{3.4.6}$$

$$w'_{n+1} = f(t,w_n) + g(t,v_n), \quad w_{n+1}(0) = x_0. \tag{3.4.7}$$

Clearly, there exist unique solutions $v_{n+1}(t),\ w_{n+1}(t)$ of the IVP (3.4.6) and (3.4.7) respectively, such that $v_{n+1},\ w_{n+1} \in C(J, E)$. We shall show that

$$v_0 \leqslant v_1 \leqslant v_2 \leqslant \ \dots \ \leqslant v_n \leqslant w_n \leqslant \ \dots \ w_2 \leqslant w_1 \leqslant w_0, \tag{3.4.8}$$

on J. Setting $n = 0$ in (3.4.6) and writing $p = \phi(v_0 - v_1)$ where $\phi \in K^*$, we obtain

$$p\ (0) = \phi(v_0(0) - v_1(0)) \leqslant 0$$

and

$$p\ ' = \phi(v_0' - v_1') \leqslant \phi(f(t, v_0) + g(t, w_0) - f(t, v_0) - g(t, w_0)) \leqslant 0.$$

This implies $p(t) \leqslant 0,\ t \in J$, which shows that $v_0 \leqslant v_1$, on J. Similarly, it can be shown that $w_1 \leqslant w_0$ on J.

Now, set $p = \phi(v_1 - w_1)$ where $\phi \in K^*$. Then by using (3.4.6), we have

$$p\ ' = \phi(v_1' - w_1') \leqslant \phi(f(t, v_0) + g(t, w_0) - f(t, w_0) - g(t, v_0)) \leqslant 0$$

because of the monotone character of $f,\ g$ and the fact that $v_0 \leqslant w_0$ on J. Thus, $p\ '(t) \leqslant 0,\ p\ (0) \leqslant 0$ which yields $p\ (t) \leqslant 0$ which implies $v_1(t) \leqslant w_1(t),\ t \in J$. Assume that for some integer $k > 1$,

$$v_{k-1} \leqslant v_k \leqslant w_k \leqslant w_{k-1} \quad \text{on} \quad J.$$

we shall show that

$$v_k \leqslant v_{k+1} \leqslant w_{k+1} \leqslant w_k \quad \text{on} \quad J.$$

Consider $p = \phi(v_k - v_{k+1})$ on J. Then by (3.4.6) and monotone nature of $f,\ g,$ we obtain

$$\begin{aligned} p' \ = \ \phi(v_k' - v_{k+1}') \ &= \ \phi(f(t, v_{k-1}) + g(t, w_{k-1}) - f(t, v_k) - g(t, w_k)) \\ &\leqslant \ \phi(f(t, v_k) + g(t, w_k) - f(t, v_k) - g(t, w_k)) \\ &\leqslant \ 0. \end{aligned}$$

Since $p\ (0) = 0$, this implies that $p\ (t) \leqslant 0$ or equivalently $v_k \leqslant v_{k+1}$ on J. Similarly, one can show that $w_{k+1} \leqslant w_k$ on J. To prove $v_{k+1} \leqslant w_{k+1}$, set

$$p = \phi(v_{k+1} - w_{k+1}), \quad \phi \in K^*$$

and see that $p\ (0) = 0$ and

$$\begin{aligned} p\ ' \ = \ \phi(v_{k+1}' - w_{k+1}') \ &= \ \phi(f(t, v_k) + g(t, w_k) - f(t, w_k) - g(t, v_k)) \\ &\leqslant \ \phi(f(t, w_k) + g(t, v_k) - f(t, w_k) - g(t, v_k)) \\ &\leqslant \ 0. \end{aligned}$$

since we have assumed $v_k \leqslant w_k$. This yields $p\ (t) \leqslant 0$, on J, i.e. $v_{k+1} \leqslant w_{k+1}$ on J. Now, by induction principle, we have (3.4.8) for all n.

From the normality of the cone, we get the uniform boundedness of $\{v_n(t)\}$, $\{w_n(t)\}$ on J. Also, the compactness assumption on f, g yields that sequences satisfy Ascoli-Arzela theorem. Since the sequences $\{v_n(t)\}, \{w_n(t)\}$ are monotone, it is clear that the entire sequences $\{v_n(t)\}, \{w_n(t)\}$ converge uniformly and monotonically to ρ, r respectively on J. It is easy to see from (3.4.6) and (3.4.7) that ρ, r are coupled solutions on J.

To show that ρ, r are coupled minimal and maximal solutions of (3.4.1), let $x(t)$ be any solution of (3.4.1) such that $v_0 \leqslant x \leqslant w_0$ on J. Suppose that for some k, $v_k \leqslant x \leqslant w_k$ on J. Setting

$$p = \phi(v_{k+1} - x), \quad \phi \in K^*$$

we get $p\,(0) = 0$, and

$$p\,' = \phi(v'_{k+1} - x') = \phi(f(t, v_k) + g(t, w_k) - f(t, x) - g(t, x)) \leqslant 0,$$

using the monotone character of f, g and the assumption $v_k \leqslant x \leqslant w_k$ on J. This implies that $p\,(t) \leqslant 0$ proving $v_{k+1} \leqslant x$ on J. Similarly, it can be shown that $x \leqslant w_{k+1}$ on J and hence, by the principle of induction, we have

$$v_n \leqslant x \leqslant w_n \quad \text{on J,} \quad \text{for all } n.$$

As $n \to \infty$, we have $\rho \leqslant x \leqslant r$ completing the proof that ρ, r are coupled extremal solutions of (3.4.1), since from (3.4.6) and (3.4.7), we get

$$\rho' = f(t, \rho) + g(t, r), \quad \rho(0) = 0,$$

$$r' = f(t, r) + g(t, \rho), \ r(0) = 0,$$

on J, respectively.

Corollary 3.4.1. *If in addition to the assumptions of Theorem 3.4.1, we suppose that for* $u_1 \geqslant u_2, \quad u_1, u_2 \in [v_0, w_0]$,

$$f(t, u_1) - f(t, u_2) \leqslant N_1(u_1 - u_2), \quad N_1 > 0$$

$$g(t, u_1) - g(t, u_2) \geqslant -N_2(u_1 - u_2), \quad N_2 > 0$$

then $\rho(t) = r(t) = x(t)$ is the unique solution of (3.4.1) on J.

Proof. Since $\rho \leqslant r$ on J, it is enough to show that $r \leqslant \rho$. Consider $p = \phi(r - \rho)$, $\phi \in K^*$. Then $p(0) = 0$ and

$$\begin{aligned} p\,' \;=\; \phi(r' - \rho') &= \phi(f(t, r) + g(t, \rho) - f(t, \rho) - g(t, r)) \\ &\leqslant \phi(N_1(r - \rho) + N_2(r - \rho)) \\ &\leqslant (N_1 + N_2)p. \end{aligned}$$

Hence $p\ (t) \leqslant 0$ on J, proving $r \leqslant \rho$. This proves $\rho(t) = r(t) = x(t)$ on J, completely the proof. □

Remark 3.4.1 Following the proof of Theorem 3.4.1, there are several interesting remarks to be made which indicate many ramifications and provide useful special cases:

(1) In Theorem 3.4.1, suppose that $g(t, x) \equiv 0$. Then $v_0(t), w_0(t)$ are natural lower and upper solutions of (3.4.1), with $f(t, x)$ nondecreasing in x for each t, we get the monotone sequences $\{v_n(t)\}, \{w_n(t)\}$ converging to minimal and maximal solutions of (3.4.1) respectively, lying in $[v_0, w_0]$.

(2) However, if $g(t, x) \equiv 0$ and $f(t, x)$ is not nondecreasing in x for each t, we can assume that $f(t, x) + Mx$ is nondecreasing in x for some $M > 0$. We still reach the same conclusion as in (1), since the IVP

$$x'(t) = \widetilde{f}(t, x) + g(t, x), \ x(t_0) = x_0,$$

with $\widetilde{f}(t, x) = f(t, x) + Mx, g(t, x) = -Mx$, now satisfies all the assumptions of Theorem 3.4.1.

(3) If $f(t, x) \equiv 0$ in Theorem 3.4.1, we obtain the result for nonincreasing $g(t, x)$, and v_0, w_0 are coupled lower and upper solutions of the IVP $x'(t) = g(t, x), x(t_0) = x_0$. In this case, the monotone iterates $\{v_n(t)\}, \{w_n(t)\}$ converge to ρ, r respectively which satisfy

$$\rho'(t) = g(t, r(t)), \ \ r'(t) = g(t, \rho(t)), \ r(t_0) = x_0 = \rho(t_0).$$

(4) If in (3) above, we suppose that $g(t, x)$ is not nonincreasing and there exists a $N > 0$ such that $\widetilde{g}(t, x) = g(t, x) - Nx$ is nonincreasing, we can consider the IVP

$$x'(t) = \widetilde{g}(t, x) + Nx(t), \ \ x(t_0) = x_0$$

which is the same as (3.4.1) with Nx in place of $f(t, x)$ and $\widetilde{g}(t, x)$ in place of $g(t, x)$. Hence the case reduces to Theorem 3.4.1 and the conclusion of Theorem 3.4.1 remains valid.

(5) Suppose $f(t, x)$ is nondecreasing but $g(t, x)$ is not nonincreasing. Then consider the IVP

$$x'(t) = \widetilde{f}(t, x) + \widetilde{g}(t, x), \ x(t_0) = x_0$$

where $\widetilde{f}(t, x) = f(t, x) + Nx, \ N > 0$ is nondecreasing and $\widetilde{g}(t, x) = g(t, x) - Nx, \ N > 0$ is nondecreasing. Note that $\widetilde{f}(t, x) + \widetilde{g}(t, x) = f(t, x) + g(t, x)$ and hence we have the same situation as in Theorem 3.4.1 and the conclusion of Theorem 3.4.1 holds.

(6) If $f(t,x)$ is not nondecreasing and $g(t,x)$ is nonincreasing, then consider the IVP
$$x'(t) = \widetilde{f}(t,x) + \widetilde{g}(t,x), \quad x(t_0) = x_0$$
with $\widetilde{f}(t,x) = f(t,x) + Mx$, $M > 0$ nondecreasing in x and $\widetilde{g}(t,x) = g(t,x) - Mx$, $M > 0$ nonincreasing in x for each t. This again results in Theorem 3.4.1 and hence the conclusion of Theorem 3.4.1 is valid.

(7) If $f(t,x)$ is not nondecreasing and $g(t,x)$ is not nonincreasing, then consider the IVP
$$x'(t) = \widetilde{f}(t,x) + \widetilde{g}(t,x), \; x(t_0) = x_0$$
where $\widetilde{f}(t,\, x) = f(t,\, x)+Mx+Nx, \widetilde{g}(t,\, x) = g(t,\, x)-Nx-Mx$, for $M, N > 0$. Now, Theorem 3.4.1 is valid.

3.5 The Method of Quasilinearization

When we utilize the technique of lower and upper solutions coupled with the method of quasi-linearization and employ the idea of Newton-Fourier, we can construct concurrently upper and lower bounding sequences whose elements are the solutions of linear initial value problems. Compared to monotone iterative technique, the method of quasilinearization has the advantage that the sequences converge quadratically to the unique solution.

We consider the IVP
$$x' = f(t,x), \; x(0) = x_0 \tag{3.5.1}$$
when $f \in C(J \times E, E)$, $J = [0,T]$. We shall prove the following simple result to bring out the ideas involved.

Theorem 3.5.1. *Assume that the cone $K \subset E$ be the distance set, K is normal and*

(*i*) $v_0, w_0 \in C^1(J,E)$ *satisfy the inequalities*
$v_0' \leqslant f(t,v_0), \quad w_0' \geqslant f(t,w_0), \quad v_0 \leqslant w_0$ *on* J;

(*ii*) *the second Frechet derivative* f_{xx} *exists and* $f_{xx}(t,x) \geqslant 0$;

(*iii*) f *is compact,* $\|f_x(t,x)\| \leqslant L$ *on* $J \times [v_0, w_0]$.

Then there exist monotone sequences $\{v_n(t)\}, \{w_n(t)\}$ *which converge quadratically to the unique solution* $x(t)$ *of* (3.5.1) *on* J.

Proof. In view of the assumptions (*ii*) and (*iii*), we obtain for $x \geqslant y$,
$$f(t,x) \geqslant f(t,y) + f_x(t,y)(x-y) \tag{3.5.2}$$
and
$$f(t,x) - f(t,y) \leqslant L(x-y), \; L > 0. \tag{3.5.3}$$

Because of (3.5.3), it follows that IVP (3.5.1) possesses the unique solution $x(t)$ on J. We define the iterates $\{v_n(t)\}, \{w_n(t)\}$ as follows:

$$v'_{n+1} = f(t, v_n) + f_x(t, v_n)(v_{n+1} - v_n), \quad v_{n+1}(0) = x_0, \tag{3.5.4}$$

$$w'_{n+1} = f(t, w_n) + f_x(t, w_n)(w_{n+1} - v_n), \quad w_{n+1}(0) = x_0 \tag{3.5.5}$$

on J, with $v_0(0) \leqslant x_0 \leqslant w_0(0)$. We shall show that

$$v_0 \leqslant v_1 \leqslant v_2 \leqslant ... \leqslant v_n \leqslant w_n \leqslant ... \leqslant w_2 \leqslant w_1 \leqslant w_0 \tag{3.5.6}$$

on J. Let us first show that

$$v_0 \leqslant v_1 \leqslant w_1 \leqslant w_0 \quad \text{on} \quad J \tag{3.5.7}$$

Set $p = v_0 - v_1$, so that $p\ (0) = 0$ and by (3.5.4),

$$\begin{aligned} p\,' &= v'_0 - v'_1 \leqslant f(t, v_0) - f(t, v_0) - f_x(t, v_0)(v_1 - v_0) \\ &\leqslant f_x(t, v_0)p \end{aligned}$$

By Theorem 2.2.5, it follows that $p\ (t) \leqslant 0$ on J, which implies $v_0 \leqslant v_1$ on J. Similarly we can prove $w_1 \leqslant w_0$ on J. Now, let us set $p = (v_1 - w_1)$, so that $p\ (0) = 0$ and by (3.5.2), (3.5.4) and (3.5.5),

$$\begin{aligned} p\,' &= f(t, v_0) + f_x(t, v_0)(v_1 - v_0) - f(t, w_0) - f_x(t, w_0)(w_1 - w_0) \\ &\leqslant f_x(t, v_0)(v_0 - w_0) + f_x(t, v_0)(v_1 - v_0) - f_x(t, w_0)(w_1 - w_0) \end{aligned}$$

which, in view of (ii), yields

$$p\,' \leqslant f_x(t, v_0)p.$$

This implies by Theorem 2.2.5, $p\ (t) \leqslant 0$ on J, which gives $v_1 \leqslant w_1$ on J. This establishes (3.5.7). Assuming that for some integer $k \geqslant 1$, the relation

$$v_{k-1} \leqslant v_k \leqslant w_k \leqslant w_{k-1} \quad \text{on} \quad J$$

is true, we can prove, by arguments similar to those in getting (3.5.7), that

$$v_k \leqslant v_{k+1} \leqslant w_{k+1} \leqslant w_k \quad \text{on} \quad J$$

and then, by induction principle, (3.5.6) is true for all n.

The iterative scheme described by (3.5.4) and (3.5.5) yields the monotone sequences $\{v_n(t)\}, \{w_n(t)\}$ satisfying (3.5.6). We now employ similar arguments as in Theorem 3.4.1 using normality of the cone K and compactness of f, that $\{v_n(t)\}, \{w_n(t)\}$ converge uniformly and monotonically to the unique solution $x(t)$ of (3.5.1) on J.

We shall next prove that the convergence is quadratic. For this purpose, consider

$$p_{n+1} = x - v_{n+1} \geqslant \theta, \quad q_{n+1} = w_{n+1} - x \geqslant \theta \text{ on J},$$

and note $p_{n+1}(0) = q_{n+1}(0) = \theta$. We then have

$$\begin{aligned} p'_{n+1} = x' - v'_{n+1} &= f(t,x) - f(t,v_n) - f_x(t,v_n)(v_{n+1} - v_n) \\ &= f_x(t,\sigma)p_n - f_x(t,v_n)(p_n - p_{n+1}) \\ &\leqslant (f_x(t,x) - f_x(t,v_n))p_n + f_x(t,v_n)p_{n+1} \end{aligned}$$

where $v_n \leqslant \sigma \leqslant x$, which further yields

$$p'_{n+1} \leqslant f_{xx}(t,\sigma)p_n^2 + f_x(t,v_n)p_{n+1}.$$

Let

$$\tilde{p}_{n+1} = \phi(x - v_{n+1}), \qquad \phi \in K^*.$$

We then get $\tilde{p}\,'_{n+1} \leqslant M\tilde{p}_{n+1} + N\tilde{p}_n^2$ where $\|f_x(t,x)\| \leqslant M$, $\|f_{xx}(t,x)\| \leqslant N$ on J. Now, applying Gronwall inequality and normality of the cone, we get

$$\theta \leqslant \tilde{p}_{n+1}(t) \leqslant N \int_0^t \tilde{p}_n^2(s)\, e^{M(t-s)}\, ds$$

which yields the desired estimate

$$\max_J \|x(t) - v_{n+1}(t)\| \leqslant \frac{N}{M}\, e^{Mt} \max_J \|x(t) - v_n(t)\|^2.$$

One can obtain, in a similar way, the estimate for $\max_J \|w_{n+1}(t) - x(t)\|$. The proof is complete. □

3.6 Extension of Quasilinearization.

We have seen in Section 3.5 that when f is convex, one can obtain lower and upper bounding sequences simultaneously that converge quadratically to the unique solution of the IVP (3.5.1). We have a similar conclusion when f is concave. The question is whether we can obtain a corresponding result when f admits a decomposition into a difference of two convex or concave functions or equivalently f admits a splitting into convex and concave parts. The answer is affirmative and we shall discuss such a general situation in this section.

Consider the IVP

$$x' = f(t,x) + g(t,x), \quad x(0) = x_0 \text{ on } J \tag{3.6.1}$$

where $f,\ g \in C(J \times E, E)$. Let us discuss the case where lower and upper solutions are natural and prove the following result.

Theorem 3.6.1. *Assume that the cone $K \subset E$ is a distance set and*

(i) $v_0,\ w_0 \in C^1(J, E),\ v_0 \leqslant w_0$ *on* J, $v_0(0) \leqslant x_0 \leqslant w_0(0)$, *satisfying the inequalities*

$$v_0' \leqslant f(t, v_0) + g(t, v_0),$$

$$w_0' \geqslant f(t, w_0) + g(t, w_0);$$

(ii) $f,\ g$ *possess Frechet derivative* $f_x,\ g_x,\ f_{xx},\ g_{xx}$ *and* $f_{xx} \geqslant 0,\ g_{xx} \leqslant 0,\ (f_x + g_x)(x)$ *is quasimonotone in* x*;*

(iii) *The cone K is also regular in addition to being normal. Then, there exist monotone sequences $\{v_n\},\ \{w_n\}$ which converge uniformly to the unique solution $x(t)$ of (3.6.1) and the convergence is quadratic.*

Proof. We note that assumption (ii) yields the following inequalities for $x \geqslant y$,

$$f(t, x) \geqslant f(t, y) + f_x(t, y)(x - y), \tag{3.6.2}$$

$$g(t, x) \geqslant g(t, y) + g_x(t, y)(x - y). \tag{3.6.3}$$

It is also clear that $f,\ g$ satisfy, for any $u_1, u_2 \in [v_0, w_0]$,

$$-L(u_1 - u_2) \leqslant f(t, u_1) - f(t, u_2) \leqslant L(u_1 - u_2) \tag{3.6.4}$$

$$-L(u_1 - u_2) \leqslant g(t, u_1) - g(t, u_2) \leqslant L(u_1 - u_2) \tag{3.6.5}$$

for some $L > 0$.

Consider the IVPs

$$\begin{aligned} v_{n+1}' &= [f(t, v_n) + f_x(t, v_n)(v_{n+1} - v_n) + g(t, v_n) \\ &\quad + g_x(t, w_n)(v_{n+1} - v_n)], \quad v_{n+1}(0) = x_0, \end{aligned} \tag{3.6.6}$$

$$\begin{aligned} w_{n+1}' &= [f(t, w_n) + f_x(t, v_n)(w_{n+1} - w_n) + g(t, w_n) \\ &\quad + g_x(t, w_n)(w_{n+1} - w_n)], \quad w_{n+1}(0) = x_0, \end{aligned} \tag{3.6.7}$$

where $v_0,\ w_0$ satisfy assumption (i). We shall show that

$$v_0 \leqslant v_1 \leqslant v_2 \leqslant ... \leqslant v_n \leqslant w_n \leqslant ... \leqslant w_2 \leqslant w_1 \leqslant w_0 \tag{3.6.8}$$

on J. Set

$$p = v_0 - v_1$$

and note that $p(0) = v_0(0) - v_1(0) \leqslant 0$. Also,

$$\begin{aligned} p' &= (v_0' - v_1') \\ &\leqslant f(t, v_0) + g(t, v_0) - f(t, v_0) - f_x(t, v_0)(v_1 - v_0) \\ &\quad -g(t, v_0) - g_x(t, w_0)(v_1 - v_0) \\ &= (f_x(t, v_0) + g_x(t, w_0))p \end{aligned}$$

which implies, by Theorem 2.2.5, that $p(t) \leqslant 0$ on J. Thus, we have $v_0 \leqslant v_1$ on J.

Similarly we can show that $w_1 \leqslant w_0$ on J. Next we shall prove that $v_1 \leqslant w_1$ on J. Set $p = v_1 - w_1$ and note $p(0) = 0$. Now, using (3.6.2), (3.6.3), (3.6.6) and (3.6.7), we have

$$\begin{aligned} p' = (v_1' - w_1') &= [f(t,v_0) + f_x(t,v_0)(v_1 - v_0) + g(t,v_0) + g_x(t,w_0)(v_1 - v_0)] \\ &\quad -[f(t,w_0) + f_x(t,v_0)(w_1 - w_0) + g(t,w_0) + g_x(t,w_0)(w_1 - w_0)] \\ &\leqslant [f_x(t,v_0)(v_0 - w_0) + g_x(t,v_0)(v_0 - w_0) + f_x(t,v_0)(v_1 - v_0) \\ &\quad -f_x(t,v_0)(w_1 - w_0) + g_x(t,w_0)(v_1 - v_0) - g_x(t,w_0)(w_1 - w_0)] \\ &\leqslant [f_x(t,v_0) + g_x(t,w_0)]p. \end{aligned}$$

This implies, by Theorem 2.2.5, that $p(t) \leqslant 0$ on J, proving $v_1 \leqslant w_1$ on J. Hence,

$$v_0 \leqslant v_1 \leqslant w_1 \leqslant w_0 \quad \text{on} \quad J. \tag{3.6.9}$$

Proceeding this way, it is easy to prove that (3.6.8) holds. The assumption (iii) now guarantees the uniform convergence of the sequences $\{v_n(t)\}, \{w_n(t)\}$ such that $v_n \to \rho$, $w_n \to r$, as $n \to \infty$, where ρ, r are the solutions of the IVP (3.6.1) on J. Because of the conditions (3.6.4) and (3.6.5), it is not difficult to show that $\rho = r = x$ is the solution of IVP (3.6.1) on J.

To prove the quadratic convergence of the sequences $\{v_n\}, \{w_n\}$, consider

$$p_{n+1} = x - v_{n+1} \geqslant \theta, \quad q_{n+1} = w_{n+1} - x \geqslant \theta$$

and note $p_{n+1}(0) = \theta = q_{n+1}(0)$. Moreover, we have

$$\begin{aligned} p_{n+1}' = x' - v_{n+1}' &= f(t,x) + g(t,x) - f(t,v_n) - f_x(t,v_n)(v_{n+1} - v_n) \\ &\quad - g(t,v_n) - g_x(t,w_n)(v_{n+1} - v_n) \\ &= [f_x(t,\sigma)p_n + f_x(t,v_n)p_{n+1} - f_x(t,v_n)p_n] \\ &\quad + [g_x(t,\eta)p_n + g_x(t,w_n)p_{n+1} - g_x(t,w_n)p_n] \\ &\leqslant [f_x(t,x) - f_x(t,v_n)]p_n + f_x(t,v_n)p_{n+1} \\ &\quad + [g_x(t,x) - g_x(t,w_n)]p_n + g_x(t,w_n)p_{n+1} \\ &\leqslant f_{xx}(t,\sigma)p_n^2 + f_x(t,v_n)p_{n+1} + g_{xx}(t,\eta)p_n^2 + g_x(t,w_n)p_{n+1}. \end{aligned}$$

It then follows that

$$\begin{aligned} p_{n+1}' &\leqslant M_1 p_n^2 + N_1 p_{n+1} + M_2 p_n^2 + N_2 p_{n+1}, \\ &\leqslant (M_1 + M_2)p_n^2 + (N_1 + N_2)p_{n+1}. \end{aligned}$$

Applying the Gronwal's inequality and normality of the cone, we obtain

$$0 \leqslant \|p_{n+1}(t)\| \leqslant (M_1 + M_2)\int_0^t \|p_n^2(s)\|(exp(M_1 + M_2)s)\, ds$$

which yields the desired estimate

$$\max_J \|x(t) - v_{n+1}(t)\| \leqslant \frac{N_1 + N_2}{M_1 + M_2} e^{(M_1+M_2)T} \max_J \|x(t) - v_n(t)\|^2$$

A similar estimate can be obtained for the difference $\|w_{n+1}(t) - x(t)\|$. The proof is complete.

3.7 Fixed Point Theorems and Quasilinearization

Let $E = (E, \leqslant, \|.\|)$ be an ordered Banach space with order cone E_+. In the first part of this section, we prove abstract fixed point results for mapping $T: E \to E$ based on the method of sub- and supersolution (also called lower and upper solution) combined with the idea of quasilinearization. We develop an iteration scheme for the $(n+1)$-st iterate involving T and its Frechet derivative T' at the n-th iterate that allows to approximate fixed points of T in a constructive and monotone way. Moreover, under certain additional assumptions on T the convergence rate can be shown to be quadratic (rapid convergence). In our first fixed point theorem, we assume a *regular* order cone E_+, i.e., all order bounded and increasing sequences of E_+ converge. However, in various applications this assumption is too strong. By imposing a compactness property on T which is often met in applications, we are able to weaken the regularity of E_+ and may instead assume merely a *normal* order cone.

In section 3.8, we provide two prototypes of applications of the abstract results to an ODE and PDE problem, which clearly demonstrate that the abstract setting given here reflects in a proper way the characteristic features of what is known as the method of quasilinearization for ODE and PDE problems.

Our first abstract fixed point results is as follows.

Theorem 3.7.1. *Let E be an ordered Banach space with regular order cone E_+. Assume that $T : E \to E$ satisfies the following hypotheses:*

(i) There exist $v_0, w_0 \in E$ such that $v_0 \leqslant Tv_0$, $Tw_0 \leqslant w_0$ and $v_0 \leq w_0$.

(ii) The Frechet derivative $T'(u)$ exists for every $u \in [v_0, w_0]$, and $u \mapsto T'(u)v$ is increasing on $[v_0, w_0]$ for all $v \in E_+$.

(iii) $[I - T'(u)]^{-1}$ exists and is a bounded and positive operator for all $u \in [v_0, w_0]$.

Then, for $n \in \mathbb{N}$, relations

$$v_{n+1} = Tv_n + T'(v_n)(v_{n+1} - v_n), \quad w_{n+1} = Tw_n + T'(v_n)(w_{n+1} - w_n), \tag{3.7.1}$$

define an increasing sequence $(v_n)_{n=0}^{\infty}$ *and a decreasing sequence* $(w_n)_{n=0}^{\infty}$ *which converge to fixed points of* T. *These fixed points are equal if*

(iv) $Tu_1 - Tu_0 < u_1 - u_0$ *whenever* $v_0 \leqslant u_0 < u_1 \leqslant w_0$.

In the proof of Theorem 3.7.1 we use the following auxiliary result.

Lemma 3.7.1. *Assume that the hypotheses (ii) of Theorem 3.7.1 holds. Then*

$$Tu_1 - Tu_0 < T'(u_1)(u_1 - u_0) \quad \textit{whenever} \quad v_0 \leqslant u_1 \leqslant u_0 \leqslant w_0.$$

Proof. According to the hypothesis (ii), the function $g : [0, 1] \to E$, defined by

$$g(t) = T'(u_0 + t(u_1 - u_0))(u_1 - u_0), \ t \in [0, 1],$$

is increasing and order bounded. Because E_+ is regular, and hence also normal, then g is regulated and norm-bounded. Thus g is Bochner integrable. Defining

$$f(t) = T(u_0 + t(u_1 - u_0)), \ t \in [0, 1],$$

we have $f'(t) = g(t)$ for all $t \in [0, 1]$. Consequently,

$$Tu_1 - Tu_0 = f(1) - f(0) = \int_0^1 f'(t)dt = \int_0^1 g(t)dt = \int_0^1 T'(u_0 + t(u_1 - u_0))(u_1 - u_0)dt.$$

Moreover, this result and the monotonicity of g, imply that

$$Tu_1 - Tu_0 = \int_0^1 g(t)dt \leqslant \int_0^1 T'(u_1)(u_1 - u_0))dt = T'(u_1)(u_1 - u_0). \quad \square$$

Proof of Theorem 3.7.1: We first show that v_n, w_n exist for all $n \in \mathbb{N}$ and satisfy

$$v_0 \leqslant v_1 \leqslant ... \leqslant v_n \leqslant w_n ... \leqslant w_1 \leqslant w_0, \quad n \in \mathbb{N}. \tag{3.7.2}$$

For this purpose consider $v_1 = Tv_0 + T'(v_0)(v_1 - v_0)$. Since $v_0 \leqslant u_0 \leqslant w_0$ we have

$$[I - T'(v_0)]v_1 = Tv_0 - T'(v_0)v_0,$$

which implies that $v_1 = [I - T'(v_0)]^{-1}(Tv_0 - T'(v_0)v_0)$. So v_1 exists. We must show that $v_0 \leqslant v_1 \leqslant w_0$. Letting $p = v_0 - v_1$ we find

$$p \leqslant Tv_0 - [Tv_0 + T'(v_0)(v_1 - v_0)] = T'(v_0)p$$

Hence $[I - T'(v_0)]p \leqslant \theta$. Thus $p \leqslant [I - T'(v_0)]^{-1}(\theta) = \theta$ by hypothesis (iii), so that $p \leqslant \theta$, or $v_0 \leqslant v_1$.

Next we show that $v_1 \leqslant w_0$. Denoting $p = v_1 - w_0$ we have by (i) and Lemma 3.7.1,

$$\begin{aligned} p &\leqslant Tv_0 + T'(v_0)(v_1 - v_0) - Tw_0 \\ &\leqslant T'(v_0)(v_0 - w_0) + T'(v_0)(v_1 - v_0) \\ &= T'(v_0)(v_0 - w_0 + v_1 - v_0) \\ &= T'(v_0)(v_1 - w_0) \\ &= T'(v_0)p. \end{aligned}$$

Arguing as before, we get $p \leqslant \theta$, or $v_1 \leqslant w_0$. In a similar way we can show that w_1 exists and $v_0 \leq w_1 \leqslant w_0$.
Next we prove that $v_1 \leqslant w_1$. Set $p = v_1 - w_1$ and apply Lemma 3.7.1 to find

$$\begin{aligned} p &= Tv_0 + T'(v_0)(v_1 - v_0) - [Tw_0 + T'(v_0)(w_1 - w_0)] \\ &\leqslant T'(v_0)(v_0 - w_0) + T'(v_0)(v_1 - v_0) - T'(v_0)(w_1 - w_0) \\ &= T'(v_0)(v_0 - w_0 - w_1 + w_0) \\ &= T'(v_0)(v_1 - w_1) \\ &= T'(v_0)p. \end{aligned}$$

The obtained inequality $p \leqslant T'(v_0)p$ implies, as before, that $p \leqslant \theta$, or $v_1 \leqslant w_1$. Thus we have

$$v_0 \leqslant v_1 \leqslant w_1 \leqslant w_0. \tag{3.7.3}$$

Suppose now that v_j, w_j exist for some $j > 0$, and that

$$v_0 \leqslant v_j \leqslant w_j \leqslant w_0. \tag{3.7.4}$$

We then prove that v_{j+1}, w_{j+1} exist, and that

$$v_0 \leqslant v_j \leqslant v_{j+1} \leqslant w_{j+1} \leqslant w_j \leqslant w_0. \tag{3.7.5}$$

For this purpose consider $v_{j+1} = Tv_j + T'(v_j)(v_{j+1} - v_j)$. It follows that $[I - T'(v_j)]v_{j+1} = Tv_j - T'(v_j)v_j$, which yields that $v_{j+1} = [I - T'(v_j)]^{-1}(Tv_j - T'(v_j)v_j)$, proving that v_{j+1} exists. To prove first that $v_j \leqslant v_{j+1} \leqslant w_j$, we prove first that $p = v_j - v_{j+1} \leqslant \theta$. Applying Lemma 3.7.1 we get

$$\begin{aligned} p &= Tv_{j-1} + T'(v_{j-1})(v_j - v_{j-1}) - [Tv_j + T'(v_j)(v_{j+1} - v_j)] \\ &\leqslant T'(v_{j-1})(v_{j-1} - v_j) + T'(v_{j-1})(v_j - v_{j-1}) - T'(v_j)(v_{j+1} - v_j) \\ &= T'(v_{j-1})(v_{j-1} - v_j + v_j - v_{j-1}) + T'(v_j)p \\ &= T'(v_j)p. \end{aligned}$$

This result and the hypothesis (iii) imply that $p \leqslant [I - T'(v_j)]^{-1}(\theta) = \theta$, so that $p \leqslant \theta$, or $v_j \leqslant v_{j+1}$. Next we prove that $p = v_{j+1} - w_j \leq \theta$. Then we get

$$\begin{aligned} p &= Tv_j + T'(v_j)(v_{j+1} - v_j) - [Tw_{j-1} + T'(v_{j-1})(w_j - w_{j-1})] \\ &\leqslant T'(v_j)(v_j - w_{j-1}) + T'(v_j)(v_{j+1} - v_j) - T'(v_{j-1})(w_j - w_{j-1}) \\ &= T'(v_j)(v_j - w_{j-1} + v_{j+1} - v_j) - T'(v_j)(w_j - w_{j-1}) \\ &= T'(v_j)(v_j - w_{j-1} + v_{j+1} - v_j + w_{j-1} - w_j) \\ &= T'(v_j)p. \end{aligned}$$

Thus arguing as before, we see that $p \leqslant \theta$, or $v_{j+1} \leqslant w_j$. Thus $v_j \leqslant v_{j+1} \leqslant w_j$. Similarly, it is easy to show that w_{j+1} exists and $v_{j+1} \leqslant w_{j+1} \leqslant w_j$. This completes the proof of (3.7.5). Now, by induction (3.7.2) is true.

Since the cone E_+ is regular, it follows from (3.7.2) that $v_n \to v$, $w_n \to w$, and that $v \leqslant w$.

To prove that v is a fixed point of T, it suffices to show that $v_{n+1} \to Tv$. Applying (3.7.1) and hypothesis (ii) it is easy to show that

$$\begin{aligned} 2T'(w_0)(v_n - v) + T'(v_0)(v - v_{n+1}) &\leqslant v_{n+1} - Tv \\ &\leqslant 2T'(v_0)(v_n - v) + T'(w_0)(v - v_{n+1}) \end{aligned} \tag{3.7.6}$$

Because E_+ is normal, and because both $2T'(w_0)(v_n - v) + T'(v_0)(v - v_{n+1}) \to 0$ and $2T'(v_0)(v_n - v) + T'(w_0)(v - v_{n+1}) \to 0$, it follows from (3.7.6) that $v_{n+1} \to Tv$. Since $v_{n+1} \to v$, then $v = Tv$. Similarly one can show that $w = Tw$.

Assume next that the hypothesis (iv) holds. If $v < w$, then

$$w - v = Tw - Tv < w - v, \quad \text{a contradiction. Thus} \quad v = w \quad \square$$

Proposition 3.7.1. *Let $T : E \to E$ satisfy the hypotheses (i) - (iv) of Theorem 3.7.1 and*

(v) $\|T'(u) - T'(v)\| \leqslant L\|u - v\|$ *for some* $L > 0$ *whenever* $v_0 \leqslant v \leqslant u \leqslant w_0$,

(vi) $M = \sup\{\|[I - T'(u)]^{-1}\| : u \in [v_0, w_0]\} < \infty$.

Then the sequences $(v_n)_{n=0}^{\infty}$ and $(w_n)_{n=0}^{\infty}$ converge quadratically to the same fixed point of T.

Proof. According to Theorem 3.7.1 both the sequences $(v_n)_{n=0}^{\infty}$ and $(w_n)_{n=0}^{\infty}$ converge to the same fixed point u of T. To prove that the convergence $v_n \to u$ is quadratic, note that $p_n = u - v_n \geqslant \theta$, for all $n \in \mathbb{N}$. Then

$$\begin{aligned} p_{n+1} &= Tu - [T(v_n) + T'(v_n)(v_{n+1} - v_n)] \\ &= \int_0^1 T'(v_n + t(u - v_n))(u - v_n)dt + T'(v_n)p_{n+1} - T'(v_n)p_n \\ &\leqslant [T'(u) - T'(v_n)]p_n - T'(v_n)p_{n+1} \end{aligned}$$

Thus $[I - T'(v_n)]p_{n+1} \leqslant [T'(u) - T'(v_n)]p_n$, so that

$$\theta \leqslant p_{n+1} \leqslant [I - T'(v_n)]^{-1}[[T'(u) - T'(v_n)]p_n \tag{3.7.7}$$

Because E_+ is normal, there exists, by definition, such a positive constant N that

$$\theta \leqslant u \leqslant v \quad \text{implies} \quad \|u\| \leqslant N\|v\|. \tag{3.7.8}$$

It then follows from (3.7.7) and (3.7.8), (v) and (vi) that

$$\|p_{n+1}\| \leqslant NML\|p_n\|^2.$$

A similar estimate can be obtained when $p_n = w_n - u$, and therefore both the sequences $(v_n)_{n=0}^{\infty}$ and $(w_n)_{n=0}^{\infty}$ converge quadratically to u. □

Corollary 3.7.1. *If $T'(u)$ is a bounded and positive operator with $\|T'(u)\| \leqslant q < 1$for all $u \in [v_0, w_0]$, then hypothesis (iii) of Theorem 3.7.1 and hypothesis (vi) of Proposition 3.7.1 are fulfilled.*

Proof. The proof is a simple consequence of Neumann series. □

In applications it may happen that the order cone E_+ is not regular. In this case one can still apply the abstract result by compensating the lack of regularity of E_+ through stronger hypotheses on the operator T. By inspection of the proof of Theorem 3.7.1 the following is obvious.

Corollary 3.7.2. *Let E be an ordered Banach space with normal order cone E_+ . Then Theorem 3.7.1 still holds true if one, in addition, requires that $T : [v_0, w_0] \to E$ is compact, and the inequality of Lemma 3.7.1 holds true, i.e.,*

$$Tu_1 - Tu_0 \leqslant T'(u_1)(u_1 - u_0) \quad \textit{whenever} \quad v_0 \leqslant u_1 \leqslant u_0 \leqslant w_0, \tag{3.7.9}$$

is fulfilled.

(Note that in the proof of Lemma 3.7.1 regularity of E_+ was used in the arguments.) Summarizing the results of Theorem 3.7.1 and Proposition 3.7.1 and Corollaries 3.7.1 and 3.7.2 leads to the following theorem.

Theorem 3.7.2. *Let $E = (E, \leqslant, \|.\|)$ be an ordered Banach space with a normal order cone E_+. Assume that $T : E \to E$ satisfies the following hypotheses:*

(a) There exist $v_0, w_0 \in E$ such that $v_0 \leqslant Tv_0$, $Tw_0 \leqslant w_0$ and $v_0 \leqslant w_0, T : [v_0, w_0] \to E$ is compact and satisfies inequality (3.7.9)

(b) The Frechet derivative $T'(u)$ exists for every $u \in [v_0, w_0]$, and $T'(u)$ is a bounded and positive operator with $\|T'(u)\| \leqslant q < 1$ for all $u \in [v_0, w_0]$,

(c) $u \mapsto T'(u)v$ is increasing on $[u_0, v_0]$ for all $v \in E_+$.

(d) The Frechet derivative T' is Lipschitz continuous on $[v_0, w_0]$, i.e.,

$$\|T'(u) - T'(v)\| \leqslant L\|u - v\| \textit{ for some } L > 0 \textit{ whenever } v_0 \leqslant v \leqslant u \leqslant w_0.$$

Then, for $n \in \mathbb{N}$, *relations*

$$v_{n+1} = Tv_n + T'(v_n)(v_{n+1} - v_n), \quad w_{n+1} = Tw_n + T'(w_n)(w_{n+1} - w_n), \quad (3.7.10)$$

define an increasing sequence $(v_n)_{n=0}^{\infty}$ *and a decreasing sequence* $(w_n)_{n=0}^{\infty}$ *which converge quadratically to fixed points of* T.

Remark 3.7.1

(i) As will be seen in the next section in various applications the assumption $\|T'(u)\| \leqslant q < 1$ for all $u \in [v_0, w_0]$ already implies uniqueness of the fixed points, and thus hypothesis (iv) of Theorem 3.7.1 can be dispensed.

(ii) In Theorem 3.7.2 the regularity of the cone E_+ has been relaxed by assuming a normal order cone. This, however, requires a stronger compactness hypothesis on the operator $T : [v_0, w_0] \to E$. The regularity of E_+ can also be dispensed with if we merely assume that T is a $\alpha-$contraction where α is the measure of noncompactness.

(iii) Note that the calculation of the approximations v_{n+1} and w_{n+1} by the linearized scheme (3.7.10) requires the knowledge of $(I - T'(u))^{-1}$ for every $u \in [v_0, w_0]$ in order to turn (3.7.10) into an explicit form. The existence of the continuous inverse $(I - T'(u))^{-1}$ is guaranteed by the assumptions made in Theorem 3.7.2. In many concrete applications, from (3.7.10) one can derive an explicit iteration scheme once the derivative T' is known. This will be demonstrated by two applications.

3.8 Applications of Fixed Point Theorems

In this section we consider the initial value problem

$$u'(t) = f(u(t)) \quad t \in J := [0, a], \quad u(0) = \eta, \quad (3.8.1)$$

under the following hypotheses:

(H1) Let $v_0, w_0 \in C^1(J)$ satisfy:

$$\begin{aligned} v_0'(t) &\leqslant f(v_0(t)) \quad \forall \ t \in J, \quad v_0(0) \leqslant \eta \\ w_0'(t) &\geqslant f(w_0(t)) \quad \forall \ t \in J, \quad w_0(0) \geqslant \eta \end{aligned}$$

(H2) For some $\delta > 0$, the function $f : [\min_{t\in J}\{v_0(t), w_0(t)\} - \delta, \max_{t\in J}\{v_0(t), w_0(t)\} + \delta] \to \mathbb{R}$ is continuously differentiable.

(i) Let $\lambda > 0$ be such that $f'(s) + \lambda \geqslant 0, \quad \forall \ s \in [\min_{t\in J}\{v_0(t), w_0(t)\}, \max_{t\in J}\{v_0(t), w_0(t)\}]$

(ii) The derivative $f' : [\min_{t\in J}\{v_0(t), w_0(t)\}, \max_{t\in J}\{v_0(t), w_0(t)\}] \to \mathbb{R}$ is increasing and Lipschitz continuous.

Remark 3.8.1 From (H2) (i) it follows, in particular, that

$$f : [\min_{t\in J}\{v_0(t), w_0(t)\}, \max_{t\in J}\{v_0(t), w_0(t)\}] \to \mathbb{R}$$

is Lipschitz continuous which together with (H1) implies that v_0 and w_0 are ordered, i.e., we have $v_0(t) \leqslant w_0(t)$ for all $t \in J$, and thus (v_0, w_0) is an ordered pair of sub-supersolution of problem (3.8.1) . By standard monotone iteration the existence of a unique solution of (3.8.1) can be ensured.

We are going to show that Theorem 3.7.2 can be applied to (3.8.1) which implies the existence of monotone sequences from above and below converging quadratically to the unique solution. To this end we consider the equivalent integral equation belonging to (3.8.1). Let $\lambda > 0$ be as in hypothesis (H2)(i). Then the initial value problem (3.8.1) is equivalent to

$$u(t) = \eta e^{-\lambda t} + \int_0^t (f(u(\tau)) + \lambda u(\tau))e^{\lambda(\tau - t)} d\tau, \quad t \in J. \tag{3.8.2}$$

Define

$$Tu(t) := \eta e^{-\lambda t} + \int_0^t (f(u(\tau)) + \lambda u(\tau))e^{\lambda(\tau - t)} d\tau, \tag{3.8.3}$$

and set $E = (C(J), \|\cdot\|_\alpha)$, where $\alpha > 0$ is specified later and the norm $\|\cdot\|_\alpha$ is given by

$$\|u\|_\alpha := \max_{t\in J} e^{-\alpha t}|u(t)|, \quad \alpha > 0.$$

Let E be equipped with the natural partial ordering. Even though the order cone E_+ is not regular, it is still normal, and the operator $T : [v_0, w_0] \to E$ is easily seen to be compact (note that due to the remark above we have $v_0 \leqslant w_0$), and by hypothesis (H1) we have $v_0 \leqslant Tv_0$ and $Tw_0 \leqslant w_0$. By linearization of T the Frechet derivative $T'(u)$ of T at $u \in [v_0, w_0]$ is given by

$$T'(u)(h)(t) = \int_0^t (f'(u(\tau)) + \lambda)h(\tau))e^{\lambda(\tau - t)} d\tau, \; h \in E. \tag{3.8.4}$$

Thus by hypothesis (H2)(ii) $u \mapsto T'(u)h$ is increasing for any $h \in E_+$, and in view of (H2)(i) $T'(u) : E \to E$ is a bounded and positive operator for any $u \in [v_0, w_0]$. Next we are going to show that $\|T'(u)\| < 1$ if α is appropriately chosen. Let $u \in [v_0, w_0]$ be fixed and let $h \in E$, then we get:

$$
\begin{aligned}
|T'(u)(h)(t)| &\leqslant \int_0^t |(f'(u(\tau)) + \lambda)h(\tau))|e^{\lambda(\tau - t)} d\tau \\
&\leqslant \int_0^t |(f'(u(\tau)) + \lambda)h(\tau))| d\tau \\
&\leqslant \int_0^t |(f'(u(\tau)) + \lambda)|e^{\alpha\tau}|h(\tau)|e^{-\alpha\tau} d\tau \\
&\leqslant \frac{K}{\alpha}(e^{\alpha t} - 1)\|h\|_\alpha \\
&\leqslant \frac{K}{\alpha}\|h\|_\alpha e^{\alpha t},
\end{aligned}
$$

which yields

$$|T'(u)(h)(t)|e^{-\alpha t} \leqslant \frac{K}{\alpha}\|h\|_\alpha, \ \forall\, t \in J,$$

and thus

$$\|T'(u)h\|_\alpha \leqslant \frac{K}{\alpha}\|h\|_\alpha, \ \forall\, h \in E, \tag{3.8.5}$$

where $K := \sup_{u \in [v_0, w_0]} \max_{t \in J} |f'(u(t)) + \lambda|$. Selecting $\alpha > K$ from (3.8.5) we see that

$$\|T'(u)\| \leqslant q < 1, \ \forall\, u \in [v_0, w_0].$$

By means of (H2)(ii) we get in a similar way the following estimate:

$$
\begin{aligned}
|T'(u) - T'(v)(h)(t)| &\leqslant \int_0^t |(f'(u(\tau)) - (f'(v(\tau))||h(\tau))|e^{\lambda(\tau - t)} d\tau \\
&\leqslant \int_0^t L|u(\tau)) - v(\tau))|e^{-\alpha\tau}|h(\tau)|e^{-\alpha\tau}e^{2\alpha\tau} d\tau \\
&\leqslant L\|u - v\|_\alpha \|h\|_\alpha \int_0^t e^{2\alpha\tau} d\tau \\
&\leqslant L\|u - v\|_\alpha \|h\|_\alpha \frac{1}{2\alpha} e^{2\alpha t},
\end{aligned}
$$

which implies

$$|T'(u) - T'(v)(h)(t)|e^{-\alpha t} \leqslant \frac{L}{2\alpha} e^{\alpha a}\|u - v\|_\alpha \|h\|_\alpha; \ \forall\, t \in J,$$

and thus

$$\|(T'(u) - T'(v))(h)\|_\alpha \leqslant \frac{L}{2\alpha}e^{\alpha a}\|u - v\|_\alpha \|h\|_\alpha;\ \forall\ h \in E.$$

The latter inequality shows that T' is Lipschitz continuous, i.e.,

$$\|T'(u) - T'(v)\| \leqslant \frac{L}{2\alpha}e^{\alpha a}L\|u - v\|_\alpha.$$

Finally, in order to apply Theorem 3.7.2 it remains to verify inequality (3.7.9). Let $u_0, u_1 \in [v_0, w_0]$ with $u_1 \leqslant u_0$. By definition of the operators T and T' and using (H2) we get

$$\begin{aligned}
(T(u_1) - T(u_0))(t) &= \int_0^t (f(u_1) - f(u_0) + \lambda(u_1 - u_0))\, e^{\lambda(\tau - t)} d\tau \\
&= \int_0^t (f'(u_1 + \theta(u_0 - u_1)) + \lambda)(u_1 - u_0)e^{\lambda(\tau - t)} d\tau \\
&\leqslant \int_0^t (f'(u_1) + \lambda)(u_1 - u_0)e^{\lambda(\tau - t)} d\tau \\
&= T'(u_1)(u_1 - u_0).
\end{aligned}$$

Now, we may apply Theorem 3.7.2 to obtain the following result.

Theorem 3.8.1. *Let the hypotheses (H1)-(H2) be satisfied. Then the unique solution u of* (3.8.1) *can be approximated by monotone sequences (v_n) and (w_n) which converge quadratically to u from below and above, respectively. The operators T and T' are given by*

$$Tv_n(t) := \eta e^{-\lambda t} + \int_0^t (f(v_n(\tau)) + \lambda v_n(\tau))\, e^{\lambda(\tau - t)} d\tau; \tag{3.8.6}$$

and

$$T'(v_n)(h)(t) = \int_0^t (f'(v_n(\tau)) + \lambda)h(\tau)e^{\lambda(\tau - t)} d\tau; \tag{3.8.7}$$

and the iterations are given by (3.7.1).

Remark 3.8.2. Note that the iterations v_{n+1} and w_{n+1} via (3.7.1) are given by an implicit iteration scheme, because they appear on both sides of the equation. To get an explicit iteration scheme one needs to know the inverse $(I - T'(u))^{-1}$ whose existence is guaranteed. In the concrete situation considered here we are able to provide the explicit scheme. By using the concrete form of the operators T and T', elementary calculations yield the following : The abstract equation

$$v_{n+1} = Tv_n + T'(v_n)(v_{n+1} - v_n)$$

is equivalent to the following linear initial value problem

$$v'_{n+1} - f'(v_n)v_{n+1} = f(v_n) - v_n f'(v_n), \quad v_{n+1}(0) = \eta, \tag{3.8.8}$$

and the abstract equation

$$w_{n+1} = Tw_n + T'(v_n)(w_{n+1} - w_n)$$

is equivalent to

$$w'_{n+1} - f'(v_n)w_{n+1} = f(w_n) - w_n f'(v_n), \quad w_{n+1}(0) = \eta, \tag{3.8.9}$$

Let us next consider the semilinear parabolic initial-boundary value problem.

Let $\Omega \subset \mathbb{R}^N$ be a bounded domain with smooth boundary $\partial\Omega \subset C^2$, let $Q = \Omega \times (0, a)$ be a parabolic cylinder, and denote by $\partial_p Q$ the parabolic boundary of Q which is given by

$$\partial_p Q = \partial Q \setminus (\Omega \times \{a\}).$$

We consider the following semilinear parabolic initial-boundary value problem

$$\frac{\partial u}{\partial t} - \Delta u = f(x, t, u) \quad \text{in } Q, \quad u|_{\partial_p Q} = \varphi,$$

where $f : \overline{Q} \times \mathbb{R} \to \mathbb{R}$ is continuous and φ is the restriction on $\partial_p Q$ of some smooth function $\Phi \in C^{2,1}(\overline{Q})$. Without loss of generality we may assume homogeneous initial - and boundary values, because otherwise homogeneous initial- and boundary values can be obtained by simple shift transformation $u \to u - \Phi$. Therefore, in what follows let us consider

$$\frac{\partial u}{\partial t} - \Delta u = f(x, t, u) \quad \text{in } Q, \quad u|_{\partial_p Q} = 0. \tag{3.8.10}$$

We assume the following hypotheses:

(P1) Let v_0 and w_0 be classical sub- and supersolutions, respectively, of (3.8.10), i.e.,

$$v_0 \in C^{2,1}(\overline{Q}) : \frac{\partial v_0}{\partial t} - \Delta v_0 \leqslant f(x, t, v_0) \quad \text{in } Q, \quad v_0|_{\partial_p Q} \leqslant 0,$$

$$w_0 \in C^{2,1}(\overline{Q}) : \frac{\partial w_0}{\partial t} - \Delta w_0 \geqslant f(x, t, w_0) \quad \text{in } Q, \quad w_0|_{\partial_p Q} \geqslant 0,$$

(P2) Assume that $f : \overline{Q} \times [m - \delta, M + \delta] \to \mathbb{R}$ is continuous and continuously differentiable with respect to u, where δ may be any positive real number, and with m and M given by

$$m := \min_{(x,t) \in \overline{Q}} \{v_0(x, t), w_0(x, t)\}, \quad M := \max_{(x,t) \in \overline{Q}} \{v_0(x, t), w_0(x, t)\}$$

(P3) The function $u \mapsto f_u(x,t,u)$ is increasing for $u \in [m,M]$ and Lipschitz continuous, i.e., there exists $L \geqslant 0$ such that for all $(x,t) \in \overline{Q}$, and for all $u, \widehat{u} \in [m,M]$,

$$|f_u(x,t,u) - f_u(x,t,\widehat{u})| \leqslant L|u - \widehat{u}|.$$

Remark 3.8.3 Hypotheses (P1) and (P2) already imply that the subsolution v_0 and the supersolution w_0 must be ordered, i.e., there exists $v_0(x,t) \leqslant w_0(x,t)$. To prove this let $y(x,t) = v_0(x,t) - w_0(x,t)$ and let $y^+ = \max\{y, 0\}$. Subtracting the corresponding inequalities from each other we get

$$\frac{\partial y}{\partial t} - \triangle y \leqslant f(x,t,v_0) - f(x,t,w_0) \quad (x,t) \in Q.$$

As y is smooth in $\overline{Q}$ it follows that y^+ possesses weak derivatives of first order that are square integrable, and we get by multiplying the last inequality by y^+ and integrating over $Q_t := \Omega \times (0,t), t \in (0,a]$ the following inequality:

$$\frac{1}{2}\|y^+(\cdot,t)\|_{L^2(\Omega)} + \|\nabla y^+\|^2_{L^2(Q_t)} \leqslant C \int\int_{Q_t} (y^+)^2 dx d\tau, \tag{3.8.11}$$

where C is bound of f_u in $\overline{Q} \times [m,M]$. Note $y^+(x,0) \equiv 0$. Setting $s(t) = \|y^+(\cdot,t)\|_{L^2(\Omega)}$, from (3.8.11) we get the following differential inequality for the scalar function s:

$$s(t) \leqslant 2C \int_0^t s(\tau) d\tau, \forall \ t \in [0,a]$$

with $s(0) = 0$. Applying Gronwall's Lemma we deduce $s(t) = 0$ for all $s \in [0,a]$ which implies $y^+(x,t) = 0$ for a.e . $(x,t) \in Q$, and since y^+ is continuous in $\overline{Q}$, we finally arrive at $y^+(x,t) \equiv 0$ and thus $v_0(x,t) \leqslant w_0(x,t)$ for all $(x,t) \in \overline{Q}$.

Let $W_p^{2,1}(Q), 1 < p < \infty$ denote the usual Sobolev space having $p-$integrable weak partial derivatives of the form

$$W_p^{2,1}(Q) = \{u \in L^p(Q) : D^s D_t^r u \in L^p(Q), \ |s| + 2r \leqslant 2\}$$

where $s \in \mathbb{N}_0^N$ is a multi-index, $r \in \{0,1\}$, D^s denote the partial derivatives with respect to the space variables, and D_t the partial derivative with respect to $t.W_p^{2,1}(Q)$ endowed with the norm

$$\|u\|_{W_p^{2,1}(Q)} = \left(\int\int_Q \sum_{|s|+2r \leqslant 2} |D^s D_t^r u|^p dx dt \right)^{1/p}$$

is a Banach space, which is continuously embedded into the Hölder space $C^{\beta,\beta/2}(\overline{Q})$ for $0 < \beta \leqslant 2 - \frac{N+2}{p}$ and $p > \frac{N+2}{2}$. Due to the compact embedding $C^{\beta,\beta/2}(\overline{Q}) \hookrightarrow\hookrightarrow C(\overline{Q})$, it follows that $W_p^{2,1}(Q) \hookrightarrow\hookrightarrow C(\overline{Q})$ is compactly embedded if $p > \frac{N+2}{2}$. In view of hypotheses (P1) and (P2) we can always assume that $p > \frac{N+2}{2}$.

We are going to make use of the theory of strong solutions from $W_p^{2,1}(Q)$ of linear initial-boundary value problems of the form

$$\frac{\partial u}{\partial t} - \Delta u + c(x,t)u = g(x,t), \quad (x,t) \in Q, \quad u|_{\partial_p Q} = 0 \tag{3.8.12}$$

with $c \in L^\infty(Q)$ and $g \in L^p(Q)$, to transform our problem (3.8.10) into a fixed point problem by using the solution operator of (3.8.12). It then will be shown that the obtained fixed point problem fits into the framework of Theorem 3.7.2. By hypothesis (P2) the derivative $f_u : \overline{Q} \times [m-\delta, M+\delta] \to \mathbb{R}$ is bounded, and thus there is a constant $\lambda \geqslant 0$ such that

$$f_u(x,t,u) + \lambda \geqslant 0, \quad \forall \ (x,t,u) \in \overline{Q} \times [m-\delta, M+\delta]. \tag{3.8.13}$$

With this constant λ the original parabolic problem (3.8.10) can equivalently be written as

$$\frac{\partial u}{\partial t} - \Delta u + \lambda u = f(x,t,u) + \lambda u, \quad (x,t) \in Q, \quad u|_{\partial_p Q} = 0 \tag{3.8.14}$$

Now we define an operator $T : u \to v = Tu$ as follows : For $u \in C(\overline{Q})$ with $u \in [v_0, u_0]$, $Tu = v$ is the unique solution of the following linear initial-boundary value problem:

$$\frac{\partial v}{\partial t} - \Delta v + \lambda v = f(x,t,u) + \lambda u, \quad (x,t) \in Q, \quad v|_{\partial_p Q} = 0 \tag{3.8.15}$$

Note that the problem (3.8.15) has a unique solution $v \in W_p^{2,1}(Q)$ for any $p > 1$, because the right-hand side $f(x,t,u) + \lambda u \in C(\overline{Q})$, and the following estimate holds

$$\|Tu\|_{W_p^{2,1}(Q)} = \|v\|_{W_p^{2,1}(Q)} \leqslant c\|f(x,t,u) + \lambda u\|_{L_p(Q)}, \tag{3.8.16}$$

(for some positive constant c) which due to the continuous embedding $C(Q) \hookrightarrow L^p(\overline{Q})$ for any $p > 1$ yields

$$\|Tu\|_{W_p^{2,1}(Q)} = \|v\|_{W_p^{2,1}(Q)} \leqslant \tilde{c}\|f(x,t,u) + \lambda u\|_{C(\overline{Q})}, \tag{3.8.17}$$

with

$$\|w\|_{C(\overline{Q})} = \sup_{(x,t)\in\overline{Q}} |w(x,t)|.$$

Thus we may choose $p > \frac{N+2}{2}$ in order to make use of the compact embedding $W_p^{2,1}(Q) \hookrightarrow\hookrightarrow C(\overline{Q})$. Hypothesis (P2) along with (3.8.17) and compact embedding $W_p^{2,1}(Q) \hookrightarrow\hookrightarrow C(\overline{Q})$ imply that

$$T : [v_0, w_0] \subset C(\overline{Q}) \to C(\overline{Q}) \quad \text{is Lipschitz continuous and compact.} \tag{3.8.18}$$

Consider Tv_0 which is the unique solution of

$$\frac{\partial Tv_0}{\partial t} - \triangle Tv_0 + \lambda Tv_0 = f(x,t,v_0) + \lambda v_0, \quad (x,t) \in Q, \quad Tv_0|_{\partial_p Q} = 0 \quad (3.8.19)$$

and compare it with v_0 which is the given subsolution, i.e.,

$$\frac{\partial v_0}{\partial t} - \triangle v_0 + \lambda v_0 \leqslant f(x,t,v_0) + \lambda v_0, \quad (x,t) \in Q, \quad v_0|_{\partial_p Q} \leqslant 0. \qquad (3.8.20)$$

Subtracting (3.8.19) from (3.8.20) results in the following inequalities for $w := v_0 - Tv_0$:

$$\frac{\partial w}{\partial t} - \triangle w + \lambda w \leqslant 0, \quad (x,t) \in Q, \quad w|_{\partial_p Q} \leqslant 0 \qquad (3.8.21)$$

which yields $w(x,t) \leqslant 0$ for all $(x,t) \in \overline{Q}$ by the (weak) maximum principle. Hence it follows $v_0 \leqslant Tv_0$. In a similar way we get $w_0 \geqslant Tw_0$.

Let $u \in [v_0, w_0]$ be fixed. For any $h \in C(\overline{Q})$ we define $z = z(h)$ as the solution of the following linear parabolic initial-boundary value problem:

$$\frac{\partial z}{\partial t} - \triangle z + \lambda z = (f_u(x,t,u) + \lambda)h(x,t), \quad (x,t) \in Q, \quad z|_{\partial_p Q} = 0 \qquad (3.8.22)$$

Note, by hypothesis (P2) we have

$$\|f_u(\cdot,\cdot,u) + \lambda\|_{C(\overline{Q})} \leqslant c, \ \forall\, u \in [v_0, w_0],$$

and therefore

$$\|z(h)\|_{C(\overline{Q})} \leqslant c\|z(h)\|_{W_p^{2,1}(\overline{Q})} \leqslant \tilde{c}\|z(h)\|_{C(\overline{Q})}. \qquad (3.8.23)$$

One readily verifies that z is linear in h, and thus by (3.8.23) we see that $z : C(\overline{Q}) \to C(\overline{Q})$ is a linear continuous mapping. We are going to show that z is the Frechet derivative of T in u, i.e., $z = T'(u)$. To this end set $w = w(h) := T(u+h) - Tu - z(h)$. Then w is the solution of the following linear parabolic problem:

$$\begin{aligned} \frac{\partial w}{\partial t} - \triangle w + \lambda w &= f(x,t,u+h) + \lambda(u+h) - f(x,t,u) - \lambda u \\ &\quad -(f_u(x,t,u) + \lambda)h \\ &= f(x,t,u+h) - f(x,t,u) - f_u(x,t,u)h \qquad (3.8.24) \end{aligned}$$

with $w|_{\partial_p Q} = 0$. In a similar way as in (3.8.23) we obtain the following estimate:

$$\begin{aligned} \|w(h)\|_{C(\overline{Q})} &\leqslant c\|w(h)\|_{W_p^{2,1}(\overline{Q})} \\ &\leqslant \tilde{c}\|f(\cdot,\cdot,u+h) - f(\cdot,\cdot,u) - f_u(\cdot,\cdot,u)h\|_{C(\overline{Q})}. \qquad (3.8.25) \end{aligned}$$

By (P2) f is continuously differentiable with respect to u which results in

$$f(x,t,u+h) = f(x,t,u) + f_u(x,t,u) + r(h)h,$$

where r satisfies $\lim_{h\to 0} r(h) = 0$. Since convergence with respect to $\|\cdot\|_{C(\overline{Q})}$ means uniform convergence, we see that

$$\frac{1}{\|h\|_{C(\overline{Q}))}} \|f(\cdot,\cdot,u+h) - f(\cdot,\cdot,u) - f_u(\cdot,\cdot,u)h\|_{C(\overline{Q}))} \leqslant \|r(h)\|_{C(\overline{Q}))} \to 0$$

as $\|h\|_{C(\overline{Q}))} \to 0$, which implies by (3.8.25) that

$$\frac{1}{\|h\|_{C(\overline{Q}))}} \|w(h)\|_{C(\overline{Q}))} \to 0, \quad \text{as} \quad \|h\|_{C(\overline{Q}))} \to 0. \tag{3.8.26}$$

Taking the definition of w into account, (3.8.26) means that z is in fact the Frechet derivative $T'(u)$. Since by hypothesis (P2) the nonlinear right-hand side $f = f(x,t,u)$ is, in particular, Lipschitz continuous with respect to $u \in [v_0, w_0]$ uniformly in $(x,t) \in \overline{Q}$, it follows that our original problem (3.8.10) (Or equivalently (3.8.14) has a unique solution in $[v_0, w_0]$. Thus the solution operator $T : [v_9, w_0] \subset C(\overline{Q})$ defined by (3.8.15) has a unique fixed point. So far we have proved the following result:

Lemma 3.8.1. *Let hypotheses (P1) and (P2) be satisfied. Then the original parabolic initial-boundary value problem* (3.8.14) *has a unique solution in* $[v_0, w_0]$, *which is a fixed point of the operator* T *defined by* (3.8.15). *The operator* $T : [v_0, w_0] \subset C(\overline{Q}) \to C(\overline{Q})$ *is Lipschitz continuous and compact, and its Frechet derivative* $T'(u)$ *exists for every* $u \in [v_0, w_0]$. *The given sub- and supersolution satisfy* $v_0 \leqslant Tv_0$ *and* $Tw_0 \leqslant w_0$.

The natural partial ordering in $C(\overline{Q})$ is generated by the order cone of nonnegative continuous functions which is a normal cone. Our goal is to show that the operator T defined by (3.8.15) fits into the general framework of Theorem 3.7.2, which allows to obtain the unique solution of the parabolic problem (3.8.14) as the limit of monotone and quadratically convergent sequences. Therefore we still need to verify hypotheses (b), (c), (d) of Theorem 3.7.2 and inequality (3.7.9). To this end we introduce first the following equivalent norm in $C(\overline{Q})$:

$$\|u\|_{\alpha\overline{Q}} = \sup_{(x,t)\in\overline{Q}} e^{-\alpha t}|u(x,t)|, \quad \alpha > 0, \tag{3.8.27}$$

where $\alpha > 0$ will be specified later. In the next three steps we are going to verify the assumptions (b) - (d) of Theorem 3.7.2.

(b): The existence of the Frechet derivative $T'(u)$ for every $u \in [v_0, w_0]$ has already been shown (see Lemma 3.8.1). In the proof of the existence of $T'(u)$ above we have seen that $z = T'(u)$ which is defined by (3.8.22). By the choice of $\lambda \geqslant 0$ according to (3.8.13), the right-hand side of (3.8.22) is nonnegative and a continuous function in $\overline{Q}$ for any $h \in C_+(\overline{Q})$, where $C_+(\overline{Q})$ is the positive cone of $C(\overline{Q})$. Thus the maximum principle implies $z(h) = T'(u)(h) \geqslant 0$ for any $h \in C_+(\overline{Q})$, which shows that $T'(u)$ is a positive operator, whose boundedness follows from (3.8.23). Next, we are going to show that $\|T'(u)\| \leqslant q < 1$ for all $u \in [v_0, w_0]$. For this purpose we equip $C(\overline{Q})$ with the equivalent norm given by (3.8.27). For any $t \in [0,a]$ set $Q_t = \Omega \times (0,t)$. Due to (3.8.22) $z(h) = T'(u)(h)$ satisfies the following linear parabolic initial-boundary value problem in the subset $Q_t \subset Q$:

$$\frac{\partial z}{\partial t} - \triangle z + \lambda z = (f_u(x,t,u) + \lambda)h(x,t), \quad (x,t) \in Q_t, \quad z|_{\partial_p Q_t} = 0 \tag{3.8.28}$$

and the following estimate holds

$$\|z\|_{W_p^{2,1}(Q_t)} \leqslant c\|(f_u(\cdot,\cdot,u) + \lambda)h\|_{L^p(Q_t)}. \tag{3.8.29}$$

Let $c > 0$ be a generic constant that may change size. By the continuous embedding $W_p^{2,1}(Q_t) \hookrightarrow C(\overline{Q}_t)$ and taking into account that

$$|f_u(x,t,u) + \lambda| \leqslant c, \quad \forall\ (x,t) \in \overline{Q}, \quad \forall\ u \in [v_0, w_0],$$

from (3.8.29) we get for every Q_t the estimate

$$\begin{aligned}
\|z\|^p_{C(\overline{Q}_t)} &\leqslant c\|z\|^p_{W_p^{2,1}(\overline{Q}_t)} \leqslant c \int\int_{Q_t} h^p(x,\tau)e^{-\alpha p\tau}e^{\alpha p\tau}\,dx d\tau \\
&\leqslant c\|h\|^p_{\alpha,\overline{Q}_t} \int_0^t \int_\Omega e^{\alpha p\tau}\,dx d\tau \\
&\leqslant c|\Omega|\|h\|^p_{\alpha,\overline{Q}_t}\left(\frac{1}{\alpha p}\right)(e^{\alpha pt} - 1) \\
&\leqslant c\|h\|^p_{\alpha,\overline{Q}_t}\left(\frac{1}{\alpha p}\right)e^{\alpha pt}.
\end{aligned} \tag{3.8.30}$$

As $|z(x,t)| \leqslant \|z\|_{C(\overline{Q}_t)}$, by means of (3.8.30) we can further estimate as follows

$$|z(x,t)|^p e^{-\alpha pt} \leqslant \frac{c}{\alpha p}\|h\|^p_{\alpha,\overline{Q}_t} \leqslant \frac{c}{\alpha p}\|h\|^p_{\alpha,\overline{Q}}$$

which yields

$$|z(x,t)|e^{-\alpha t} \leqslant \left(\frac{c}{\alpha p}\right)^{1/p}\|h\|_{\alpha,\overline{Q}}, \quad \forall\ x \in \overline{\Omega}, \quad \forall\ t \in [0,a],$$

and thus

$$\|z\|_{\alpha,\overline{Q}} \leqslant \left(\frac{c}{\alpha p}\right)^{1/p}\|h\|_{\alpha,\overline{Q}}. \tag{3.8.31}$$

As $z = z(h) = T'(u)(h)$ and because the inequality holds true for every $u \in [v_0, w_0]$, it follows from (3.8.31) that

$$\|T'(u)\| \leqslant q < 1, \quad \forall\, u \in [v_0, w_0],$$

by choosing $\alpha > 0$ large enough such that

$$\left(\frac{c}{\alpha p}\right)^{1/p} < 1.$$

This proves that (b) is satisfied.

(c): To prove that $u \mapsto T'(u)h$ is increasing on $[v_0, w_0]$ for any fixed $h \in C_+(\overline{Q})$, we make use of hypothesis (P3). If $u_1 \leqslant u_2$ then $(f_u(x,t,u_1) + \lambda)h(x,t) \leqslant (f_u(x,t,u_2)+\lambda)h(x,t)$. As $T'(u_i)h$ is the unique solution of (3.8.22) with the right-hand side $(f_u(x,t,u_i)+\lambda)h(x,t)$, $i = 1, 2$, we obtain by applying the maximum principle $T'(u_1) \leqslant T'(u_2)h$, which proves the assertion.

(d): For arbitrary $h \in C(\overline{Q}), T'(u)h$ and $T'(v)h$ with $u, v \in [v_0, w_0]$ are the unique solution of (3.8.22) with the right-hand side $(f_u(x,t,u) + \lambda)h(x,t)$ and $(f_u(x,t,v) + \lambda)h(x,t)$, respectively. Similarly as in the proof of (b) we consider these solutions in Q_t. By subtracting the corresponding equations we get the following estimate

$$\|(T'(u) - T'(v))h\|_{W_p^{2,1}(Q_t)} \leqslant c\|(f_u(\cdot,\cdot,u) - f_u(\cdot,\cdot,v))h\|_{L^p(Q(t))}, \tag{3.8.32}$$

which by means of the uniform Lipschitz continuity given in (P3) and the continuous embedding $W_p^{2,1}(Q) \hookrightarrow C(\overline{Q})$ yields

$$\|(T'(u) - T'(v))h\|^p_{C(\overline{Q})} \leqslant c \int\!\!\int_Q L^p|u - v|^p h^p \, dx d\tau,$$

and thus

$$\|(T'(u) - T'(v))h\|_{C(\overline{Q})} \leqslant c\|u - v\|_{C(\overline{Q})}\|h\|_{C(\overline{Q})}, \quad \forall\, h \in C(\overline{Q}). \tag{3.8.33}$$

Since the norms $\|\cdot\|_{C(\overline{Q})}$ and $\|\cdot\|_{\alpha,\overline{Q}}$ are equivalent, it also follows from (3.8.33)

$$\|(T'(u) - T'(v))h\|_{\alpha,\overline{Q}} \leqslant c\|u - v\|_{\alpha,\overline{Q}}\|h\|_{\alpha,\overline{Q}}, \quad \forall\, h \in C(\overline{Q}), \tag{3.8.34}$$

and hence it follows

$$\|T'(u) - T'(v)\| \leqslant c\|u - v\|_{\alpha,\overline{Q}}, \quad \forall\, u, v \in [v_0, w_0],$$

i.e., the Lipschitz continuity of T'. To show that (3.7.9) is satisfied, consider $u_1, u_0 \in [v_0, w_0]$ with $u_1 \leqslant u_0$ and set $w = Tu_1 - Tu_0$, which by definition of the operator T(see (3.8.15)) satisfies the following initial-boundary value problem:

$$\frac{\partial w}{\partial t} - \Delta w + \lambda w = f(x,t,u_1) + \lambda u_1 - f(x,t,u_0) - \lambda u_0, \quad \text{in } Q, \tag{3.8.35}$$

with $w|_{\partial_p Q} = 0$. Using (P3) and (3.8.13) the right-hand side can be estimated above as

$$f(x,t,u_1) + \lambda u_1 - f(x,t,u_0) - \lambda u_0 \leqslant (f_u(x,t,u_1) + \lambda)(u_1 - u_0). \tag{3.8.36}$$

Since $z = T'(u_1)(u_1 - u_0)$ is the unique solution of

$$\frac{\partial z}{\partial t} - \Delta z + \lambda z = (f_u(x,t,u_1) + \lambda)(u_1 - u_0), \quad \text{in } Q \tag{3.8.37}$$

with $z|_{\partial_p Q} = 0$ we obtain from (3.8.35) - (3.8.37) by comparison based on the maximum principle that $w \leqslant z$ which prove (3.7.9).

Now, all the assumptions of Theorem 3.7.2 are verified which finally leads to the following result for our original initial-boundary value problem (3.8.10).

Theorem 3.8.2. *Let the hypotheses (P1) - (P3) be satisfied. Then* (3.8.10) *has a unique solution u within the ordered interval $[v_0, w_0]$ formed by the sub- and supersolution which can be approximated by monotone sequences (v_n) and (w_n) converging quadratically to u from below and above, respectively. The operators T and T' are defined by* (3.8.15) *and* (3.8.22), *respectively, which allow to derive from the implicitly defined iterations*

$$v_{n+1} = Tv_n + T'(v_n)(v_{n+1} - v_n)$$

of (3.7.1) *the following explicit iteration*

$$\frac{\partial v_{n+1}}{\partial t} - \Delta v_{n+1} - f_u(x,t,v_n)v_{n+1} = f(x,t,v_n) - f_u(x,t,v_n)v_n \tag{3.8.38}$$

with $v_{n+1}|_{\partial_p Q} = 0$, and the abstract equation

$$w_{n+1} = Tw_n + T'(v_n)(w_{n+1} - w_n)$$

is equivalent to the linear parabolic problem

$$\frac{\partial w_{n+1}}{\partial t} - \Delta w_{n+1} - f_u(x,t,v_n)w_{n+1} = f(x,t,w_n) - f_u(x,t,v_n)w_n \tag{3.8.39}$$

with $w_{n+1}|_{\partial_p Q} = 0$

Remark 3.8.4 *The equivalence of the abstract equation* (3.7.1) *with* (3.8.38) *and* (3.8.39), *respectively, can easily be shown by applying the definition of T and T' given in* (3.8.15) *and* (3.8.22), *respectively.*

3.9 Notes and Comments

The results of Section 3.2 are from Guo and Lakshmikantham [1] and Lakshmikantham and Leela [1]. Section 3.3 contains the well known monotone iterative techniques and the absractization of the ideas of Krasnoselskii[1]. These are adopted from Guo and Lakshmikantham [1]. All the results of Section 3.4 – 3.6 are new and are modelled on the corresponding results in R^n. See Ladde, Lakshmikantham and Vatsala [1] and Lakshmikantham and Vatsala [1] The contents of Sections 3.7 and 3.8 are from Carl, Heikkila and Lakshmikantham [1].

Chapter 4

Stability Analysis

4.1 Introduction

This chapter is dedicated to the study of stability theory in terms of Lyapunov functions in cones. We begin by tracing the trends in the basic theory of Lyapunov in Section 4.2. Necessary comparison results in terms of Lyapunov functions and some refinements of Lyapunov theory are described in Section 4.3. Section 4.4 deals with the results on boundedness and practical stability.

In Section 4.5, the method of vector Lyapunov functions is described. The stability concepts in terms of two different measures are given in Section 4.6, providing suitable stability results in terms of two measures. Practical stability and boundedness results in terms of two measures are proved in Section 4.7. Section 4.8 deals with the perturbed systems and some results in perturbation theory.

Large scale dynamic systems are considered in Section 4.9, where vector Lyapunov functions appear naturally in the method of vector Lyapunov functions. Section 4.10 develops a technique in perturbation theory which is a blending of two different methods, namely, the method of variation of parameters and the Lyapunov method. This technique is very fruitful in discussing the perturbation theory. Section 4.11 deals with the notion of quasi-solutions that is useful when, in the method of Lyapunov functions, needed quasimonotone property is not available. Also, in this case, the method of cone-valued Lyapunov functions is very useful, which is developed in Section 4.12. In Section 4.13, some new directions in the method of Lyapunov functions are described. The method of conevalued Lyapunov functions is described in a general setup in section 4.14 where several results are included. Finally, in section 4.15, notes and comments are provided.

4.2 Trends in Basic Lyapunov Theory

It is well known that Lyapunov's second method is an interesting and fruitful technique that has gained increasing significance and has given decisive impetus for modern development of stability theory of differential equations. A manifest advantage of this method is that it does not require the knowledge of solutions and thus has exhibited a great power in applications. There are several books available expounding the main ideas of Lyapunov's second method including some extensions and generalizations.

It is now recognized that the concept of Lyapunov function can be utilized to study various qualitative and quantitative properties of nonlinear differential equations. Lyapunov function serves as a vehicle to transform a given complicated differential system into a relatively simpler system, and therefore, it is enough to investigate the properties of this simpler system. It is also being realized that the same versatile tools are adaptable to study entirely different nonlinear systems, and these effective methods offer an exciting prospect for further advancement. In this section, we shall recall the main trends in the basic theory of Lyapunov method.

Consider the differential system

$$x' = f(t,x),\ x(t_0) = x_0,\ t_0 \geq 0, \tag{4.2.1}$$

where $f \in C(R_+ \times R^n, R^n)$. Assume, for convenience, that the solutions $x(t) = x(t, t_0, x_0)$ of (4.2.1) exist, are unique for $t \geq t_0$ and $f(t,0) = 0$ so that we have the trivial solution $x = 0$.

Let us state the well known original theorems of Lyapunov for stability and asymptotic stability in a suitable form.

Theorem 4.2.1. *Assume (A) $V \in C^1(R_+ \times S(\rho), R_+), V$ is positive definite and $V(t,0) \equiv 0$, where $S(\rho) = x \in R^n : \|x\| < \rho$. If $V'(t,x) \equiv V_t + V_x \cdot f(t,x) \leq 0$ on $R_+ \times S(\rho)$, then $x = 0$ of (4.2.1) is stable.*

Theorem 4.2.2. *Suppose that condition (A) holds. Assume further that V is decrescent and $V'(t,x) \leq -c(\|x\|)$ on $R_+ \times S(\rho)$, where $c \in \kappa$. Then $x = 0$ of (4.2.1) is uniformly asymptotically stable.*

In the forgoing results $\kappa = [\sigma \in C[[0,\rho), R_+]$ such that $\sigma(t)$ is strictly increasing and $\sigma(0) = 0]$. Also V is positive definite (decrescent) means there exists a function $a \in \kappa$ such that $V(t,x) \geq a(|x|)$, $(V(t,x) \leq a(|x|))$.

These two theorems have been modified, extended and generalized in various aspects. We shall discuss some important trends that have occurred in recent years.

4.2.1 Loss of Decrescentness

If we omit the requirement that V is decrescent in Theorem 4.2.2 and suppose that $V(t,0) \equiv 0$, we still get the stability of trivial solution. To examine closely the asymptotic behavior of solutions, we first observe that it is easy to show that

$\liminf_{t\to\infty} \|x(t)\| = 0$. If $\limsup_{t\to\infty} = \gamma \geq 0$, then there exist a sequence of intervals (t'_n, t''_n) such that $\|x(t)\| \geq \gamma' > 0$, for $t'_n < t < t''_n$. We therefore conclude from $V'(t,x) \leq -c(\|x\|)$ that

$$V(t, x(t)) \leq V(t_0, x_0) - c(\gamma') \sum_{n=1}^{\infty} (t''_n - t'_n), \tag{4.2.2}$$

and this is compatible with positive definiteness of V only if the series $\sum_{n=1}^{\infty}(t''_n - t'_n)$ converges. For large time $\|x(t)\|$ is thus almost everywhere arbitrarily small. Its graph is a sawtooth curve. An example of this kind of behavior is due to Massera which is $x' = \dfrac{g'(t)}{g(t)}x$ where $g(t) = \sum_{n=1}^{\infty} \dfrac{1}{1 + n^4(t - n^2)}$.

If we suppose that f is bounded by M say, then $\|\dfrac{dx(t)}{dt}\| \leq M$ and this yields $t''_n - t'_n \geq \frac{\gamma'}{M}$. Consequently, assuming that $x = 0$ is not asymptotically stable leads to a contradiction because of (4.2.2) and hence one can conclude that $x = 0$ is asymptotically stable if f is bounded when we delete the assumption of decrescence of V. This is Marachkov's Theorem.

Theorem 4.2.3. *Suppose that condition (A) holds and f is bounded on $R_+ \times S(\rho)$. Then the trivial solution of (4.2.1) is asymptotically stable if $V'(t, x) \leq -c(\|x\|)$ on $R_+ \times S(\rho)$, where $c \in \kappa$.*

An interesting generalization of Marachkov's result is due to Salvadori which uses two Lyapunov functions. The first Lyapunov function V serves to obtain stability and the second Lyapunov function W relates suitably to the first one. The advantage is that one can utilize the monotone character of $V(t, x(t))$.

Theorem 4.2.4. *Assume that*

(i) $V \in C^1(R_+ \times S(\rho), R_+)$, V is positive definite, $V(t, 0) \equiv 0$ and $V'(t, x) \leq -C(W(t, x))$ on $R_+ \times S(\rho)$, where $C \in \kappa$;

(ii) $W \in C^1(R_+ \times S(\rho), R_+)$, W is positive definite and $W'(t, x)$ is bounded from above or from below on $R_+ \times S(\rho)$.

Then $x = 0$ of (4.2.1) is asymptotically stable.

4.2.2 Loss of Positive Definiteness

The positive definiteness of $V(t, x)$ in Marachkov's Theorem can be weakened.

Theorem 4.2.5. *In Theorem 4.2.3, assume that, instead of positive definiteness of $V(t, x)$, a weaker condition, namely, $V(t, 0) \equiv 0$ and $V(t, x) \geq 0$. Then the conclusion of Theorem 4.2.3 holds.*

Proof. It is enough to prove that $x = 0$ is stable. Let $0 < \varepsilon < \rho$ and $t_0 \in R_+$ be given. Let $\|f(t,x)\| \leq M$ on $R_+ \times S(\rho)$. Choose $\delta = \delta(t_0, \varepsilon)$ so that $\delta \in (0, \frac{\varepsilon}{2})$ and $V(t_0, x_0) < C(\frac{\varepsilon}{2})\frac{\varepsilon}{2M}$ for $\|x_0\| < \delta$ and suppose that $t_2, t_1 > t_0$ such that

$$\|x(t_2)\| = \varepsilon, \quad \|x(t_1)\| = \frac{\varepsilon}{2} \text{ and } \|x(t)\| < \frac{\varepsilon}{2}, \; t \in [t_0, t_1).$$

Then we have

$$\begin{aligned} 0 \leq V(t_2, x(t_2)) &\leq V(t_0, x_0) + V(t_2, x(t_2)) - V(t_1, x(t_1)) \\ &\leq V(t_0, x_0) - C(\frac{\varepsilon}{2})(t_2 - t_1) \\ &\leq V(t_0, x_0) - C(\frac{\varepsilon}{2})\frac{\varepsilon}{2M} < 0 \end{aligned}$$

which is a contradiction. Hence the proof is complete.

4.2.3 Loss of Negative Definiteness of $V'(t, x)$

When the hypothesis on $V'(t, x)$ is weakened in Theorem 4.2.2, we need to impose compensating condition since otherwise the result is not true. In this direction, we have the following first result due to Krasovskii which concerns periodic systems.

Theorem 4.2.6. *Assume that*

(i) $V \in C^1(R_+ \times S(\rho), R_+)$, *$V$ is periodic in t, positive definite,* $V(t, 0) \equiv 0$ *and* $V'(t, x) \leq 0$ *on* $R_+ \times S(\rho)$*;*

(ii) The set $A = \{(t, x) \in R_+ \times S(\rho) \setminus \{0\} : \; V'(t, x) = 0\}$ *does not contain any noncontinuable positive trajectory.*

Then $x = 0$ *of* (4.2.1) *is uniformly asymptotically stable.*

For autonomous systems LaSalle has introduced a notion called invariance principle which we state below.

Theorem 4.2.7. *Assume that* $f(t, x) \equiv f(x)$ *in* (4.2.1) *and* $V \in C^1(R^n, R_+)$. *Let* $\Omega = \{x \in R^n : V(x) < l\}$. *Suppose that* Ω *is bounded and that within* Ω, $V(x) > 0$ *for* $x \neq 0$, $V'(x) \leq 0$. *Let* $E = \{x \in \Omega : V'(x) = 0\}$ *and* M *be the largest invariant set in* E. *Then every solution* $x(t)$ *in* Ω *tends to* M *as* $t \to \infty$. *If, in addition,* $V'(x) < 0$ *for all* $x \neq 0$ *in* Ω, *then* $x = 0$ *of* (4.2.1) *is asymptotically stable*

4.2.4 Comparison Principle

The concept of Lyapunov function together with the theory of differential inequalities provides a very general comparison principle under much less restrictive assumptions. In this set up, Lyapunov function may be viewed as a transformation which reduces the study of a given complicated differential system to the study of relatively simpler scalar differential equation. Not only stability but also other properties of solutions can be discussed with this general comparison principle. We shall merely be content in stating the following simple result.

Theorem 4.2.8. *Suppose that $V \in C^1(R_+ \times S(\rho), R_+)$, V is positive definite and decrescent and $V'(t,x) \le g(t, V(t,x))$ on $R_+ \times S(\rho)$, where $g \in C[R_+, R]$ and $g(t,0) \equiv 0$. Then the stability properties of the trivial solution of*

$$u' = g(t,u), \;\; u(t_0) = u_0,$$

imply the corresponding stability properties of $x = 0$ of (4.2.1).

4.2.5 Instability

Original Lyapunov instability results may be summed up in the following theorem.

Theorem 4.2.9. *Assume that there exists a $t_0 \in R_+$ and an open set $U \subset S(\rho)$ such that $V \in C^1([t_0,\infty) \times S(\rho), R_+)$ and for $[t_0,\infty) \times U$,*

(i) $0 < V(t,x) \le a(\|x\|)$, $a \in \kappa$;

(ii) either $V'(t,x) \ge b(\|x\|)$, $b \in \kappa$ or $V'(t,x) = CV(t,x) + w(t,x)$, where $C > 0$ and $w \in C[[t_0,\infty) \times U, R_+]$;

(iii) $V(t,x) = 0$ on $[t_0,\infty) \times (\partial U \cap S(\rho))$, where ∂U denotes the boundary of U and $0 \in \partial U$.

Then $x = 0$ of (4.2.1) is unstable.

We now list various definitions of stability.

Definition 4.2.1. *The trivial solution of (4.2.1) is said to be*

(S_1) equi-stable, if, for each $\epsilon > 0$, $t_0 \in R_+$, there exists a positive function $\delta = \delta(t_0, \epsilon)$ which is continuous in t_0 for each ϵ such that $\|x_0\| < \delta$ implies $\|x(t)\| < \epsilon, \;\; t \ge t_0$;

(S_2) uniformly stable if the δ in (S_1) is independent of t_0;

(S_3) quasi-equi asymptotically stable, if for each $\epsilon > 0$, $t_0 \in R_+$, there exist positive numbers $\delta_0 = \delta_0(t_0)$ and $T = T(t_0, \epsilon)$ such that $\|x_0\| < \delta_0$ implies $\|x(t)\| < \epsilon$, $t \ge t_0 + T$;

(S_4) quasi-uniformly asymptotically stable if the numbers δ_0 and T in (S_3) are independent of t_0;

(S_5) equi-asymptotically stable if (S_1) and (S_3) hold together;

(S_6) uniformly asymptotically stable if (S_2) and (S_4) hold together;

(S_7) *quasi-equi asymptotically stable in the large if for each* $\epsilon > 0$, $\alpha > 0$, $t_0 \in R_+$, *there exists a positive number* $T = T(t_0, \epsilon, \alpha)$ *such that* $|x_0| \leq \alpha$ *implies* $\|x(t)\| < \epsilon$, $t \geq t_0 + T$;

(S_8) *quasi-uniformly asymptotically stable (in the large) if the number* T *in* (S_7) *is independent of* t_0;

(S_9) *completely stable if* (S_1) *holds and* (S_7) *is satisfied for all* α, $0 \leq \alpha < \infty$;

(S_{10}) *uniformly completely stable if* (S_2) *holds and* (S_8) *is verified for all* α, $0 \leq \alpha < \infty$;

(S_{11}) *unstable if* (S_1) *fails to hold.*

Remark 4.2.1. *We note that the existence of the trivial solution of* (4.2.1) *is not necessary for the notions* (S_3), (S_4), (S_7) *and* (S_8) *to hold. Further more, even when the trivial solution does not exist, we may have stability eventually which is a generalization of Lyapunov stability. We shall define such a concept below.*

Definition 4.2.2. *The system* (4.2.1) *is said to be*

(E_1) *eventually stable, if for each* $\epsilon > 0$, *there exists two positive numbers* $\delta = \delta(\epsilon)$ *and* $\tau = \tau(\epsilon)$ *such that* $\|x_0\| < \delta$ *implies* $\|x(t)\| < \epsilon, \quad t \geq t_0 \geq \tau$;

(E_2) *eventually asymptotically stable if* (E_1) *and* (S_4) *hold simultaneously.*

It is clear that, in applications, asymptotic stability is more important than stability. It is therefore necessary to know the size of the domain of asymptotic stability so that based on estimates of conditions under which the system will actually operate, requirements on its performance etc., we can judge whether or not the system is sufficiently stable to function properly and may be able to see how to improve its stability. Thus for practical purposes, complete stability seems desirable.

Corresponding to different types of stability, we can define concepts of boundedness. To do this, we do not need the existence of the trivial solution.

Definition 4.2.3. *The differential system* (4.2.1) *is said to be*

(B_1) *equi-bounded if, for each* $\alpha \geq 0$, $t_0 \in R_+$, *there exists a positive function* $\beta = \beta(t_0, \alpha)$ *that is continuous in* t_0 *for each* α *such* $\|x_0\| \leq \alpha$ *implies* $\|x(t)\| < \beta$, $t \geq t_0$;

(B_2) *uniformly bounded if* β *in* (B_1) *is independent of* t_0;

(B_3) *quasi-equi ultimately bounded if, for each* $\alpha \geq 0$, $t_0 \in R_+$, *there exist positive numbers* N *and* $T = T(t_0, \alpha)$ *such that* $\|x_0\| \leq \alpha$ *implies* $\|x(t)\| < N$, $t \geq t_0 + T$;

(B_4) *quasi-unifrom ultimately bounded if the* T *in* (B_3) *is independent of* t_0;

(B_5) *equi-ultimately bounded if* (B_1) *and* (B_3) *hold together;*

(B_6) *uniform-ultimately bounded if* (B_2) *and* (B_4) *hold simultaneously;*

(B_7) *equi-Lagrange stable if* (B_1) *and* (S_7) *are satisfied;*

(B_8) *uniform-Lagrange stable* (B_2) *and* (S_8) *are satisfied;*

(B_9) *eventually bounded if, for each* $\alpha \geq 0$ *there exist two positive numbers* $\tau = \tau(\alpha)$, $\beta = \beta(\alpha)$ *such that* $|x_0| \leq \alpha$ *implies* $\|x(t)\| < \beta$, $t \geq t_0 \geq \tau$;

(B_{10}) *eventually Lagrange stable if* (B_9) *and* (S_8) *hold together.*

We note that if $f(t, 0) \equiv 0$ and β occurring in (B_1), (B_2) has the property that $\beta \to 0$ as $\alpha \to 0$, then the definitions (B_1), (B_2) imply (S_1), (S_2) respectively.

Let us illustrate the definitions given above with some examples.

Example 4.2.1. *Consider the equations*

$$\begin{cases} x'(t) = n(t)y + m(t)x(x^2 + y^2), \ x(t_0) = x_0, \\ y'(t) = -n(t)x + m(t)y(x^2 + y^2), \ y(t_0) = y_0, \end{cases} \tag{4.2.3}$$

where $n, m \in C[R_+, R]$. *The general solution of* (4.2.3) *is given by*

$$x(t) = \frac{x_0 \cos\left(\int_{t_0}^t n(t)\mathrm{d}t\right) + y_0 \sin\left(\int_{t_0}^t n(t)\mathrm{d}t\right)}{\left(1 - 2(x_0^2 + y_0^2)\int_{t_0}^t m(t)\mathrm{d}t\right)^{\frac{1}{2}}}$$

$$y(t) = \frac{y_0 \cos\left(\int_{t_0}^t n(t)\mathrm{d}t\right) - x_0 \sin\left(\int_{t_0}^t n(t)\mathrm{d}t\right)}{\left(1 - 2(x_0^2 + y_0^2)\int_{t_0}^t m(t)\mathrm{d}t\right)^{\frac{1}{2}}}$$

which reduces to

$$r^2(t) = x^2(t) + y^2(t) = r_0^2\left(1 - 2r_0^2 \int_{t_0}^t m(t)\mathrm{d}t\right)^{-1}, \tag{4.2.4}$$

where $r_0^2 = x_0^2 + y_0^2$. *It follows from* (4.2.4) *that the trivial solution of* (4.2.3) *is stable if* $m(t) \leq 0, t \geq t_0$. *If* $m(t) > 0$, $t \geq t_0$, *then the trivial solution of* (4.2.3) *is stable when the integral*

$$\int_{t_0}^t m(t)\mathrm{d}t \tag{4.2.5}$$

is bounded and unstable when (4.2.5) *is unbounded.*

Example 4.2.2. *Consider the differential system*

$$\begin{cases} x'(t) = -x - y + k(x-y)(x^2+y^2), \ x(t_0) = x_0, \\ y'(t) = x - y + k(x+y)(x^2+y^2), \ y(t_0) = y_0, \end{cases} \tag{4.2.6}$$

where $k > 0$ is a constant. The general solution of (4.2.6) is given by

$$x(t) = \frac{1}{\sqrt{k\mu}}(x_0 \cos\theta - y_0 \sin\theta), \ y(t) = \frac{1}{\sqrt{k\mu}}(x_0 \sin\theta + y_0 \cos\theta)$$

with $\theta = 2(t-t_0) - \frac{1}{2}\ln\mu$ and $\mu = r_0^2 + \left(\frac{1}{k} - r_0^2\right)\exp(2(t-t_0))$.

This reduces to

$$(r(t))^2 = \frac{1}{k\mu} r_0^2. \tag{4.2.7}$$

It is clear that if $r_0^2 = x_0^2 + y_0^2 < \dfrac{1}{k}$, then the trivial solution of (4.2.6) is asymptotically stable.

Example 4.2.3. *Consider the equation of perturbed motion of a mechanical system with one degree of freedom given by*

$$\frac{d^2x}{dt^2} - 2h\frac{dx}{dt} + gx = 0, \ x(0) = x_0, \ x'(0) = y_0,$$

which can be rewritten as the system

$$\begin{cases} \dfrac{dx}{dt} = y, \ x(0) = x_0 \\ \dfrac{dy}{dt} = 2hy - gx, \ y(0) = y_0 \end{cases} \tag{4.2.8}$$

with $g - h^2 > 0$. Since the characteristic roots of $\lambda^2 - 2h\lambda + g = 0$ are complex conjugate, let $\lambda_1 = \alpha + i\beta$, $\lambda_2 = \alpha - i\beta$ and the general solution of (4.2.8) is of the form

$$x(t) = \left[x_0\cos(\beta t) + \frac{y_0 - \alpha x_0}{\beta}\sin(\beta t)\right]\exp(\alpha t).$$

Consequently, the trivial solution of (4.2.8) is unstable if $\alpha > 0$, $h < 0$ and $g > 0$.

Example 4.2.4. *Consider the differential equation*

$$x' = -\lambda'(t), \ x(t_0) = x_0 \tag{4.2.9}$$

where $\lambda \in C^1[R_+, R_+]$ with $\lambda'(t) \geq 0$. Then $x(t) = x_0 + \lambda(t_0) - \lambda(t)$ and hence, we have

$$\|x(t)\| < \|x_0\| + \lambda(t_0), \ t \geq t_0.$$

It is easy to see that (B_1) holds with $\beta = \alpha + \lambda(t_0)$ and that β does not tend to zero as $\alpha \to 0$. If $\lambda(t)$ is decreasing to zero as $t \to \infty$, then, given $\epsilon > 0$, there exists a $\tau(\epsilon)$ such that $\lambda(t_0) < \dfrac{\epsilon}{2}$ if $t_0 \geq \tau(\epsilon)$. As a result, (E_1) holds with $\delta = \dfrac{\epsilon}{2}$ and $\tau(\epsilon)$ defined above.

Example 4.2.5. *Consider the equation*

$$x' = \frac{-x}{1+t}, \; x(t_0) = x_0 \tag{4.2.10}$$

whose solution $x(t) = \dfrac{x_0(1+t_0)}{(1+t)}$ *does not tend to zero uniformly with respect to* t_0.

We have seen that complete stability is a more desirable feature in applications than asymptotic stability. Sometimes even instability may be good enough. Since the desired state of system may be mathematically unstable but the system may oscillate sufficiently near this state so that its performance is considered acceptable. For example, an aircraft or a missile may oscillate around a mathematically unstable course yet its performance may be acceptable. Many problems fall into this category including the travel of a space vehicle between two points and the problem, in a chemical process, of keeping the temperature within ceratin bounds. To deal with such situations, the notion of practical stability is more useful, which we define below. In fact, it is easy to see that the trivial solution of (4.2.10) is uniformly stable and quasi-equi-asymptotically stable which implies that we have (S_2) and(S_5). This example suggests that it may be fruitful to study other combinations than the usual ones and for such an investigation several Lyapunov functions are more appropriate.

Definition 4.2.4. *The system* (4.2.1) *is said to be*

(PS_1) *practically stable if, given* (λ, A) *with* $0 < \lambda < A$, *we have* $\|x_0\| < \lambda$ *implies* $\|x(t)\| < A$, $t \geq t_0$ *for some* $t_0 \in R_+$;

(PS_2) *uniformly practically stable if* (PS_1) *holds for every* $t_0 \in R_+$;

(PS_3) *practically quasi-stable if given* $(\lambda, B, T) > 0$ *and some* $t_0 \in R_+$, *we have* $\|x_0\| < \lambda$ *implies* $\|x(t)\| < B, t \geq t_0 + T$;

(PS_4) *uniformly practically quasi stable if* (PS_3) *holds for all* $t_0 \in R_+$;

(PS_5) *strongly practically stable if* (PS_1) *and* (PS_3) *hold simultaneously;*

(PS_6) *strongly uniformly practically stable if* (PS_2) *and* (PS_4) *hold together;*

(PS_7) *practically asymptotically stable if* (PS_1) *and* (S_7) *hold with* $\alpha = \lambda$;

(PS_8) *uniformly practically asymptotically stable if* (PS_2) *and* S_8 *hold at the same time with* $\alpha = \lambda$;

(PS_9) *practically unstable if* (PS_1) *does not hold;*

(PS_{10}) *eventually practically stable if given* (λ, A), $0 < \lambda < A$, *there exists a* $\tau = \tau(\lambda, A)$ *such that* $\|x_0\| < \lambda$ *implies* $\|x(t)\| < A$, $t \geq t_0 \geq \tau$;

(PS_{11}) *eventually strongly practically stable if* (PS_{10}) *and* (PS_3) *hold;*

(PS_{12}) *eventually uniformly strongly practically stable if* (PS_{10}) *and* (PS_4) *are satisfied simultaneously.*

In (PS_5) and (PS_6), if $0 < B < \lambda < A$, then we say that the system (4.2.1) is contractively practically stable, while for $0 < \lambda < B < A$, the system is said to be expansively practically stable.

Sometimes in physical problems, one is interested in the behavior of systems within specified bounds during a fixed time interval. The concept of "finite time stability" is appropriate to cover such situations. For example, the notion (PS_5) would translate into the following: the differential system (4.2.1) is strongly practically stable if, given positive numbers λ, A, B and T we have $\|x_0\| < \lambda$ implies $\|x(t)\| < A$, $t_0 \le t \le t_0 + T$ and $\|x(t_0 + T)\| < B$.

Let us consider some examples to illustrate the notions of practical stability.

Example 4.2.6. *Consider the example* (4.2.1) *and let* $A = 2\lambda$. *Suppose that* $\int_{t_0}^{\infty} m(s) = \beta > 0$. *Then we get from* (4.2.3)

$$\lim_{t\to\infty} r(t) = r_0^2(1 - 2r_0^2\beta)^{-1}. \tag{4.2.11}$$

It therefore follows from (4.2.11) *that the system* (4.2.3) *is practically stable if* $\beta \le \dfrac{3}{8\lambda^2}$ *and practically unstable if* $\beta > \dfrac{3}{8\lambda^2}$.

Example 4.2.7. *Consider the example 4.2.2. Recall that* $r_0^2 = x_0^2 + y_0^2 < \dfrac{1}{k}$ *is the region of asymptotic stability. Let* λ, A *be such that* $\dfrac{1}{\sqrt{k}} < \lambda < A$. *Then for the initial value* (x_0, y_0) *such that* $\dfrac{1}{k} \le r_0^2 < \lambda^2$, *the system* (4.2.6) *is not practically stable which shows that asymptotic stability is not sufficient for practical stability to hold. In other words, the presence or absence of practical stability in a system does not depend on asymptotic stability.*

Example 4.2.8. *Consider the system* (4.2.8) *of Example 4.2.3. Recall that the system* (4.2.8) *is unstable if* $\alpha > 0$, $h < 0$ *and* $g > 0$. *Define the sets*

$$S_0 = \left\{ (x, y) : x^2 + \left(\frac{y - hx}{\beta}\right)^2 < \lambda^2 \right\}$$

$$S = \left\{ (x, y) : x^2 + \left(\frac{y - hx}{\beta}\right)^2 < \delta\lambda^2 \right\}$$

with $\delta > 0$ *and* $\beta = \sqrt{g - h^2}$. *Take* $T < \dfrac{\ln \delta}{2\alpha}$. *Then, it is easy to check that the system* (4.2.8) *is practically stable with respect to the sets* S_0, S *and the finite time interval* $[t_0, t_0 + T]$.

Example 4.2.8 indicates that it is more natural to define practical stability in terms of arbitrary sets instead of neighborhoods of the origin. We have seen that practical stability is neither weaker nor stronger than Lyapunov stability. Before we speak of practical stability we must decide on

(i) how near the desired state (that is, set S) it is necessary to have the system operate; and

(ii) how well the initial set (that is set S_0) can be controlled.

Also, practical stability is somewhat similar to uniform boundedness. It is, however, not merely that a bound exists but that the bound be pre-assigned. Note that Lagrange stability is somewhat similar to practical asymptotic stability and ultimate boundedness is a necessary condition for the system to possess strong practical stability.

Sometimes, the interdependence of (λ, A, B, T) may be useful in practice. For example, (PS_3) may be weakened as follows. The system (4.2.1) is said to be

(PS_3^*) practically quasi-stable if given $(\lambda, B) > 0$ and $t_0 \in R_+$, there exists a $T = T(t_0, \lambda, B)$ such that $\|x_0\| < \lambda$ implies $\|x(t)\| < B$, $t \geq t_0 + T$.

If (PS_1) and (PS_3^*) hold together, we can identify that as (PS_5^*) and other similar concepts may be introduced. Occasionally, it is advantageous to restrict the initial times t_0 to a given set $T_0 \subset R_+$, instead of allowing the initial set (set T_0) to be the whole real line.

4.3 Comparison Results and Refinements

Consider the differential system

$$x' = f(t, x), \; x(t_0) = x_0, \; t_0 \in R_+ \tag{4.3.1}$$

where $f \in C(R_+ \times R^n, R^n)$. For any function $V \in C(R_+ \times R^n, R^N)$, we define the generalized derivative

$$D^+V(t, x) = \limsup_{h \to 0^+} \frac{1}{h}[V(t + h, x + hf(t, x)) - V(t, x)] \tag{4.3.2}$$

for $(t, x) \in R_+ \times R^n$. One could also utilize the other generalized derivative

$$D_-V(t, x) = \limsup_{h \to 0^-} \frac{1}{h}[V(t + h, x + hf(t, x)) - V(t, x)]. \tag{4.3.3}$$

Occasionally, we shall denote (4.3.2) by $D^+V(t, x)_{(4.3.1)}$ to emphasize the definition of D^+V with respect to the system (4.3.1). We note that if $V \in C^1(R_+ \times R^n, R_+^N)$, then $D^+V = D_-V = V'(t, x)$ where $V'(t, x) = V_t(t, x) + V_x(t, x)f(t, x)$.

A function $F \in C(R^n, R^n)$ is said to be quasimonotone nondecreasing in x, if $x \leq y$ and $x_i = y_i$ for $1 \leq i \leq n$ implies $F_i(x) \leq F_i(y)$ for all i. The inequalities between vectors are understood to be componentwise inequalities.

Theorem 4.3.1. *Let $V \in C(R_+ \times R^n, R_+^N)$ be locally Lipschitzian in x. Assume that*

$$D^+V(t,x) \leq g(t, V(t,x)), \ (t,x) \in R_+ \times R^n, \tag{4.3.4}$$

where $g \in C(R_+ \times R_+^N, R^N)$ and $g(t,u)$ is quasimonotone nondecreasing in u. Let $r(t) = r(t, t_0, u_0)$ be the maximal solution of

$$u' = g(t,u), \ u(t_0) = u_0 \geq 0, \ t_0 \in R_+, \tag{4.3.5}$$

existing for $t \geq t_0$. Then

$$V(t, x(t)) \leq r(t), \ t \geq t_0, \tag{4.3.6}$$

where $x(t) = x(t, t_0, x_0)$ is any solution of (4.3.1) existing on $[t_0, \infty)$, provided that $V(t_0, x_0) \leq u_0$.

Proof. Let $x(t) = x(t, t_0, x_0)$ be any solution of (4.3.1) existing for $t \geq t_0$ such that $V(t_0, x_0) \leq u_0$. Define $m(t) = V(t, x(t))$. For sufficiently small $h > 0$, we have

$$\begin{aligned} m(t+h) - m(t) &= V(t+h, x(t+h)) \\ &\quad - V(t+h, x(t) + hf(t, x(t))) \\ &\quad + V(t+h, x(t) + hf(t, x(t))) - V(t, x(t)). \end{aligned}$$

Since $V(t,x)$ is locally Lipschitzian in x, we obtain, using (4.3.4) the differential inequality

$$D^+m(t) \leq g(t, m(t)), \ m(t_0) \leq u_0, \ t \geq t_0,$$

and Theorem 1.7.1 now gives the desired result (4.3.6).

Corollary 4.3.1. *Suppose that in Theorem 4.3.1, we have $g(t,u) \equiv 0$. Then $V(t, x(t))$ is nonincreasing in t and $V(t, x(t)) \leq V(t_0, x_0), t \geq t_0$. The case $N = 1$ in Theorem 4.3.1 is a well known basic comparison result which is often utilized in the study of nonlinear systems.*

A comparison result analogous to Theorem 4.3.1 which yields lower bounds is the following. We merely state this result since its proof is similar.

Theorem 4.3.2. *If in Theorem 4.3.1, assumption (4.3.4) is reversed to*

$$D^+V(t,x) \geq g(t, V(t,x)), \ (t,x) \in R_+ \times R^n,$$

and $\rho = \rho(t, t_0, u_0)$ is the minimal solution of (4.3.5) existing for $t \geq t_0$, then

$$V(t, x(t)) \geq \rho(t), \ t \geq t_0,$$

whenever $V(t_0, x_0) \geq u_0$.

In some situations, estimating $D^+V(t,x)$ as a function of t, x and $V(t,x)$ is more natural. The next comparison result is in that direction whose proof is similar to that of Theorem 4.3.1.

Theorem 4.3.3. *Let $V \in C(R_+ \times R^n, R_+^N)$ and be locally Lipschitzian in x. Assume that $g \in C(R_+ \times R^n \times R_+^N, R^N)$ and for $(t,x) \in R_+ \times R^n$,*

$$D^+V(t,x) \le g(t,x,V(t,x))$$

where $g(t,x,u)$ is quasimonotone nondecreasing in u. If $x(t) = x(t,t_0,x_0)$ is any solution of (4.3.1) existing for $t \ge t_0$ and $r(t,t_0,u_0,x_0)$ is the maximal solution of

$$u' = g(t,x(t),u), u(t_0) = u_0 \ge 0,$$

existing for $t \ge t_0$, then $V(t_0,x_0) \le u_0$ implies

$$V(t,x(t)) \le r(t,t_0,u_0,x_0),\ t \ge t_0.$$

As mentioned earlier, the inequalities in the foregoing considerations are componentwise. Instead of considering componentwise inequalities between vectors, we could utilize the notion of a cone to introduce partial order on R^n and prove comparison results in that framework. Naturally, this approach is more general and is useful when we deal with conevalued functions. We shall only prove a result corresponding to Theorem 4.3.1 in arbitrary cones. Other results can be proved similarly.

A proper subset $K \subset R^n$ is called a cone if the following properties hold:

$$\begin{cases} \lambda K \subset K,\ \lambda \ge 0,\ K + K \subset K,\ K = \bar{K}, \\ K \cap \{-K\} = \{0\} \text{ and } K^\circ \ne \phi, \end{cases} \tag{4.3.7}$$

where $\bar{K}$ denotes the closure of K, K° is the interior of K. We shall denote by ∂K the boundary of K. The cone K induces the order relations on R^n defined by

$$x \le_{\bar{K}} y \text{ iff } y - x \in K \text{ and } x <_K y \text{ iff } y - x \in K^\circ. \tag{4.3.8}$$

The set K^* defined by $K^* = \{\psi \in R^n : \psi(x) \ge 0 \text{ for all } x \in K\}$, where $\psi(x)$ denotes the scalar product $< \psi, x >$, is called the adjoint cone and satisfies the properties (4.3.7). We note that $K = (K^*)^*$, $x \in K^\circ$ iff $\psi(x) > 0$ for all $\psi \in K_0^*$ and $x \in \partial K$ iff $\psi(x) = 0$ for some $\psi \in K_0^*$ where $K_0 = K - \{0\}$.

We can now define quasimonotone property of a function relative to the cone K. A function $f \in C(R^n, R^n)$ is said to be quasimonotone nondecreasing relative to K if $x \le_K y$ and $\psi(x-y) = 0$ for some $\psi \in K_0^*$ imply $\psi(f(x) - f(y)) \le 0$. If f is linear, that is, $f(x) = Ax$ where A is an n by n matrix, the quasimonotone property of f means the following: $x \ge 0$ and $\psi(x) = 0$ for some $\psi \in K_0^*$ imply $\psi(Ax) \ge 0$. If $K = R_+^n$, the quasimonotonicity of f reduces precisely to what we defined before.

In the general setting of a cone, Theorem 4.3.1 is true.

Theorem 4.3.4. *Assume that $V \in C(R_+ \times R^n, K)$ and $V(t,x)$ is locally Lipschitzian in x relative to the cone $K \subset R^N$ and for $(t,x) \in R_+ \times R^n$,*

$$D^+V(t,x) \le_K g(t,V(t,x)).$$

Let $g \in C(R_+ \times K, R^N)$, $g(t,u)$ be quasimonotone nondecreasing in u with respect to K and $r(t) = r(t, t_0, u_0)$ be the maximal solution of (4.3.5) existing for $t \geq t_0$. Then, any solution $x(t) = x(t, t_0, u_0)$ of (4.3.1) existing for $t \geq t_0$ satisfies the estimate

$$V(t, x(t)) \leq_K r(t), \; t \geq t_0,$$

provided $V(t_0, x_0) \leq_{\bar{K}} u_0$.

Proof. Proceeding as in Theorem 4.3.1 with suitable modifications, we arrive at the differential inequality

$$D^+ m(t) \leq_K g(t, m(t)), \; m(t_0) \leq u_0, \; t \geq t_0.$$

Then, Theorem 1.7.3 yields the stated result.

The next theorem which is a variant of Theorem 4.3.4 is more flexible in applications. We merely state the result.

Theorem 4.3.5. *Let P and Q be cones in R^N such that $P \subset Q$. Suppose that $V \in C(R_+ \times R^n, Q)$, $V(t,x)$ satisfies a local Lipschitz condition relative to P and*

$$D^+ V(t,x) \leq_P g(t, V(t,x)), (t,x) \in R_+ \times R^n.$$

Assume further that $g \in C(R_+ \times Q, R^N)$ and $g(t,u)$ is quasimonotone nondecreasing in u relative to P and $x(t) = x(t, t_0, x_0)$ is any solution of (4.3.1) existing for $t \geq t_0$ such that $V(t_0, x_0) \leq_P u_0$. Then,

$$V(t, x(t)) \leq_Q r(t), t \geq t_0, \tag{4.3.9}$$

where $r(t) = r(t, t_0, u_0)$ is the maximal solution of (4.3.5) relative to P. In particular, if $Q = R^N_+$, then (4.3.9) implies the componentwise estimate $V(t, x(t)) \leq r(t)$, $t \geq t_0$.

We shall next discuss some stability results by employing two Lyapunov like functions and obtain refinements of the results considered in section 4.2. We begin with the following result which gives nonuniform stability under weaker assumptions and also includes several results.

Theorem 4.3.6. *Assume that*

(i) $V_1 \in C(R_+ \times S(\rho), R_+)$, $V_1(t,x)$ is locally Lipschitzian in x, $V_1(t,0) \equiv 0$ and

$$D^+ V_1(t,x) \leq g_1(t, V_1(t,x)), (t,x) \in R_+ \times S(\rho),$$

where $g_1 \in C(R^2_+, R)$ and $g_1(t,0) \equiv 0$;

(ii) for every $\eta > 0$, there exists a $V_{2,\eta} \in C(R_+ \times S(\rho) \cap S^c(\eta), R_+)$, $V_{2,\eta}$ is locally Lipschitzian in x and for $(t,x) \in R_+ \times S(\rho) \cap S^c(\eta)$,

$$b(\|x\|) \leq V_{2,\eta}(t,x) \leq a(\|x\|), \; a, b \in \mathcal{K}$$

and

$$D^+ V_1(t,x) + D^+ V_{2,\eta}(t,x) \leq g_2(t, V_1(t,x) + V_{2,\eta}(t,x)),$$

where $g_2 \in C(R^2_+, R)$, $g_2(t,0) \equiv 0$, and $S^c(\eta)$ is the complement of $S(\eta)$;

(iii) the trivial solution of

$$u' = g_1(t, u), \ u(t_0) = u_0 \geq 0 \tag{4.3.10}$$

is equistable and the trivial solution of

$$v' = g_2(t, v), \ v(t_0) = v_0 \geq 0 \tag{4.3.11}$$

is uniformly stable.

Then, the trivial solution of the system (4.3.1) *is equistable.*

Proof. Let $0 < \epsilon < \rho$ and $t_0 \in R_+$ be given. Since the trivial solution of (4.3.11) is uniformly stable, given $b(\epsilon) > 0$ and $t_0 \in R_+$, there exists a $\delta_0 = \delta_0(\epsilon) > 0$ such that

$$v(t, t_0, v_0) < b(\epsilon), \ t \geq t_0, \tag{4.3.12}$$

provided $v_0 < \delta_0$, where $v(t, t_0, v_0)$ is any solution of (4.3.11). In view of $a \in \kappa$ there is a $\delta_2 = \delta_2(\epsilon) > 0$ such that

$$a(\delta_2) < \frac{\delta_0}{2}. \tag{4.3.13}$$

By the equistability of $u \equiv 0$ relative to (4.3.10), given $\dfrac{\delta_0}{2} > 0$ and $t_0 \in R_+$, there exists a $\delta^* = \delta^*(t_0, \epsilon)$ such that

$$u(t, t_0, u_0) < \frac{\delta_0}{2}, \ t \geq t_0, \tag{4.3.14}$$

whenever $u_0 < \delta^*$, $u(t, t_0, u_0)$ being any solution of (4.3.10).

Choose $u_0 = V_1(t_0, x_0)$. Since $V_1(t, x)$ is continuous and $V_1(t, 0) \equiv 0$, there exists a $\delta_1 > 0$ such that

$$\|x_0\| < \delta_1 \text{ and } V_1(t_0, x_0) < \delta^* \tag{4.3.15}$$

hold simultaneously. Set $\delta = \min(\delta_1, \delta_2)$. Then we claim that $\|x_0\| < \delta$ implies $\|x(t, t_0, x_0)\| < \epsilon$ for $t \geq t_0$. If this were false, there would exist a solution $x(t, t_0, x_0)$ of (4.3.1) with $\|x_0\| < \delta$ and $t_1, t_2 > t_0$ such that

$$\|x(t_1, t_0, x_0)\| = \delta_2, \ \|x(t_2, t_0, x_0)\| = \epsilon \tag{4.3.16}$$

and $x(t, t_0, x_0) \in \overline{S(\epsilon) \bigcap S(\delta_2)}$ on $[t_1, t_2]$. Let $\delta_2 = \eta$ so that the existence of $V_{2,\eta}$ satisfying hypothesis (ii) is assured. Hence setting,

$$m(t) = V_1(t, x(t, t_0, x_0)) + V_{2,\eta}(t, x(t, t_0, x_0)), \ t \in [t_1, t_2],$$

we obtain by Theorem 4.3.1 with $N = 1$, the estimate

$$m(t_2) \leq r_2(t_2, t_1, m(t_1)),$$

$r_2(t, t_1, v_0)$ being the maximal solution of (4.3.11) such that $r_2(t_1, t_1, v_0) = v_0$. We also have similarly,

$$V_1(t_1, x(t_1, t_0, x_0)) \leq r_1(t_1, t_0, V_1(t_0, x_0)),$$

where $r_1(t, t_0, u_0)$ is the maximal solution of (4.3.10). By (4.3.14) and (4.3.15), we get

$$V_1(t_1, x(t_1, t_0, x_0)) < \frac{\delta_0}{2}. \tag{4.3.17}$$

Also, by (4.3.13), (4.4.16) and the assumptions on $V_{2,\eta}$, we have

$$V_{2,\eta}(t_1, x(t_1, t_0, x_0)) \leq a(\delta_2) < \frac{\delta_0}{2}. \tag{4.3.18}$$

The inequalities (4.3.17), (4.3.18) together with (4.3.12), (4.3.16), $V_1 \geq 0$ and $V_{2,\eta} \geq b(\|x\|)$ lead to the contradiction $b(\epsilon) < b(\epsilon)$. Hence the proof of the theorem is complete.

Remark 4.3.1. *Theorem 4.3.6 is a refinement of Theorem 4.2.8 relative to nonuniform stability. When* $g_1(t, u) \equiv g_2(t, u) \equiv 0$, *Theorem 4.3.5 improves Theorem 4.2.1 by relaxing conditions. If* $V_1(t, x) \equiv 0$, $\quad g_1(t, u) \equiv 0$, *then Theorem 4.3.5 yields uniform stability under weaker assumptions. If, on the other hand,* $V_{2,\eta}(t, x) \equiv 0$ *and* $g_2(t, u) \equiv 0$ *and* $V_1(t, x)$ *is assumed to be positive definite, Theorem 4.3.5 reduces to Theorem 4.2.1*

We shall consider a result on asymptotic stability which is in the spirit of Theorem 4.3.6

Theorem 4.3.7. *In addition to the assumptions of Theorem 4.3.6, suppose that*

(a) *either*

(a_1) $\mathrm{D}^+V_1(t, x) + C(\|x\|) \leq g_1(t, V_1(t, x))$ *where* $C \in k$ *and* $g_1(t, u)$ *is nondecreasing in* u, *or*

(a_2) $\mathrm{D}^+V_1(t, x) + \mathrm{D}^+V_{2,\eta}(t, x) + C(\|x\|) \leq g_2(t, V_1(t, x) + V_{2,\eta}(t, x))$ *where* $C \in k$ *and* $g_2(t, u)$ *is nondecreasing in* u;

(b) *there exists a sufficiently small* $\rho_0 > 0$ *such that*

$$V_1(t, x) \leq a_0(\|x\|) \quad for \quad (t, x) \in R_+ \times S(\rho_0), a_0 \in k;$$

(c) *the trivial solution of* (4.3.10) *is uniformly stable for* $0 \leq u_0 < \rho_0$.

Then the trivial solution of (4.3.1) *is equi-asymptotically stable.*

Proof. Suppose that (a_1) holds. Then since the assumptions of Theorem 4.3.6. hold we get taking $\epsilon = \rho$,

$$\|x_0\| < \delta(t_0, \rho) \quad implies \quad \|x(t)\| < \rho, t \geq t_0.$$

Now let $0 < \epsilon < \rho_0$ and $t_0 \in R_+$ be given and $\delta, \delta_1, \delta_2$ be chosen as in Theorem 4.3.6. Note that in view of (c), δ^* is independent of t_0 and hence by (b),

$\delta_1 = a_0^{-1}(\delta^*)$. As a result δ is independent of t_0. Choose a $T = T(\epsilon) > 0$ such that

$$T > \frac{\delta_0(\epsilon)}{2C(\delta(\epsilon))}. \tag{4.3.19}$$

We claim that there exists a $t^* \in [t_0, t_0 + T]$ such that $\|x(t^*)\| < \delta$ for any solution $x(t)$ of (4.3.1) with $|x_0| < \delta(t_0, \rho)$. If this is not true, let us suppose the $\delta \leq |x(t)|$ for $t \in [t_0, t_0 + T]$. Then setting

$$m(t) = V_1(t, x(t)) + \int_{t_0}^{t} C(\|x(s)\|)\mathrm{d}s,$$

using condition (a_1) and the monotonic character of g_1, we get

$$\mathrm{D}^+ m(t) \leq g_1(t, m(t)), \quad t \in [t_0, t_0 + T]$$

using arguments similar to the proof of Theorem 4.3.1. Hence by Theorem 1.7.1, we obtain, choosing $u_0 = V_1(t_0, x_0)$, the estimate

$$V_1(t, x(t)) + \int_{t_0}^{t} C(\|x(s)\|)\mathrm{d}s \leq r_1(t, t_0, V_1(t_0, x_0)), \quad t \in [t_0, t_0 + T],$$

where $r_1(t, t_0, u_0)$ is the maximal solution of (4.3.10). This then implies that

$$C(\delta(\epsilon))T \leq \frac{\delta_0(\epsilon)}{2},$$

which contradicts (4.3.19). It therefore follows that there exists a $t^* \in [t_0, t_0 + T]$ such that $\|x(t^*)\| < \delta(\epsilon)$ whenever $\|x_0\| < \delta(t_0, \rho)$. We now claim that $|x(t)| < \epsilon, \quad t \geq t_0 + T$. If this is not true, we proceed as in the proof of Theorem 1.4.1 replacing t_0 by t^* to arrive at a contradiction. Hence it follows that $\|x(t)\| < \epsilon, \quad t \geq t_0 + T$ whenever $\|x_0\| < \delta(t_0, \rho)$.

If (a_2) holds, we proceed as before with suitable modification to get a contradiction. Hence the proof is complete.

Remark 4.3.2. *The function $g_1(t, u) \equiv g_2(t, u) \equiv 0$ are admissible in Theorem 4.3.7. In this case, we can assume, instead of (b), that $f(t, x)$ is bounded on $R_+ \times S(\rho)$ which is an improvement of Theorem 4.2.3. Other conclusions, similar to Remark 4.3.1 can be drawn from Theorem 4.3.6*

Theorem 4.2.6 and 4.2.7 cover particular situations of $f(t, x)$ when $\mathrm{D}^+V(t, x)$ loses negative definiteness. The result that follows takes care of the general case of $f(t, x)$ and requires two Lyapunov functions.

Theorem 4.3.8. *Assume that*

(i) $f(t, x)$ is bounded on $R_+ \times S(\rho)$;

(ii) $V_1 \in C(R_+ \times S(\rho), R_+)$, $V_1(t,x)$ *is positive definite, decrescent, locally Lipschitzian in* x*, and*

$$\mathrm{D}^+ V_1(t,x) \le w(x) \le 0, \quad (t,x) \in R_+ \times S(\rho),$$

where $w(x)$ *is continuous for* $x \in S(\rho)$*;*

(iii) $V_2 \in C(R_+ \times S(\rho), R_+)$*, and* $V_2(t,x)$ *is bounded on* $R_+ \times S(\rho)$ *and is locally Lipschitzian in* x*. Furthermore, given any number* α*,* $0 < \alpha < \rho$*, there exist positive numbers* $\xi = \xi(\alpha) > 0$*,* $\eta = \eta(\alpha) > 0$*,* $\eta < \alpha$*, such that*

$$\mathrm{D}^+ V_2(t,x) > \xi$$

for $\alpha < \|x\| < \rho$ *and* $\mathrm{d}(x,E) < \eta$*,* $t \ge 0$*, where*

$$E = [x \in S(\rho): \quad w(x) = 0]$$

and $\mathrm{d}(x,E)$ *is the distance between the point* x *and the set* E*.*

Then, the trivial solution of (4.3.1) *is uniformly asymptotically stable.*

Proof. Let $\epsilon > 0$ and $t_0 \in R_+$ be given. Since $V_1(t,x)$ is positive definite and decrescent, there exist functions $a, b \in k$ such that

$$b(\|x\|) \le V_1(t,x) \le a(\|x\|), \quad (t,x) \in R_+ \times S(\rho).$$

We choose $\delta = \delta(\epsilon)$ so that

$$b(\epsilon) > a(\delta).$$

Then, it is easy to conclude that the trivial solution of (4.3.1) is uniformly stable.

Let us now fix $\epsilon = \rho$ and define $\delta_0 = \delta(\rho)$. Let $0 < \epsilon < \rho$, $t_0 \in R_+$ and $\delta = \delta(\epsilon)$ be the same δ obtained above for uniform stability . Assume that $\|x\| < \delta_0$. To prove uniform asymptotic stability of the solution $x = 0$, it is enough to show that there exists a $T = T(\epsilon)$ such that, for some $t^* \in [t_0, t_0 + T]$, we have

$$\|x(t^*, t_0, x_0)\| < \delta.$$

This we achieve in a number of stages:

(1) If $\mathrm{d}[x(t_1, x(t_2)] > r > 0$. $t_2 > t_1$, then

$$r \le M n^{1/2}(t_2 - t_1),$$

where $\|f(t,x)\| \le M$, $(t,x) \in R_+ \times S(\rho)$. For, consider

$$\begin{aligned}\|x_i(t_1) - x_i(t_2)\| &\le \int_{t_1}^{t_2} \|x_i'(s)\|\mathrm{d}s \le \int_{t_1}^{t_2} \|f_i(s, x(s))\|\mathrm{d}s \\ &\le M(t_2 - t_1) \quad (i = 1, 2, \ldots, n),\end{aligned}$$

and therefore

$$\begin{aligned} r &< d[x(t_1), x(t_2)] \\ &= \left\{[x_1(t_1) - x_1(t_2)]^2 + \cdots + [x_n(t_1) - x_n(t_2)]^2\right\}^{1/2} \\ &\le M n^{1/2}(t_2 - t_1).\end{aligned}$$

(2) By assumption (iii), given $\delta = \delta(\epsilon)$, $0 < \delta < \rho$, there exists $\xi = \xi(\epsilon)$, $\eta = \eta(\epsilon)$, $\eta < \delta$ such that

$$D^+V(t,x) > \xi, \quad \delta < \|x\| < \rho, \quad \mathrm{d}(x,E) < \eta, t \geq 0.$$

Let us consider the set

$$U = \big[x \in S_\rho : \delta < \|x\| < \rho, \, \mathrm{d}(x,E) < \eta\big],$$

and let

$$\sup_{\substack{\|x\|<\rho \\ t\geq 0}} V_2(t,x) = L.$$

Assume that, at $t = t_1$, $x(t_1) = x(t_1, t_0, x_0) \in U$. Then for $t > t_1$, we have, letting $m(t) = V_2(t, x(t))$,

$$D^+m(t) \geq D^+V_2(t, x(t)) > \xi,$$

because of condition (iii) and the fact that $V_2(t,x)$ satisfies a Lipschitz condition in x locally. Thus

$$m(t) - m(t_1) = \int_{t_1}^{t} D^+m(s)\mathrm{d}s,$$

and hence

$$\begin{aligned} m(t) + m(t_1) \geq \int_{t_1}^{t} D^+m(s)\mathrm{d}s &\geq \int_{t_1}^{t} D^+V_2(t, x(s))\mathrm{d}s \\ &> \xi(t - t_1) \end{aligned}$$

as long as $x(t)$ remains in U. This inequality can simultaneously be realized with $m(t) \leq L$ only if

$$t < t_1 + 2L/\xi.$$

It therefore follows that there exists a t_2, $t_1 < t_2 \leq t_1 + 2L/\xi$ such that $x(t_2)$ is on the boundary of set U. In other words, $x(t)$ cannot stay permanently in the set U.

(3) Consider the sequence $\{t_k\}$ such that

$$t_k = t_0 + k\frac{2L}{\xi} \quad (k = 0, 1, 2, \dots).$$

Set $n(t) = V_1(t, x(t))$. Then, by assumption (ii), we have

$$D^+n(t) \leq D^+V_1(t, x(t)) \leq 0.$$

We let

$$\lambda = \inf\big[|w(x)|, \delta < \|x\| < \rho, \, \mathrm{d}(x,E) \geq \eta/2\big],$$

and

$$\lambda_1 = \frac{\lambda\eta}{2Mn^{1/2}}.$$

Suppose that $x(t)$ is such that, for $t_k \leq t \leq t_{k+2}$, $\delta < \|x(t)\| < \rho$. If for $t_k \leq t \leq t_{k+1}$, we have $\delta < \|x(t)\| < \rho$ and $\mathrm{d}(x, E) \geq \frac{1}{2}\eta$, then using assumption (ii) together with the definition of the set E, we obtain

$$\begin{aligned} n(t_{k+2}) - n(t_k) &= \int_{t_k}^{t_{k+2}} D^+ n(s)\mathrm{d}s \\ &\leq \int_{t_k}^{t_{k+2}} D^+ V_1(s, x(s))\mathrm{d}s \\ &\leq \int_{t_k}^{t_{k+1}} D^+ V_1(s, x(s))\mathrm{d}s + \int_{t_{k+1}}^{t_{k+2}} D^+ V_1(s, x(s))\mathrm{d}s \\ &\leq -\lambda(t_{k+1} - t_k) = -\lambda\frac{2L}{\xi}. \end{aligned} \tag{4.3.20}$$

On the other hand, if it happens that, for $t_k \leq t \leq t_{k+1}$,

$$\delta < \|x(t_1)\| < \rho, \quad \mathrm{d}\big[x(t_1), E\big] < \frac{1}{2}\eta,$$

then there exists a $t_3, t_1 \leq t_3 \leq t_1 + 2L/\xi$ such that $\mathrm{d}\big[x(t_3), E\big] = \eta$, in view of (2). It follows that there also exists a $t_4, t_1 \leq t_4 \leq t_3$ satisfying $\mathrm{d}\big[x(t_4), E\big] = \frac{1}{2}\eta$. These considerations lead to $\mathrm{d}\big[x(t_3), x(t_4)\big] \geq \frac{1}{2}\eta$, and hence we obtain, because of (1),

$$\frac{1}{2}\eta \leq Mn^{1/2}(t_3 - t_4),$$

which implies

$$\frac{\eta}{2Mn^{1/2}} \leq t_3 - t_4 \leq \frac{2L}{\xi}. \tag{4.3.21}$$

Moreover,

$$\begin{aligned} n(t_3) - n(t_1) &\leq \int_{t_1}^{t_4} D^+ V_1(s, x(s))\mathrm{d}s + \int_{t_4}^{t_3} D^+ V_1(s, x(s))\mathrm{d}s \\ &\leq -\lambda(t_3 - t_4) \leq \frac{-\lambda\eta}{2Mn^{1/2}} = -\lambda_1. \end{aligned}$$

Since $n(t)$ is a nondecreasing function, we have

$$\begin{aligned} n(t_{k+2}) &\leq n(t_3) \leq n(t_1) - \lambda_1 \\ &\leq n(t_k) - \lambda_1. \end{aligned}$$

Also, on the basis of (4.3.21), we obtain from (4.3.20) that

$$n(t_{k+2}) \leq n(t_k) - \lambda_1.$$

Thus, in any case,

$$V_1(t_{k+2}, x(t_{k+2})) \leq V_1(t_k, x(t_k)) - \lambda_1.$$

Choose an integer k^* such that $\lambda_1 k^* > a(\delta_0)$ and $T = T(\epsilon) = 4k^* L/\xi(\epsilon)$. Assume that, for $t_0 \leq t \leq t_0 + T$,

$$\|x(t, t_0, x_0)\| \geq \delta.$$

It then results from the preceding considerations that

$$\begin{aligned} V_1(t_0 + T, x(t_0 + T)) &\leq V_1(t_0, x_0) - k^* \lambda_1 \\ &\leq a(\delta_0) - k^* \lambda_1 \\ &\leq 0 \end{aligned}$$

which is incompatible with the positive definiteness of $V_1(t, x)$. Thus, there exists a $t^* \in [t_0, t_0 + T]$ satisfying

$$\|x(t^*, t_0, x_0)\| < \delta,$$

and the proof is complete.

We give below an instability theorem in which two Lyapunov functions are used which is an improvement of Theorem 4.2.9.

Theorem 4.3.9. *Suppose that the following conditions hold:*

(i) *$f \in C(R_+ \times S(\rho), R^n), f(t, 0) \equiv 0$, and $f(t, x)$ is bounded on $R_+ \times S(\rho)$;*

(ii) *$V_1 \in C(R_+ \times S(\rho), R_+), V_1(t, x)$ is locally Lipschitzian in x, decrescent, and, for any $t \geq 0$, it is possible to find points x lying in any given small neighborhood of the origin such that $V_1(t, x) > 0$;*

(iii) *$D^+V_1(t, x) \geq 0$, $(t, x) \in R_+ \times S(\rho)$, and, in each domain $t \geq 0$, $\alpha < \|x\| < \rho$, $D^+V_1(t, x) \geq \phi_\alpha(t)\omega(x)$, where $\omega(x) \geq 0$ is continuous for $x \in S(\rho)$ and $\phi_\alpha(t) \geq 0$ is continuous in t such that, for any infinite system S of closed, nonintersecting intervals of R_+ of an identical fixed interval, we have*

$$\int_s \phi_\alpha(s) ds = \infty; \tag{4.3.22}$$

(iv) *$V_2 \in C(R_+ \times S(\rho), R_+)$, and $V_2(t, x)$ is bounded on $R_+ \times S(\rho)$ and is locally Lipschitizian in x. Furthermore, given any number α, $0 < \alpha < \rho$, it is possible to find $\eta = \eta(\alpha), \eta < \alpha$, and a continuous function $\xi_\alpha(t) > 0$ such that*

$$\int_t^\infty \xi_\alpha(t) dt = \infty, \tag{4.3.23}$$

and, in the set $\alpha < \|x\| < \rho$, $d(x, E) < \eta$, $t \in R_+$,

$$D^+V_2(t, x) \geq \xi_\alpha(t), \tag{4.3.24}$$

where

$$E = [x \in S(\rho) : \quad \omega(x) = 0].$$

Then, the solution of (4.3.1) *is unstable.*

Proof. The proof of this theorem closely resembles that of Theorem 4.3.8 and hence we shall be brief. Suppose that, under the conditions of the theorem, the trivial solution is stable. That is, given $0 < \epsilon < \rho, \quad t_0 \in R_+$ there exists a $\delta > 0$ such that $\|x_0\| < \delta$ implies $\|x(t, t_0, x_0)\| < \epsilon, \quad t \geq t_0$.

According to assumption (ii), a point (t_0, x_0^*) can be found such that $\|x_0^*\| < \delta$ and $V_1(t_0, x_0^*) > 0$. We shall consider the motion $x(t) = x(t, t_0, x_0^*)$ and its properties:

(1) $\mathrm{d}(x(t), x(\tau)) \geq \eta, \quad t > \tau$; then $t - \tau \geq \eta / M n^{1/2}$. This is clean from (1) in the proof of Theorem 4.3.7.

(2) For every $t \geq t_0$, there will be a positive number α such that

$$\alpha < \|x(t)\| < \epsilon < \rho. \tag{4.3.25}$$

This is compatible with the assumption of stability, that is,$\|x(t)\| < \epsilon, t \geq t_0$. However, since $\mathrm{D}^+ V_1(t, x) \geq 0$, it follows that

$$V_1(t, x(t)) \geq V_1(t_0, x_0^*) > 0.$$

Since $V_1(t, x)$ is decrescent, for numbers $V_1(t_0, x_0^*) > 0$, a number $\alpha > 0$ can be found such that, for all $t \geq t_0$, $\|x\| \leq \alpha$, we shall have

$$V_1(t, x) < V_1(t_0, x_0^*).$$

Consequently, $\|x\| < \alpha$ is not possible. According to (iv), there exists a number $\eta = \eta(\alpha, \epsilon), \eta < \alpha$, and a continuous function $\xi_\alpha(t) > 0$ such that (4.3.23) and (4.3.24) hold.

(3) If $\mathrm{d}(x(\tau), E) < \eta$, then a $t^* > \tau$, can be found such that

$$\mathrm{d}(x(t^*), E) = \eta. \tag{4.3.26}$$

Suppose that $\mathrm{d}(x(t), E) < \eta$ for all $t \geq \tau$. Letting $m(t) = V_2(t, x(t))$, we obtain, using the Lipschitzian character of $V_2(t, x)$ in x, the inequality

$$\mathrm{D}^+ m(t) \geq \mathrm{D}^+ V_2(t, x(t)) \geq \xi_\alpha((t),$$

and hence

$$m(t) + m(\tau) \geq \int_\tau^t \mathrm{D}^+ m(s) \mathrm{d}s \geq \int_\tau^t \xi_\alpha(s) \mathrm{d}s.$$

Since $V_2(t, x)$ is assumed to be bounded, the relation (4.3.23) shows that $\mathrm{d}(x(t), E) < \eta$ can not hold for all $t \geq \tau$. Hence, there exists a $t^* \geq \tau$ such that (4.3.26) is satisfied.

(4) if $\mathrm{d}(x(t), E) < \eta/2$, then, for $t = t^*$, when $\mathrm{d}(x(t^*), E) = \eta$, we have

$$V_1(t^*, x(t^*)) \geq V_1(\tau, x(\tau)) + \epsilon \int_{t^{**}}^{t^*} \phi_\alpha(s)\mathrm{d}s,$$

where

$$\tau \leq t^{**} = t^* - \frac{\eta}{2Mn^{1/2}}$$

and

$$\epsilon = \inf[\omega(x), \alpha < \|x\| < \rho,\ \mathrm{d}(x, E) \geq \frac{1}{2}\eta] > 0.$$

In fact, under the given condition, $\tau < t_* < t^*$ can be found such that

$$\mathrm{d}(x(t_*), E) = \frac{1}{2}\eta,$$

and, for $t_* \leq t \leq t^*$, we shall

$$\frac{1}{2}\eta \leq \mathrm{d}(x(t), E) \leq \eta.$$

Hence, by (iii), it follows that

$$\mathrm{D}^+ V_1(t, x(t)) \geq \phi_\alpha(t)\, \omega(x(t)) \geq \epsilon\phi_\alpha(t),$$

using the fact that $V_1(t, x)$ is locally Lipschitzian in x, and, consequently,

$$V_1(t^*, x(t^*)) \geq V_1(\tau, x(\tau)) + \int_{t_*}^{t^*} \phi_\alpha(s)\mathrm{d}s.$$

Observing, however, that $\mathrm{d}(x(t^*), x(t_*)) \geq \frac{1}{2}\eta$, we get, in view of (1), that

$$t^* - \tau \geq t^* - t_* \geq \frac{\eta}{2Mn^{1/2}}.$$

(5) There is no number $t_1 \geq t_0$ such that, for all $t > t_1$, we would have

$$\mathrm{d}(x(t), E) \geq \frac{1}{2}\eta.$$

Indeed, if such a t_1 exists, then, for all $t > t_1$, we would have

$$\begin{aligned} V_1(t, x(t)) &= V_1(t_1, x(t_1)) + \int_{t_1}^{t} \mathrm{D}^+ V_1(s, x(s))\mathrm{d}s \\ &\geq V_1(t_1, x(t_1)) + \epsilon \int_{t_1}^{t} \phi_\alpha(s)\mathrm{d}s. \end{aligned}$$

By (4.3.22), this implies that $V_1(t, x(t)) \to \infty$ as $t \to \infty$, which is absurd because of the relation (4.3.25) and the fact that $V_1(t, x)$ is decrescent. Thus it follows that, for any t_i^*, a $\tau_{i+1} > t_i^*$ can be found such that

$$\mathrm{d}(x(\tau_{i+1}), E) < \frac{1}{2}\eta,$$

and, according, to (3), there corresponds a $t_{i+1}^* > \tau_{i+1}$ satisfying

$$\mathrm{d}(x(t_{i+1}^*), E) = \eta.$$

Let us consider the infinite sequence of numbers

$$t_0 < \tau_1 < t_1^* < \cdots < \tau_i < t_i^* < \cdots .$$

In view of assumption (iii) and (4), we have

$$V_1(t_i^*, x(t_i^*)) \geq V_1(t_0, x_0) + \epsilon \sum_{j=1}^{i} \int_{t_j^{**}}^{t_j^*} \phi_\alpha(s)\mathrm{d}s,$$

where $\tau_j \leq t_j^{**} = t_j^* - \eta/2Mn^{1/2}$. The infinite system of segments $[t_j^{**}, t_j^*]$ satisfies condition (iii), and therefore the last sum increases indefinitely with i. In other words, $V_1(t_i^*, x(t_i^*)) \to \infty$ as $i \to \infty$. This is not compatible with the boundedness of $V_1(t, x(t))$. This contradiction shows that the assumption of stability is wrong, and the proof is complete.

4.4 Boundedness and Practical Stability

It is well known that in proving uniform boundedness of a differential system by means of Lyapunov functions, it is sufficient to impose conditions in the complement of a compact set in R^n, whereas, in the case of equiboundedness, the proofs demand that the assumptions hold everywhere in R^n.

For any set $E \subset R^n$, we denote by $\overline{E}$, E^c and ∂E, the closure, the complement and the boundary of E respectively. Then we have the following result which employs two Lyapunov-like functions and only assumes conditions outside a compact set.

Theorem 4.4.1. *Assume that*

(i) $E \subset R^n$ *is compact,* $V_1 \in C(R_+ \times \overline{E}^c, R_+)$, $V_1(t, x)$ *is locally Lipschitzian in* x, *bounded for* $(t, x) \in R_+ \times \partial E$, *and*

$$D^+V_1(t, x) \leq g_1(t, V_1(t, x)), (t, x) \in R_+ \times \overline{E}^c, \tag{4.4.1}$$

where $g_1 \in C(R_+ \times R_+, R)$;

(ii) $V_2 \in C(R_+ \times S^c(\rho), R_+)$, $V_2(t, x)$ *is locally Lipschitzian in* x,

$$b(\|x\|) \leq V_2(t, x) \leq a(\|x\|), (t, x) \in R_+ \times S^c(\rho), \tag{4.4.2}$$

where $a, b \in C([\rho, \infty), R_+)$ *such that* $b(u)$ *is nondecreasing in* u *and* $b(u) \to \infty$ *as* $u \to \infty$ *(*ρ *may be sufficiently large) and for* $(t, x) \in R_+ \times S^c(\rho)$,

$$D^+V_1(t, x) + D^+V_2(t, x) \leq g_2(t, V_1(t, x) + V_2(t, x)) \tag{4.4.3}$$

where $g_2 \in C(R_+ \times R_+, R)$;

(iii) the scalar differential equations

$$u' = g_1(t,u), u(t_0) = u_0 \geq 0, \tag{4.4.4}$$

and

$$v' = g_2(t,v), v(t_0) = v_0 \geq 0, \tag{4.4.5}$$

are equibounded and uniformly bounded respectively. Then, the differential system (4.3.1) is equibounded.

Proof. Since E is compact, there exists ρ (may be sufficiently large) such that $S(\rho) \supset S(E, \rho_0)$ for some $\rho_0 > 0$. Here

$$S(E, \rho_0) = [x \in R^n : d(x, E) \leq \rho_0],$$

where $d(x, E) = \inf_{y \in E} \|x - y\|$. Let $t_0 \in R_+$ and $\alpha \geq \rho$ be given. Let $\alpha_1 = \alpha_1(t_0, \alpha) = \max(\alpha_0, \alpha^*)$ where $\alpha_0 = \max[V_1(t_0, x_0) : x_0 \in \overline{S(\alpha) \cap E^c}]$ and $\alpha^* \geq V_1(t, x)$ for $(t, x) \in R_+ \times \partial E$. Since the equation (4.4.4) is equibounded, given $\alpha_1 > 0$ and $t_0 \in R_+$, there exists a $\beta_0 = \beta_0(t_0, \alpha_1)$ such that

$$u(t, t_0, u_0) < \beta_0, t \geq t_0, \tag{4.4.6}$$

provided $u_0 < \alpha_1$, where $u(t, t_0, u_0)$ is any solution of (4.4.4). Also, uniform boundedness of the equation (4.4.5) yields that

$$v(t, t_0, v_0) < \beta_1(\alpha_2), t \geq t_0, \tag{4.4.7}$$

provided $v_0 < \alpha_2$, where $v(t, t_0, v_0)$ is any solution of (4.4.5). We set $u_0 = V_1(t_0, x_0)$ and $\alpha_2 = a(\alpha) + \beta_0$. As $b(u) \to \infty$ with $u \to \infty$, we can choose a $\beta = \beta(t_0, \alpha)$ such that

$$b(\beta) > \beta_1(\alpha_2). \tag{4.4.8}$$

We now claim that $x_0 \in S(\alpha)$ implies that any solution $x(t, t_0, x_0)$ satisfies $x(t, t_0, x_0) \in S(\beta)$, for $t \geq t_0$. If this is not true, there exists a solution $x(t, t_0, x_0)$ of (4.3.1) with $x_0 \in S(\alpha)$ such that for some $t^* > t_0$, $\|x(t^*, t_0, x_0)\| = \beta$. Since $S(E, \rho_0) \subset S(\alpha)$, there are two possibilities to consider:

(i) $x(t, t_0, x_0) \in E^c$ for $t \in [t_0, t^*]$;

(ii) there exists a $\tilde{t} \geq t_0$ such that $x(\tilde{t}, t_0, x_0) \in \partial E$ and $x(t, t_0, x_0) \in \overline{E}^c$ for $t \in [\tilde{t}, t^*]$.

If case (i) holds, we can find $t_1 > t_0$ such that

$$\begin{cases} x(t_1, t_0, x_0) \in \partial S(\alpha), \\ x(t^*, t_0, x_0) \in \partial S(\beta), \\ x(t, t_0, x_0) \in S^c(\alpha), t \in [t_1, t^*]. \end{cases} \tag{4.4.9}$$

Setting $m(t) = V_1(t, x(t, t_0, x_0)) + V_2(t, x(t, t_0, x_0))$ for $t \in [t_1, t^*]$, it is easy to obtain by Theorem 4.3.1 with $N = 1$, the relation

$$m(t) \leq r_2(t, t_1, m(t_1)), t \in [t_1, t^*],$$

where $r_2(t, t_1, v_0)$ is the maximum solution of (4.4.5) such that $r_2(t_1, t_1, v_0) = v_0$. Thus,

$$\begin{aligned} &V_1(t^*, x(t^*, t_0, x_0)) + V_2(t^*, x(t^*, t_0, x_0)) \\ &\qquad \leq r_2(t^*, t_1, V_1(t_1, x(t_1, t_0, x_0)) + V_2(t_1, x(t_1, t_0, x_0))) \end{aligned} \tag{4.4.10}$$

Similarly, because of (4.4.4), we also have by Theorem 4.3.1

$$V_1(t_1, x(t_1, t_0, x_0)) \leq r_1(t_1, t_0, V_1(t_0, x_0)) \tag{4.4.11}$$

where $r_1(t, t_0, u_0)$ is the maximal solution of (4.4.4). In view of the fact that $u_0 = V_1(t_0, x_0) < \alpha_1$, (4.4.6) yields

$$r_1(t_1, t_0, V_1(t_0, x_0)) < \beta_0.$$

Furthermore, $V_2(t_1, x(t_1, t_0, x_0)) \leq a(\alpha)$ because of (4.4.2) and (4.4.9). Consequently, we have

$$v_0 = V_1(t_1, x(t_1, t_0, x_0)) + V_2(t_1, x(t_1, t_0, x_0)) < \beta_0 + a(\alpha) = \alpha_2. \tag{4.4.12}$$

Hence the inequality (4.4.10) gives, because of the relations (4.4.2), (4.4.7), (4.4.8), (4.4.9), (4.4.12) and the fact that $V_1 \geq 0$,

$$b(\beta) \leq \beta_1(\alpha_2) < b(\beta) \tag{4.4.13}$$

which is a contradiction.

If case (ii) holds we again arrive at the inequality (4.4.6), where $t_1 > \tilde{t}$ satisfies (4.4.9). We now have, in place of (4.4.11) the relation

$$V_1(t_1, x(t_1, t_0, x_0)) \leq r_1(t_1, \tilde{t}, V_1(\tilde{t}, x(\tilde{t}, t_0, x_0))).$$

Since $x(\tilde{t}, t_0, x_0) \in \partial E$ and $V_1(\tilde{t}, x(\tilde{t}, t_0, x_0)) \leq \alpha^* \leq \alpha_1$, arguing as before, we arrive at the contradiction (4.4.13). This proves that if $x_0 \in S(\alpha)$, $\alpha \geq \rho$, $x(t, t_0, x_0) \in S(\beta)$, for $t \geq t_0$. For $\alpha < \rho$ we set $\beta(t_0, \alpha) = \beta(t_0, \rho)$ and hence the proof is complete.

We shall next consider a result concerning equi-ultimate boundedness.

Theorem 4.4.2. *In addition to the assumption of Theorem 4.4.1, we suppose that*

(a) $D^+V_1(t,x) + C(\|x\|) \leq g_1(t, V_1(t,x))$, $(t,x) \in R_+ \times \bar{E}^c$, *where* $C \in ([\rho, \infty), R_+)$, $C(u)$ *is nondecreasing in* u, $C(u) \to \infty$ *as* $u \to \infty$ *and* $g_1(t,u)$ *is nondecreasing in* u;

(b) $V_1(t,x) \le \alpha_1(\rho^*)$ *for* $t \in R_+$ *and* $\|x\| = \rho^*$ *for some* $\rho^* \ge \rho$*;*

(c) The differential equation (4.4.4) is uniformly bounded only for $u_0 \le \alpha_1(\rho^*)$*.*

Then the differential system 4.3.1 is equi-ultimately bounded.

Proof. Set $\alpha = \rho^*$. Then because of (b) and (c), we get $\alpha_1 = \alpha_1(\rho^*)$, $\beta_0 = \beta_0(\rho^*)$ and $\beta = \beta(\rho^*)$ following the proof of Theorem 4.4.1. We designate $B = \beta(\rho^*)$ so that we have

$$\|x_0\| \le \rho^*; \quad \text{implies} \quad \|x(t)\| < B, t \ge t_0. \tag{4.4.14}$$

Now let $\|x_0\| \le \alpha$ for any $\alpha > \rho^*$. Then Theorem 4.4.1 also yields $\|x(t)\| < \beta(t_0,\alpha), t \ge t_0$, where $x(t) = x(t,t_0,x_0)$ is any solution of (4.3.1) with $\|x_0\| \le \alpha$.

We claim that there exists a $t^* \in [t_0, t_0 + T]$ where

$$T = T(t_0,\alpha) \ge \frac{\beta_0(t_0,\alpha)}{C(\rho^*)}, \tag{4.4.15}$$

such that $\|x(t^*)\| < \rho^*$ whenever $\rho^* \le \|x_0\| \le \alpha$. If this is not true, there would exists a solution $x(t)$ of (4.3.1) with $\rho^* \le \|x_0\| \le \alpha$ such that $\rho^* \le \|x(t)\| \le \beta$, $t \in [t_0, t_0 + T]$. Hence, setting

$$m(t) = V_1(t,x(t)) + \int_{t_0}^{t} C(\|x(s)\|)ds,$$

and using condition (a) together with monotonic nature of $g_1(t,u)$, we get $D^+m(t) \le g_1(t,m(t)), t_0 \le t \le t_0 + T$. It then follows by Theorem 1.7.1, the relation

$$V_1(t,x(t)) + \int_{t_0}^{t} C(\|x(s)\|)ds \le r_1(t,t_0,V_1(t_0,x_0))$$

for $t \in [t_0, t_0 + T]$. This implies,

$$C(\rho^*)T \le r_1(t_0 + T, t_0, V_1(t_0,x_0)) < \beta_0(t_0,\alpha),$$

since $V_1(t_0,x_0) \le \alpha_1$. Thus $T < \frac{\beta_0(t_0,\alpha)}{C(\rho^*)}$ which contradicts (4.4.15). Hence, there exists a $t^* \in [t_0, t_0 + T]$ such that $\|x(t^*)\| < \rho^*$. It then follows from (4.4.15) that $\|x(t)\| < B$ for $t \ge t_0 + T$ whenever $\|x_0\| \le \alpha$. Hence the proof is complete.

Remark 4.4.1. *Theorems 4.4.1 and 4.4.2, which are in the spirit of Theorems 4.3.5 and 4.3.6, contain several refinements of boundedness results. In order to avoid monotony, we leave it to the reader to ponder over these special cases.*

We shall next discuss nonuniform practical stability of (4.3.11) in the same spirit. We denote by $C_\kappa = [a \in C[R_+ \times [0,A), R_+] : a(t,u) \in \kappa \text{ for each } t \in R_+]$.

Theorem 4.4.3. *Assume that*

(i) $0 < \lambda < A$;

(ii) $V_1 \in C(R_+ \times S(a), R_+)$, *$V_1(t,x)$ is locally Lipschitzian in x and for* $(t,x) \in R_+ \times S(A)$, $V_1(t,x) \leq a_1(t, \|x\|)$, $a_1 \in C_k$ *and*

$$D^+V_1(t,x) \leq g_1(t, V_1(t,x)),$$

where $g_1 \in C(R_+^2, R)$;

(iii) $V_2 \in C(R_+ \times S(A) \cap S^c(\lambda), R_+)$, *$V_2(t,x)$ is locally Lipschitzian in x and for* $(t,x) \in R_+ \times S(A) \cap S^c(\lambda)$,

$$b(\|x\|) \leq V_2(t,x) \leq a_2(\|x\|), b, a_2 \in \kappa,$$

$$D^+V_1(t,x) + D^+V_2(t,x) \leq g_2(t, V_1(t,x) + V_2(t,x))$$

where $g_2 \in C(R_+^2, R)$;

(iv) $a_1(t_0, \lambda) + a_2(\lambda) < b(A)$ *for some* $t_0 \in R_+$;

(v) $u_0 < a_1(t_0, \lambda)$ *implies* $u(t, t_0, u_0) < a_1(t_0, \lambda)$ *for* $t \geq t_0$, *where* $u(t, t_0, u_0)$ *is any solution of* (4.4.4) *and* $v_0 < a_1(t_0, \lambda) + a_2(\lambda)$ *implies* $v(t, t_0, v_0) < b(A), t \geq t_0$, *for every* $t_0 \in R_+$, *where* $v(t, t_0, v_0)$ *is any solution of* (4.4.5).

Then, the system (4.3.1) is practically stable.

Proof. We claim that if $\|x_0\| < \lambda$ we have $\|x(t)\| < A, t \geq t_0$ where $x(t) = x(t, t_0, x_0)$ is any solution of (4.3.1). If this is not true, there would exists a $t_2 > t_1 > t_0$ and solution $x(t) = x(t, t_0, x_0)$ of (4.3.1) such that

$$\|x(t_1)\| = \lambda, \|x(t_2)\| = A \text{ and } \lambda \leq \|x(t)\| \leq A, t_1 \leq t \leq t_2. \tag{4.4.16}$$

Hence, we get by Theorem 4.3.1 with N=1, using (iii), the estimate

$$V_1(t, x(t)) + V_2(t, x(t)) \leq r_2(t, t_1, V_1(t_1, x(t_1)) + V_2(t_1, x(t_1))) \tag{4.4.17}$$

for $t_1 \leq t \leq t_2$, where $r_2(t, t_1, v_0)$ is the maximal solution of (4.4.5) through (t_1, v_0). Similarly, condition (ii) gives the estimate

$$V_1(t, x(t)) \leq r_1(t, t_0, V_1(t_0, x_0)), t_0 \leq t \leq t_1,$$

where $r_1(t, t_0, u_0)$ is the maximal solution of (4.4.4). Since $\|x_0\| < \lambda$, we have, because of (ii),

$$V_1(t_0, x_0) \leq a_1(t_0, \|x_0\|) < a_1(t_0, \lambda)$$

and hence (v) shows that

$$V_1(t_1, x(t_1)) < a_1(t_0, \lambda).$$

Also, $V_2(t_1, x(t_1)) \leq a_2(\|x(t_1)\|) = a_2(\lambda)$, because of (ii) and consequently

$$V_1(t_1, x(t_1)) + V_2(t_1, x(t_1)) \leq a_1(t_0, \lambda) + a_2(\lambda).$$

Now, using (4.4.17) and (v), we obtain

$$V_1(t_2, x(t_2)) + V_2(t_2, x(t_2)) \le r_2(t_2, t_1, a_1(t_0, \lambda) + a_2(\lambda)) < b(A.) \tag{4.4.18}$$

But, in view of ((4.4.17)), (ii) and (iii), we have

$$V_1(t_2, x(t_2)) + V_2(t_2, x(t_2)) \ge V_2(t_2, x(t_2)) \ge b(\|x(t_2)\|) = b(A)$$

which contradicts (4.4.18). This proves the claim.

We shall consider a simple result which yields practical asymptotic stability.

Theorem 4.4.4. *Suppose that hypotheses of Theorem 4.4.3 hold except that the estimate on $D^+V_1(t, x)$ is strengthened to*

$$D^+V_1(t, x) \le -C(\|x\|), C \in \kappa, (t, x) \in R_+ \times S(A). \tag{4.4.19}$$

Suppose further that $f(t, x)$ is bounded on $R_+ \times S(A)$. Then the system (4.4.4) is practically asymptotically stable.

Proof. It is clear from (4.4.19) that $D^+V_1(t, x) \le 0$ which means that $g_1(t, u) \equiv 0$ so that condition (v) corresponding to (4.4.4) is automatically satisfied. As a result, we have by Theorem 4.4.3, practical stability of the system (4.3.1). It is now easy to show that $\lim_{t\to\infty} x(t) = 0$. Hence the proof is complete.

4.5 Method of Vector Lyapunov Functions

It is well known that using a single Lyapunov function, it is possible to investigate a variety of problems in a unified way. However, as we have seen, employing two Lyapunov functions is more advantageous in improving several results. It is therefore natural to ask whether it might be more fruitful to use several Lyapunov functions. The answer is positive and this approach offers a more flexible mechanism since each function can satisfy less rigid requirements.

Let us first consider the method of vector Lyapunov functions. Naturally, Theorem 4.3.1 plays an important role whenever we employ vector Lyapunov functions. As a typical result, we shall prove a theorem that gives sufficient conditions in terms of vector Lyapunov functions for the stability properties of the trivial solution of (4.3.1) which is an extension of Theorem 4.2.8.

Theorem 4.5.1. *Assume that*

(i) $g \in C(R_+ \times R^N, R^N)$,$g(t, 0) \equiv 0$ and $g(t, u)$ is quasimonotone nondecreasing in u for each $t \in R_+$;

(ii) $V \in C(R_+ \times S(\rho), R^N_+)$,$V(t, x)$ is locally Lipschitzian in x and the function

$$V_0(t, x) = \sum_{i=1}^{N} V_i(t, x) \tag{4.5.1}$$

is positive definite and decrescent;

(iii) $f \in C(R_+ \times S(\rho), R^n), f(t,0) \equiv 0$ *and*

$$D^+V(t,x) \leq g(t, V(t,x)), (t,x) \in R_+ \times S(\rho).$$

Then, the stability properties of the trivial solution of

$$u' = g(t,u), u(t_0) = u_0 \geq 0, \tag{4.5.2}$$

imply the corresponding stability properties of the trivial solution of (4.3.1).

Note that in condition (ii) of Theorem 4.5.1, we have used the measure $V_0(t,x)$ defined by (4.5.1). We could use other convenient measures such as

$$V_0(t,x) = \max_{1 \leq i \leq N} V_i(t,x), \qquad V_0(t,x) = \sum_{i=1}^{N} d_i V_i(t,x)$$

for a positive vector $d > 0$, or, $V_0(t,x) = Q(V(t,x))$ where $Q \in C(R_+^N, R_+)$, $Q(u)$ is nondecreasing in u and $Q(0) = 0$.

Corresponding to the stability and the boundedness notions given in Section 4.2, we need similar notions relative to the comparison system (4.5.2). We merely state one of the concepts.

Definition 4.5.1. *The trivial solution of (4.5.2) is set to be equistable if given* $\epsilon > 0$ *and* $t_0 \in R_+$, *there exists a* $\delta = \delta(t_0, \varepsilon) > 0$ *such that*

$$\sum_{i=1}^{N} u_{0i} < \delta \text{ implies } \sum_{i=1}^{N} u_i(t, t_0, u_0) < \epsilon, t \geq t_0,$$

where $u(t, t_0, u_0)$ *is any solution of (4.5.2).*

We note that one need to use the same measure in Definition 4.5.1 which is adopted for the measure of $V(t,x)$.

Proof of Theorem 4.5.1. We shall only prove equiasymptotic stability of the trivial solution of (4.3.1). For this purpose, let us first prove equistability.

Since V_0 is positive definite and decrescent there exist $a, b \in \kappa$ such that

$$b(\|x\|) \leq V_0(t,x) \leq a(\|x\|), (t,x) \in R_+ \times S(\rho). \tag{4.5.3}$$

Let $0 < \epsilon < \rho$ and $t_0 \in R_+$ be given and suppose that the trivial solution of (4.5.2) is equistable. Then, given $b(\epsilon) > 0$ and $t_0 \in R_+$, there exists a $\delta_1 = \delta_1(t_0, \epsilon) > 0$ such that

$$\sum_{i=1}^{N} u_{0i} < \delta_1 \text{ implies } \sum_{i=1}^{N} u_i(t, t_0, u_0) < b(\epsilon), t \geq t_0, \tag{4.5.4}$$

where $u(t, t_0, u_0)$ is any solution of (4.5.2). Choose $u_0 = V(t_0, x_0)$ and a $\delta = \delta(t_0, \epsilon) > 0$ satisfying

$$a(\delta) < \delta_1 \tag{4.5.5}$$

Let $\|x_0\| < \delta$. Then we claim that $\|x(t)\| < \epsilon$, $t \geq t_0$ for any solution $x(t)$ of (4.3.1). If this is not true, there would exist a solution $x(t)$ of (4.3.1) with $\|x_0\| < \delta$ and a $t_1 > t_0$ such that

$$\|x(t_1)\| = \epsilon \text{ and } \|x(t)\| \leq \epsilon \text{ for } t \in [t_0, t_1], \tag{4.5.6}$$

Hence by Theorem 4.3.1 we have

$$V(t, x(t)) \leq r(t, t_0, u_0), t \in [t_0, t_1], \tag{4.5.7}$$

where $r(t, t_0, u_0)$ is the maximal solution of (4.5.2). Since

$$V_0(t_0, x_0) \leq a(\|x_0\|) \leq a(\delta) < \delta_1,$$

the relations (4.5.4), (4.5.5), (4.5.6) and (4.5.7) yield

$$b(\epsilon) \leq V_0(t_1, x(t_1)) \leq r_0(t_1, t_0, u_0) < b(\epsilon),$$

where $r_0(t, t_0, u_0) = \sum_{i=1}^{N} r_i(t, t_0, u_0)$. This contradiction proves that the trivial solution of (4.3.1) is equistable.

Suppose next that the trivial solution of (4.5.2) is quasiequiasymptotically stable. Set $\epsilon = \rho$ and $\hat{\delta}_0(t_0) = \delta(t_0, \rho)$. Let $0 < \eta < \rho$. Then, given $b(\eta) > 0$ and $t_0 \in R_+$, there exist $\delta_1^* = \delta_1(t_0) > 0$ and $T = T(T_0, \eta) > 0$ satisfying

$$\sum_{i=1}^{N} u_{0i} < \delta_1^* \text{ implies } \sum_{i=1}^{N} u_i(t, t_0, u_0) < b(\eta), t \geq t_0 + T. \tag{4.5.8}$$

Choosing $u_0 = V(t_0, x_0)$ as before, we find a $\delta_0^* = \delta_0(t_0) > 0$ such that $a(\delta_0^*) < \delta_1^*$. Let $\delta_0 = \min(\delta_1^*, \delta_0^*)$ and $\|x_0\| < \delta_0$. This implies that $\|x(t)\| < \rho, t \geq t_0$ and therefore, the estimate (4.5.7) is valid for all $t \geq t_0$. Suppose now that there is a sequence $\{t_k\}$,$t_k \geq t_0 + T$,$t_k \to \infty$ as $k \to \infty$, and $\eta \leq \|x(t_k)\|$, where $x(t)$ is any solution of (4.3.1) with $\|x_0\| < \delta_0$. In view of (4.5.7) and (4.5.8), this leads to the contradiction

$$b(\eta) \leq V_0(t_k, x(t_k)) \leq r_0(t_k, t_0, u_0) < b(\eta).$$

Hence the trivial solution of (4.3.1) is equiasymptotically stable and the proof is complete.

To exhibit the advantage in using vector Lyapunov functions, consider the following example.

Example 4.5.1. *Let us consider the systems*

$$\begin{cases} x' = e^{-t}x + y\sin t - (x^3 + xy^2)\sin^2 t, \\ y' = x\sin t + e^{-t}y - (x^2y + y^3)\sin^2 t. \end{cases} \tag{4.5.9}$$

Suppose we choose a single Lyapunov function V given by

$$V(t,x) = x^2 + y^2.$$

Then, it is evident that

$$D^+V(t,x) \leq 2(e^{-t} + |\sin t|)V(t,x)$$

using the inequality $2|ab| \leq a^2 + b^2$ and observing that $[x^2 + y^2]^2 \sin^2 t \geq 0$. Clearly, the trivial solution of the scalar differential equation

$$u' = 2(e^{-t} + |\sin t|)u, u(t_0) = u_0 \geq 0$$

is not stable and so we cannot deduce any information about the stability of the trivial solution of (4.5.9) from Theorem 4.5.1 although it is easy to check that it is stable. On the other hand, let us seek a Lyapunov function as a quadratic form with constant coefficients given by

$$V(t,x) = \frac{1}{2}[x^2 + 2Bxy + Ay^2]. \tag{4.5.10}$$

Then, the function $D^+V(t,x)$ with respect to (4.5.9) is equal to the sum of two functions $w_1(t,x)$, $w_2(t,x)$ where

$$w_1(t,x) = x^2[e^{-t} + B\sin t] + xy[2Be^{-t} + (A+1)\sin t]$$

$$+y^2[Ae^{-t} + B\sin t],$$

$$w_2(t,x) = -\sin^2 t[x^2 + y^2](x^2 + 2Bxy + Ay^2)$$

For arbitrary A and B, the function $V(t,x)$ defined in (4.5.10) does not satisfy Lyapunov's theorem on the stability of motion. Let us try to satisfy the condition of Theorem 4.2.8 by assuming $w_1(t,x) = \lambda(t)V(t,x)$. This equality can occur in two cases:

(i) $A_1 = 1, B_1 = 1, \lambda_1(t) = 2[e^{-t} + \sin t]$ when $V_1(t,x) = \frac{1}{2}(x+y)^2$;

(ii) $A_2 = 1, B_2 = -1, \lambda_2(t) = 2[e^{-t} - \sin t]$ when $V_2(t,x) = \frac{1}{2}(x-y)^2$.

The functions V_1, V_2 are not positive definite hence do not satisfy Theorem 4.2.8. However they do fulfill the conditions of Theorem 4.5.1. In fact,

(a) the functions $V_1(t,x) \geq 0, V_2(t,x) \geq 0$ and $\sum_{i=1}^{2} V_i(t,x) = x^2 + y^2$ and therefore $V_0(t,x) = \sum_{i=1}^{2} V_i(t,x)$ is positive definite and decrescent;

(b) the vectorial inequality $D^+V(t,x) \leq g(t, V(t,x))$ is satisfied with the functions

$$g_1(t, u_1, u_2) = 2(e^{-t} + \sin t)u_1,$$

$$g_2(t, u_1, u_2) = 2(e^{-t} - \sin t)u_2,$$

It is clear that $g(t,u)$ is quasimonotone nondecreasing in u, and the null solution of $u' = g(t,u)$ is stable. Consequently, the trivial solution of (4.5.9) is stable by Theorem 4.5.1.

Let us next extend Theorem 4.2.7 in term of vector Lyapunov functions. For this purpose, let us consider the autonomous differential system

$$x' = f(x), x(0) = x_0, \tag{4.5.11}$$

where $f \in C(G^*, R^n)$, G^* being an open set in R^n. Let G be an arbitrary set in R^n such that $\bar{G} \subset G^*$. Suppose also that $f(0) = 0$. We then have a variant of Theorem 4.2.7 in a more convenient form.

Theorem 4.5.2. *Assume that $V \in C^1(G^*, R_+)$ and $V'(x) = V_x(x)f(x) \leq 0, x \in G$. Let $E = [x : V'(x) = 0, x \in \bar{G} \cap G^*]$ and M be the largest invariant set in E. Then every bounded solution for $t \geq 0$ of (4.5.11) that remains in G tends to M as $t \to \infty$.*

Definition 4.5.2. *A solution $x(t)$ of (4.5.11) is said to be compact if $x(t)$ is contained in a compact set relative to G^*, that is, $x(t)$ is bounded for $t \geq 0$ and has no positive limit points on the boundary of G^*. (p is said to be positive limit point of $x(t)$ if there exists a sequence $\{t_n\}$ such that $t_n \to \infty$ and $x(t_n) \to p$ as $n \to \infty$).*

An immediate extension of Theorem 4.5.2 in terms of vector Lyapunov functions is as follows.

Theorem 4.5.3. *Let $V \in C^1(G^*, R^N)$ and $V_i'(x) \leq 0$ on G for each $i = 1, 2, \ldots, N$. If $x(t)$ is a compact solution of (4.5.11) that remains in G for $t \geq 0$, then $x(t) \to M = \bigcap_{i=1}^N M_i$, where M_i is the largest invariant set of $E_i = [x : V_i'(x) = 0, x \in \bar{G} \cap G^*]$.*

Taking $V_0(x) = \sum_{i=1}^N V_i(x)$, we see that V_0 satisfies Theorem 4.5.2.

Another interesting extension of Theorem 4.5.2 is as follows.

Theorem 4.5.4. *Let $V \in C^1(G^*, R^N)$, $V_j'(x) \leq 0$ for each $j \in N(x)$ and $x \in G$ where $N(x) = [j : V_j(x) = V_0(x)]$ and $V_0(x) = \max_j V_j(x)$. Let $E = [x : V_i'(x) = 0$ for some $i \in N(x), x \in \bar{G} \cap G^*]$ and M be the largest invariant set in E. Then every solution of (4.5.11) that is compact and remains in G for $t \geq 0$ approaches M as $t \to \infty$.*

Proof. It is easy to see that $V_0(x)$ is continuous and $V_0'(x) = \max_{j \in N(x)} V_j'(x) \leq 0$, $x \in G$. Hence V_0 satisfies the conditions of Theorem 4.5.2 and so the conclusion follows.

A sufficient condition for the vector Lyapunov function V to satisfy the assumptions of Theorem 4.5.4 is as follows which is useful in applications.

Theorem 4.5.5. *Let $V \in C^1(G^*, R^N)$. If for each $x \in G$, there exists a vector $c(x)$ with $c(x) \neq 0$, such that for each $i \in N(x)$, $V_i(x) \geq V_j(x), j = 1, 2, \ldots, N$ implies* *(i)* $\frac{\partial V_i}{\partial x_j}(x)c_j(x) \geq 0, i \neq j;$

(ii) $\left(\frac{\partial V}{\partial x}(x)c(x)\right)_i \leq 0;$

(iii) $0 \leq \frac{f_i(x)}{c_i(x)}$ *and* $\frac{f_i(x)}{c_i(x)} \geq \frac{f_j(x)}{c_j(x)}, j = 1, 2, \ldots, N.$

Then V is the desired function in Theorem 4.5.4.

Proof. For each $i \in N(x)$ if $V_i(x) \geq V_j(x)$,$j = 1, 2, \ldots, N$, we obtain successively using (i),(ii) and (iii),

$$V_i'(x) = \frac{\partial V_i}{\partial x_1} f_1(x) + \ldots + \frac{\partial V_i(x)}{\partial x_i} f_i(x) + \ldots$$

$$= \left(\frac{\partial V_i(x)}{\partial x_1} c_i(x)\right)\left(\frac{f_1(x)}{c_1(x)}\right) + \ldots + \left(\frac{\partial V_i(x)}{\partial x_i} c_i(x)\right)\left(\frac{f_i(x)}{c_i(x)}\right) + \ldots$$

$$\leq \frac{f_i(x)}{c_i(x)}\left[\frac{\partial V_i(x)}{\partial x_1} c_1(x) + \ldots + \frac{\partial V_i(x)}{\partial x_i} c_i(x) + \ldots\right]$$

$$\leq 0, x \in G.$$

This implies that V satisfies the assumptions of Theorem 4.5.2

4.6 Stability Concepts in Terms of Two Measures

Let us begin by defining the following classes of functions for future use:

$$\kappa = \{a \in C[R_+, R_+] : a(u) \text{ is strictly increasing in } u \text{ and } a(0) = 0\},$$

$$\mathcal{L} = \{\sigma \in C[R_+, R_+] : \sigma(u) \text{ is strictly decreasing in } u \text{ and } \lim_{u \to \infty} \sigma(u) = 0\},$$

$$\kappa\mathcal{L} = \{a \in C[R_+^2, R_+] : a(t, s) \in \kappa \text{ for each } s \text{ and } a(t, s) \in \mathcal{L} \text{ for each } t\},$$

$$C\kappa = \{a \in C[R_+^2, R_+] : a(t, s) \in \kappa \text{ for each } t\},$$

$$\Gamma = \{h \in C[R_+ \times R^n, R_+] : \inf h(t, x) = 0\},$$

$$\Gamma_0 = \{h \in \Gamma : \inf_x h(t, x) = 0 \text{ for each } t \in R_+\}.$$

We note that the class κ has already been utilized earlier.

We shall now define the equistability concept for the system (4.3.1) in terms of two measures $h_0, h \in \Gamma$.

Definition 4.6.1. *The differential system* (4.3.1) *is said to be* (h_0, h)*-equistable, if for each* $\epsilon > 0$ *and* $t_0 \in R_+$*, there exists a function* $\delta = \delta(t_0, \epsilon) > 0$ *which is continuous in* t_0 *for each* ϵ *such that*

$$h_0(t_0, x_0) < \delta \text{ implies } h(t, x(t)) < \epsilon, t \geq t_0,$$

where $x(t) = x(t, t_0, x_0)$ *is any solution of* (4.3.1).

On the basis of this definition, it is easy to formulate various notions of stability, boundedness and practical stability in terms of two measures (h_0, h) corresponding to the deinitions given in Section 4.2.

A few choices of the two measures (h_0, h) given below will demonstrate the generality of the Definition 4.6.1. Furthermore, the concepts in terms of the two measures (h_0, h) enable us to unify a variety of stability notions found in the literature. It is easy to see that Definition 4.6.1 reduces to

(1) the well known stability of the trivial solution $x(t) \equiv 0$ of (4.3.1) or equivalently, of the invariant set $\{0\}$, if $h(t,x) = h_0(t,x) = \|x\|$;

(2) the stability of the prescribed motion $x_0(t)$ of (4.3.1) if $h(t,x) = h_0(t,x) = \|x - x_0(t)\|$;

(3) the partial stability of the trivial solution of (4.3.1) if $h(t,x) = \|x\|_s$, $1 \leq s \leq n$ and $h_0(t,x) = \|x\|$;

(4) the stability of asymptotically invariant set $\{0\}$, if $h(t,x) = h_0(t,x) = \|x\| + \sigma(t)$ where $\sigma \in \mathcal{L}$;

(5) the stability of the invariant set $A \in R^n$ if $h(t,x) = h_0(t,x) = \mathrm{d}(x,A)$, where $\mathrm{d}(x,A)$ is the distance of x from the set A;

(6) the stability of conditionally invariant set B with respect to A, where $A \subset B \subset R^n$, if $h(t,x) = \mathrm{d}(x,B)$ and $h_0(t,x) = \mathrm{d}(x,A)$;

(7) the conditional stability of the trivial solution of (4.3.1) if $h(t,x) = \|x\|$ and $h_0(t,x) = \|x\| + \mathrm{d}(x,M)$ where M is the k-dimensional manifold containing the origin.

We recall that the set $\{0\}$ is said to be asymptotically invariant relative to (4.3.1) if given $\epsilon > 0$, there exists a $\tau(\epsilon) > 0$ such that $x_0 = 0$ implies $\|x(t, t_0, 0)\| < \epsilon$ for $t \geq t_0 \geq \tau(\epsilon)$. Recall also that, $x = 0$ is said to be conditionally equistable, if given $\epsilon > 0$ and $t_0 \in R_+$ there exists a $\delta = \delta(t_0, \epsilon) > 0$ such that

$$x_0 \in S(\delta) \cap M \text{ implies } x(t) \in S(\epsilon),\ t \geq t_0.$$

We remark that when we wish to discuss the notion indicated in (4), we need to restrict the initial time t_0 to a suitable subset of R_+ so that it is possible to have $h_0(t_0, x_0) < \delta$. Similarly, when we intend to consider the concept defined in (7), we choose the initial data x_0 to be in the manifold M in order that $h_0(t_0, x_0) < \delta$ implies $x_0 \in S(h_0, \delta) \cap M$.

We note further that several other combinations of choices are possible for h_0, h in addition to those given in (1) to (7). Moreover, similar comments can be made to practical stability and boundedness concepts.

Definition 4.6.2. *Let $h_0, h \in \Gamma$. Then, we say that*

(i) h_0 is finer than h if there exists a $\rho > 0$ and a function $\phi \in C\mathcal{K}$ such that $h_0(t,x) < \rho$ implies $h(t,x) \leq \phi(t, h_0(t,x))$;

(ii) h_0 *is uniformly finer than* h *if in (i)* ϕ *is independent of* t*;*

(iii) h_0 *is asymptotically finer than* h *if there exists a* $\rho > 0$ *and a function* $\phi \in \kappa\mathcal{L}$ *such that* $h_0(t,x) < \rho$ *implies* $h(t,x) \leq \phi(h_0(t,x),t)$.

Definition 4.6.3. *Let* $V \in C(R_+ \times R^n, R_+^N)$ *and* $V_0(t,x) = \sum_{i=1}^{N} V_i(t,x)$. *Then* V *is said to be*

(i) h-positive definite if there exists a $\rho > 0$ *and a function* $b \in \kappa$ *such that* $b(h(t,x)) \leq V_0(t,x)$ *whenever* $h(t,x) < \rho$*;*

(ii) h-decrescent if there exists a $\rho > 0$ *and a function* $a \in \kappa$ *such that* $V_0(t,x) \leq a(h(t,x))$ *whenever* $h(t,x) < \rho$*;*

(iii) h-weakly decrescent if there exists a $\rho > 0$ *and a function* $a \in C\kappa$ *such that* $V_0(t,x) \leq a(t,(h(t,x))$ *whenever* $h(t,x) < \rho$*;*

(iv) h-asymptotically decrescent if there exists a $\rho > 0$ *and a function* $a \in \kappa\mathcal{L}$ *such that* $V_0(t,x) \leq a((h(t,x),t)$ *whenever* $h(t,x) < \rho$.

Let us now establish some sufficient conditions for the (h_0, h) stability properties of the differential system (4.3.1).

Theorem 4.6.1. *Assume that*

(A_0) $h, h_0 \in \Gamma$ *and* h_0 *is uniformly finer than* h*;*

(A_1) $V \in C(R_+ \times R^n, R_+^N)$, $V(t,x)$ *is locally Lipschitzian in* x, V *is h-positive definite and* h_0*-descrescent;*

(A_2) $g \in C(R_+ \times R_+^N, R^N)$, $g(t,u)$ *is quasimonotone nondecreasing in* u*;*

(A_3) $D^+V(t,x) \leq g(t,V(t,x))$ *for* $(t,x) \in S(h,\rho)$ *for some* $\rho > 0$, *where* $S(h,\rho) = \{(t,x) \in R_+ \times R^n;\ h(t,x) < \rho\}$.

Then the stability properties of the trivial solution of (4.5.2) imply the corresponding (h_0,h)*-stability properties of (4.3.1).*

Proof. We shall only prove (h_0,h)-equiasymptotic stability of (4.3.1). For this purpose, let us first prove (h_0,h) equistability.
Since V is h-positive definite, there exists a $\lambda \in (0,\rho]$ and $b \in \kappa$ such that

$$b(h(t,x)) \leq V_0(t,x),\ (t,x) \in S(h,\lambda). \tag{4.6.1}$$

Let $0 < \epsilon < \lambda$ and $t_0 \in R_+$ be given and suppose that the trivial solution of (4.5.2) is equistable. Then, given $b(\epsilon) > 0$ and $t_0 \in R_+$, there exists a function $\delta_1 = \delta_1(t_0,\epsilon)$ that is continuous in t_0 such that

$$\sum_{i=1}^{N} u_{0i} < \delta_1 \text{ implies } \sum_{i=1}^{N} u_i(t,t_0,u_0) < b(\epsilon), \quad t \geq t_0, \tag{4.6.2}$$

where $u(t, t_0, u_0)$ is any solution of (4.5.2). We choose $u_0 = V(t_0, x_0)$. Since V is h_0-descrescent and h_0 is uniformly finer than h, there exists a $\lambda_0 > 0$ and a function $a \in \kappa$ such that for $(t_0, x_0) \in S(h_0, \lambda_0)$,

$$h(t_0, x_0) < \lambda \text{ and } V_0(t_0, x_0) \leq a(h_0(t_0, x_0)). \tag{4.6.3}$$

It then follows from (4.6.1) that

$$b(h(t_0, x_0)) \leq V_0(t_0, x_0) \leq a(h_0(t_0, x_0)), \ (t_0, x_0) \in S(h_0, \lambda_0). \tag{4.6.4}$$

Choose $\delta = \delta(t_0, \epsilon)$ such that $\delta \in (0, \lambda_0]$, $a(\delta) < \delta_1$ and let $h_0(t_0, x_0) < \delta$. Then (4.6.4) shows that $h(t_0, x_0) < \epsilon$ since $\delta_1 < b(\epsilon)$. We claim that

$$h(t, x(t)) < \epsilon, \ t \geq t_0 \text{ whenever } h_0(t_0, x_0) < \delta$$

where $x(t) = x(t, t_0, x_0)$ is any solution of (4.3.1) with $h_0(t_0, x_0) < \delta$. If this is not true, then there exists a $t_1 > t_0$ and a solution $x(t)$ of (4.3.1) such that

$$h(t_1, x(t_1)) = \epsilon, \text{ and } h(t, x(t)) < \epsilon, \ t_0 \leq t < t_1, \tag{4.6.5}$$

in view of the fact that $h(t_0, x_0) < \epsilon$ whenever $h_0(t_0, x_0) < \delta$. This means that $x(t) \in S(h, \lambda)$ for $[t_0, t_1]$ and hence by Theorem 4.3.1, we have

$$V(t, x(t)) \leq r(t, t_0, u_0), \quad t_0 \leq t < t_1. \tag{4.6.6}$$

where $r(t, t_0, u_0))$ is the maximal solution of (4.5.2). Now the relations (4.6.1), (4.6.2), (4.6.5) and (4.6.6) yield

$$b(\epsilon) \leq V_0(t_1, x(t_1)) \leq r_0(t_1, t_0, u_0) < b(\epsilon),$$

a contradiction proving (h_0, h)-equistability of (4.3.1) where

$$r_0(t, t_0, u_0) = \sum_{i=1}^{N} r_i(t, t_0, u_0).$$

Suppose next that the trivial solution of (4.5.2) is quasi-equiasymptotically stable. From the (h_0, h)-equistability, we set $\epsilon = \lambda$ so that $\hat{\delta}_0 = \delta(t_0, \lambda)$. Now, let $0 < \eta < \lambda$. Then, by quasi-equiasymptotic stability of (4.5.2), we have that, given $b(\eta) > 0$ and $t_0 \in R_+$, there exist positive numbers $\delta_1^* = \delta_1^*(t_0)$ and $T = T(t_0, \eta) > 0$ such that

$$\sum_{i=1}^{N} u_{0i} < \delta_1^* \quad \text{implies} \quad \sum_{i=1}^{N} u_i(t, t_0, u_0) < b(\eta), \quad t \geq t_0 + T. \tag{4.6.7}$$

Choosing $u_0 = V(t_0, x_0)$ as before, we find a $\delta_0^* = \delta_0^*(t_0) > 0$ such that $\delta_0^* \in (0, \lambda_0]$ and $a(\delta_0^*) < \delta_1^*$. Let $\delta_0 = \min(\delta_0^*, \hat{\delta}_0)$ and $h_0(t_0, x_0) < \delta_0$. This implies that $h(t, x(t)) < \lambda$, $t \geq t_0$ and hence the estimate (4.6.6) is valid for all $t \geq t_0$. Suppose now that there exists a sequence $\{t_k\}$, $t_k \geq t_0 + T$, $t_k \to \infty$ as $k \to$

∞. such that $\eta \leq h(t_k, x(t_k))$ where $x(t)$ is any solution of (4.3.1) such that $h_0(t_0, x_0) < \delta_0$. This leads to a contradiction

$$b(\eta) \leq V_0(t_k, x(t_k)) \leq r_0(t_k, t_0, u_0) < b(\eta)$$

because of (4.6.6) and (4.6.7). Hence the system (4.3.1) is (h_0, h)-equi-asymptotically stable and the proof is complete.

We have assumed in Theorem 4.6.1 stronger requirements on V, h, h_0 only to unify all the stability criteria in one theorem. This obviously puts burden on the comparison equation (4.5.2). However, to obtain only non-uniform stability criteria, we could weaken certain assumptions of Theorem 4.3.1 as in the next result. The details of proof are omitted.

Theorem 4.6.2. *Assume that conditions $(A_0) - (A_3)$ hold with the following changes:*

(i) $h_0, h \in \Gamma_0$ and h_0 is finer than h;

(ii) V is h_0-weakly decrescent.

Then, the equi or uniform stability properties of the trivial solution of (4.5.2) imply the corresponding equi (h_0, h)-stability properties of (4.3.1).

We shall next consider a result on (h_0, h)-asymptotic stability which generalizes classical results.

Theorem 4.6.3. *Assume that*

(i) $h_0, h \in \Gamma_0$ and h_0 is finer than h;

(ii) $V \in C(R_+ \times R^n, R_+^N)$, $V(t, x)$ is locally Lipschitzian in x, V is h-positive definite and h_0-weakly decrescent;

(iii) $W \in C(R_+ \times R^n, R_+)$, $W(t, x)$ is locally Lipschitzian in x, W is h-positive definite, $D^+W(t, x)$ is bounded from above or from below on $S(h, \rho)$ and for $(t, x) \in S(h, \rho)$, $1 \leq p \leq N$, $D^+V_p(t, x) \leq -C(W(t, x))$, $C \in \kappa$ and $D^+V_i(t, x) \leq g_i(t, V(t, x))$; $i \neq p$.

Then, equi or uniform stability of the trivial solution of (4.5.2) implies that the system (4.3.1) is (h_0, h)-asymptotically stable.

Proof. By Theorem 4.6.2 with $g_p(t, u) \equiv 0$, it follows that the system (4.3.1) is (h_0, h)-equistable. Hence it is enough to prove that given $t_0 \in R_+$, there exists a $\delta_0 = \delta_0(t_0) > 0$ such that

$$h_0(t_0, x_0) < \delta_0 \quad \text{implies} \quad h(t, x(t)) \to 0 \quad \text{as} \quad t \to \infty.$$

For $\epsilon = \lambda$, let $\delta_0 = \delta(t_0, \lambda)$ be associated with (h_0, h)-equistability. We suppose that $h_0(t_0, x_0) < \delta_0$. Since, $W(t, x)$ is h-positive definite, it is enough to prove that $\lim_{t \to \infty} W(t, x(t)) = 0$ for any solution $x(t)$ of (4.3.1) with $h_0(t_0, x_0) <$

δ_0. We first note that $\liminf_{t\to\infty} W(t, x(t)) = 0$. For otherwise, in view of (iii), we get $V_p(t, x(t)) \to -\infty$ as $t \to \infty$.

Suppose that $\lim_{t\to\infty} W(t, x(t)) \neq 0$. Then, for any $\epsilon > 0$, there exists divergent sequences $\{t_n\}, \{t_n^*\}$ such that $t_i < t_i^* < t_{i+1}$, $i = 1, 2, \ldots$, and

$$\begin{cases} W(t_i, x(t_i)) = \frac{\epsilon}{2}, \ W(t_i^*, x(t_i^*)) = \epsilon, \quad \text{and} \\ \frac{\epsilon}{2} < W(t, x(t)) < \epsilon, \quad t \in (t_i, t_i^*). \end{cases} \tag{4.6.8}$$

Of course, we could also have, instead of (4.6.8),

$$W(t_i, x(t_i)) = \epsilon, \ W(t_i^*, x(t_i^*)) = \frac{\epsilon}{2}, \ W(t, x(t)) \in (\frac{\epsilon}{2}, \epsilon). \tag{4.6.9}$$

Suppose that $D^+W(t, x) \leq M$. Then, it is easy to obtain, using (4.6.8), the relation $t_i^* - t_i > \frac{\epsilon}{2M}$. In view of (iii), we have for large n,

$$\begin{aligned} 0 \leq V_p(t_n^*, x(t_n^*)) &\leq V_p(t_0, x_0) \\ &+ \sum_{1\leq i\leq n} \int_{t_i}^{t_i^*} D^+V_p(s, x(s))\mathrm{d}s \\ &\leq V_p(t_0, x_0) - nC(\frac{\epsilon}{2})\frac{\epsilon}{2M} \\ &< 0, \end{aligned}$$

which is a contradiction. Thus, $W(t, x(t)) \to 0$ as $t \to \infty$ and hence $h(t, x(t)) \to 0$ as $t \to \infty$. The argument is similar when D^+W is bounded from below and we use (4.6.9). The proof is therefore complete.

If we desire only uniform stability properties, we can relax the assumptions of Theorems 4.6.1 and 4.6.2 by employing a vector Lyapunov function which satisfies less restrictive conditions.

Theorem 4.6.4. *Assume (A_0) and (A_2) of Theorem 4.6.1. Suppose further* (A_1^*) *for each $\eta \in (0, \rho)$, $\rho > 0$ there exists a function*

$$V \in C(S(h, \rho) \cap S^C(h_0, \eta), R_+^N)$$

such that V is locally Lipschitzian in x, V is h-positive definite and h_0-descrescent, where $S^C(h_0, \eta)$ is the complement of $S(h_0, \eta)$; and

$$(A_3^*) \quad D^+V(t, x) \leq g(t, V(t, x)) \quad \textit{for} \quad (t, x) \in S(h, \rho) \cap S^C(h_0, \eta).$$

Then, the uniform stability of trivial solution of (4.5.2) implies (h_0, h)-uniform stability of (4.3.1).

Proof. Suppose that the trivial solution of (4.5.2) is uniformly stable. Because of (A_0) and (A_1^*), there exists a $\lambda \in (0, \rho]$ such that the relations (4.6.1) and (4.6.4) hold for $(t, x) \in S(h, \lambda) \cap S^C(h_0, \eta)$ and λ_0 is independent of t_0. Also, (4.6.2) holds with δ_1 independent to t_0. We choose $\delta = \delta(\epsilon) > 0$ such that $\delta \in (0, \lambda_0]$ and $a(\delta) < \delta_1$. We let $h_0(t_0, x_0) < \delta$ and note that $h(t_0, x_0) < \epsilon$ as before. If (h_0, h)-uniform stability of (4.3.1) does not hold, then there will exist a solution $x(t)$ of (4.3.1) and $t_1, t_2 > t_0$ such that

$$\begin{cases} h_0(t_1, x(t_1)) = \delta, \ h(t_2, x(t_2)) = \epsilon, \quad \text{and} \\ x(t) \in S(h, \epsilon) \cap S^C(h_0, \delta) \quad \text{for} \quad t \in [t_1, t_2]. \end{cases} \tag{4.6.10}$$

Hence, choosing $\eta = \delta$ and using Theorem 1.3.1, we have

$$V(t, x(t)) \le r(t, t_1, u_0), \quad t \in [t_1, t_2], \tag{4.6.11}$$

where $r(t, t_1, u_0)$ is the maximal solution of (4.5.2) through (t_1, u_0). The relations (4.6.1), (4.6.2), (4.6.10) and (4.6.11) lead us to the contradiction.

$$b(\epsilon) \le V_0(t_2, x(t_2)) \le r_0(t_2, t_1, u_0) < b(\epsilon),$$

proving (h_0, h)-uniform stability of (4.3.1), V_0 and r_0 being the same as in the proof of Theorem 4.6.1.

4.7 Practical Stability in Terms of Two Measures

In this section, we shall continue to consider concepts in terms of two measures and prove some typical results relative to practical stability and boundedness.

Theorem 4.7.1. *Assume that*

(A_0) $0 < \lambda < A$, $h, h_0 \in \Gamma$ *and* $h(t, x) \le \phi(h_0(t, x))$ *if* $h_0(t, x) < \lambda$;

(A_1) $V \in C(R_+ \times R^n, R_+^N)$, $V(t, x)$ *is locally Lipschitzian in* x *and*

$$D^+V(t, x) \le g(t, V(t, x)) \quad \textit{for} \quad (t, x) \in R_+ \times S(A),$$

where $g \in C(R_+ \times R_+^N, R^N)$ *and* $g(t, u)$ *is quasimonotone nondecreasing in* u;

(A_2) $b(h(t, x)) \le V_0(t, x)$ *if* $h(t, x) < A$ *and*

$$V_0(t, x) \le a(h_0(t, x)) \quad \textit{if} \quad h_0(t, x) < \lambda,$$

where $a, b \in \kappa$ *and* $V_0 = \sum_{i=1}^{N} V_i(t, x)$;

(A_3) $\phi(\lambda) < A$ *and* $a(\lambda) < b(A)$.

Then the practical stability properties of (4.5.2) *implies the corresponding* (h_0, h)*-practical stability properties of the system* (4.3.1).

Proof. Let us first suppose that (4.5.2) is practically stable. Then, given $(a(\lambda), b(A))$, it follows, because of (A_3), that

$$\sum_{i=1}^{N} u_{0i} < a(\lambda) \text{ implies } \sum_{i=1}^{N} u_i(t, t_0, u_0) < b(A), \quad t \geq t_0. \tag{4.7.1}$$

Let $h_0(t_0, x_0) < \lambda$. Then by (A_0) and (A_3), it follows that

$$h(t_0, x_0) \leq \phi(h_0(t_0, x_0)) < \phi(\lambda) < A. \tag{4.7.2}$$

We claim that $h(t, x(t)) < A$, $t \geq t_0$, where $x(t) = x(t, t_0, x_0)$ is any solution of (4.3.1). If this is not true because of (4.7.2), then there would exist a solution $x(t)$ of (4.3.1) with $h_0(t_0, x_0) < \lambda$ and a $t_1 > t_0$ such that

$$h(t_1, x(t_1)) = A \quad \text{and} \quad h(t, x(t)) \leq A, \quad t_0 \leq t \leq t_1.$$

By (A_2) this yields

$$b(A) \leq V_0(t_1, x(t_1)). \tag{4.7.3}$$

Choose $u_0 = V(t_0, x_0)$. Using (A_1), we obtain by Theorem 4.3.1, the estimate

$$V(t, x(t)) \leq r(t, t_0, u_0), \quad t_0 \leq t \leq t_1, \tag{4.7.4}$$

where $r(t, t_0, u_0)$ is the maximal solution of (4.5.2). The relation (4.7.1), (4.7.3) and (4.7.4) imply

$$b(A) \leq V_0(t_1, x(t_1)) \leq r_0(t_1, t_0, u_0) < b(A),$$

since $V_0(t_0, x_0) \leq a(h_0(t_0, x_0)) < a(\lambda)$ because of (A_2). This is a contradiction which proves (h_0, h)-practical stability of the system (4.3.1).

We shall next prove that the system (4.3.1) is (h_0, h)-strongly practically stable for $(\lambda, A, B, T) > 0$. To do this, let us suppose that (4.5.2) is strongly practically stable for $(a(\lambda), b(A), b(B), T) > 0$. This means we need to prove only (h_0, h)-practical stability of the system (4.3.1). The practical quasi-stability of (4.5.2) means that

$$\sum_{i=1}^{N} u_{0i} < a(\lambda) \text{ implies } \sum_{i=1}^{N} u_i(t, t_0, u_0) < b(B), \quad t \geq t_0 + T. \tag{4.7.5}$$

Suppose that $h_0(t_0, x_0) < \lambda$ so that by (h_0, h)-practical stability of (4.3.1), we have $h(t, x(t)) < A$, $t \geq t_0$. Consequently, the relation (4.7.4) holds for all $t \geq t_0$, that is,

$$V(t, x(t)) \leq r(t, t_0, u_0), \quad t \geq t_0, \tag{4.7.6}$$

which yields because of (4.7.5), (4.7.6), (A_2) and (A_3),

$$b(h(t,x(t)) \leq V_0(t,x(t)) \leq r_0(t,t_0,u_0) < b(B), \quad t \geq t_0 + T.$$

Thus we have, whenever $h_0(t_0,x_0) < \lambda$, $\quad h(t,x(t)) < B$, $\quad t \geq t_0 + T$ and hence the system (4.3.1) is (h_0,h) strongly practically stable.

One can prove similarly other $(h_0,h)-$ practical stability properties of (4.3.1) and hence the proof is complete.

If we desire only uniform practical stability properties, we can relax the assumptions of Theorem 4.7.1 considerably following Theorem 4.6.4. We shall not discuss such results to avoid monotony.

We shall next prove (h_0,h) boundedness results in this framework. As we have seen in Section 4.4, when we investigate boundedness properties, we also utilize class k functions. However, in this situation, the class k is defined by $k = [a \in [\rho,\infty), R_+]$, $a(u)$ is strictly increasing in u and $a(u) \to \infty$ as $u \to \infty$. We shall employ the same symbol k for both situations, with this understanding.

Theorem 4.7.2. *Assume that*

(B_0) $h_0, h \in \Gamma$ and $h(t,x) \leq \phi(h_0(t,x))$, if $h_0(t,x) < \rho_0$ where $\phi \in k$;

(B_1) $V \in C(R_+ \times S^C(h,\rho), R_+^N)$, $V(t,x)$ is locally Lipschitzian in x and for $(t,x) \in R_+ \times S^C(h,\rho)$,

$$D^+V(t,x) \leq g(t,V(t,x)),$$

where $\phi(\rho_0) \leq \rho, g \in C(R_+ \times R_+^N, R^N)$ and $g(t,u)$ is quasimonotone nondecreasing in u;

(B_2) $b(h(t,x)) \leq v_0(t,x) = \sum_{i=1}^N V_i(t,x)$ if $h(t,x) < \rho$ and

$$V_0(t,x) \leq a(h_0(t,x)), \quad \textit{if} \quad h_0(t,x) < \rho_0.$$

Then boundedness properties of (4.5.2) *imply the corresponding* (h_0,h) *boundedness properties of* (4.3.1)

Proof. Let $\alpha \geq \rho_0$ and $t_0 \in R_+$ be given and let $\alpha_1 = a(\alpha)$. Suppose that (4.5.2) is uniform bounded. Then, given $\alpha_1 > 0$ and $t_0 \in R_+$, there exists a $\beta_1 = \beta_1(\alpha) > 0$ such that

$$\sum_{i=1}^N u_{0_i} < \alpha_1 \text{ implies } \sum_{i=1}^N u_i(t,t_0,u_0) < \beta, t \geq t_0, \tag{4.7.7}$$

where $u(t,t_0,u_0)$ is any solution of (4.5.2). Choose $u_0 = V(t_0,x_0)$ and let $h_0(t_0,x_0) < \alpha$. By (B_0), we have

$$h(t_0,x_0) \leq \phi(h_0(t_0,x_0)) < \phi(\alpha) \equiv \alpha_0.$$

Let $\beta = \beta(\alpha) > 0$ be chosen such that $\beta_1(\alpha) < b(\beta)$ and $a(\alpha) < \beta$. We then claim that $h(t,x(t)) < \beta, t \geq t_0$, where $x(t) = x(t,t_0,x_0)$ is any solution of (4.3.1). If this is not true, there would exist a solution $x(t) = x(t, t_0, x_0)$ of (4.3.1) with $h_0(t_0, x_0) < \alpha$ and $t_1, t_2 > t_0$ satisfying

$$\begin{cases} h(t_1, x(t_1)) = \alpha_0, h(t_2, x(t_2)) = \beta \quad \text{and} \\ \rho \leq \alpha_0 \leq h(t, x(t)) \leq \beta, \quad t_1 \leq t \leq t_2. \end{cases} \tag{4.7.8}$$

By Theorem 4.3.1, we have because of (B_1), the estimate

$$V(t, x(t)) \leq r(t, t_1, V(t_1, x(t_1))), \quad t_1 \leq t \leq t_2 \tag{4.7.9}$$

where $r(t, t_1, u_0)$ is the maximal solution of (4.5.2). It then follows using (4.7.7), (4.7.8), (4.7.9) and (B_2), that

$$b(\beta) \leq V_0(t_2, x(t_2)) \leq r_0(t_2, t_1, V(t_1, x(t_1))) < b(\beta),$$

since $V_0(t_1, x(t_1)) \leq a(h_0(t_1, x(t_1))) = a(\alpha) = \alpha_1$. This contradiction proves (h_0, h) uniform boundedness of (4.3.1)

We can prove analogously other (h_0, h)- boundedness properties and hence the proof is complete.

We shall finally consider a result on $(h_0, h)-$ uniform ultimate boundedness under a different set of conditions.

Theorem 4.7.3. *Let the assumptions of Theorem 4.7.2 hold except that* (B_1) *is strengthened to*

(B_1^*) $D^+V_p(t, x) \leq -C(h_0(t, x)), C \in \kappa$ *and*

$$D^+V_i(t, x) \leq g_i(t, V(t, x)), \quad i \neq p, \ \textit{for} \ (t, x) \in R_+ \times S^C(h, \rho).$$

Then uniform boundedness of (4.5.2) *implies* $(h_0, h)-$ *uniform ultimate boundedness of* (4.3.1)

Proof. By Theorem 4.7.2 with $g_p(t, u) \equiv 0$, it follows that (4.3.1) is $(h_0.h)-$ uniform bounded. Setting $\alpha = \rho_0$ and $B = \beta(\rho)$, we have

$$h_0(t_0, x_0) < \rho_0 \quad \text{implies} \quad h(t, x(t)) < B, t \geq t_0.$$

Now let $h_0(t_0, x_0) < \alpha$ for any $\alpha > \rho_0$. Then we claim that there exists a $t^* \in [t_0, t_0 + T]$, where $T = T(\alpha) > \dfrac{a(\alpha)}{C(\rho_0)}$, such that $h_0(t^*, x(t^*)) < \rho_0$. If this does not hold, then we would have $\rho_0 \leq h_0(t, x(t)), t \in [t_0, t_0 + T]$, for a solution $x(t)$ of (4.3.1) with $h_0(t_0, x_0) < \alpha$. We then obtain using (B_1^*),

$$0 \leq V_p(t, x(t)) \leq V_p(t_0, x_0) - C(\rho_0)T \leq a(\alpha) - C(\rho_0)T < 0,$$

by the choice of T. This contradiction proves the existence of a $t^* \in [t_0, t_0 + T]$ with $h_0(t^*, x(t^*)) < \rho_0$, which yields that $h(t, x(t)) < B, \quad t \geq t^* \geq t_0 + T$. The proof is therefore complete

4.8 Perturbed Systems

In order to unify the investigation of stability and boundedness properties of perturbed systems, it is fruitful to employ coupled comparison systems as in Theorem 4.3.3. Naturally, the use of coupled comparison systems is also beneficial in the discussion of unperturbed systems, since estimating $D^+V(t,x)$ by a function of $(t,x,V(t,x))$ is more advantageous than by a function of $(t,V(t,x))$ only. To avoid technical complications, we shall, hereafter, restrict ourselves to one measure $h(t,x) = h_0(t,x) = \|x\|$ so that we get the usual stability and boundedness concepts.

Let us consider the perturbed system

$$x' = F(t,x), \quad x(t_0) = x_0, \tag{4.8.1}$$

where $F \in C(R_+ \times R^n, R^n)$. If $F(t,x) = f(t,x) + R(t,x)$ where $R(t,x)$ is a perturbation term then (4.8.1) is a perturbed system relative to unperturbed system (4.3.1). Of course, perturbation may enter the system (4.3.1) in a variety of ways; for example

$$F(t,x) = f(t,x,R(t,x)),$$
$$F(t,x) = f(t,x)R_0(t,x) + R_1(t,x),$$

where $R_0(t,x)$ is $n \times n$ matrix function and $R_1(t,x)$ is a vector function.

As indicated earlier, we consider the coupled comparison system

$$u' = g(t,x,u), \quad u(t_0) = u_0 \geq 0, \tag{4.8.2}$$

where $g \in C(R_+ \times R^n \times R_+^N, R^N)$. We require suitable definitions of stability and boundedness properties relative to (4.8.2).

Definition 4.8.1. *Let $z \in C(R_+, R^n)$ with $z(t_0) = x_0$ and $u(t,t_0,x_0,u_0)$ be any solution of the system*

$$u' = g(t,z(t),u), \quad u(t_0) = u_0 \geq 0. \tag{4.8.3}$$

Then

(a) *the trivial solution of (4.8.3) is said to be*

(i) *equistable, if given $0 < \epsilon < \rho$, $b \in \kappa$ and $t_0 \in R_+$, there exists $\delta_1(t_0,\epsilon)$, $\delta_2(t_0,\epsilon) > 0$ such that*

$$\|x_0\| < \delta_2(t_0,\epsilon) \quad \text{and} \quad \sum_{i=1}^{N} u_{0i} < \delta_1(t_0,\epsilon) \quad \text{imply} \quad \sum_{i=1}^{N} u_i(t,t_0,x_0,u_0) < b(\epsilon)$$

on any interval $t_0 \leq t \leq t_1$ on which $|z(t)| \leq \epsilon$;

(ii) *quasi-equiasymptotically stable, if given $0 < \epsilon < \rho$, $b \in \kappa$ and $t_0 \in R_+$, there exists $\delta_{10}(t_0)$, $\delta_{20}(t_0) > 0$ and a $T(t_0,\epsilon) > 0$ such that*

$$\|x_0\| < \delta_{20}(t_0) \quad \textit{and} \quad \sum_{i=1}^{N} u_{0i} < \delta_{10}(t_0) \quad \textit{imply} \quad \sum_{i=1}^{N} u_i(t, t_0, x_0, u_0) < b(\epsilon)$$

for $t \geq t_0 + T(t_0, \epsilon)$ *whenever* $\|z(t)\| < \rho$, $t \geq t_0$;

(b) *the system* (4.8.3) *is said be*

(iii) *practically stable, if given* $0 < \lambda < A$, a, $b \in \kappa$ *with* $a(\lambda) < b(A)$ *we have*

$$\|x_0\| < \lambda \quad \textit{and} \quad \sum_{i=1}^{N} u_{0i} < a(\lambda) \quad \textit{imply} \quad \sum_{i=1}^{N} u_i(t, t_0, x_0, u_0) < b(A)$$

on any interval $t_0 \leq t \leq t_1$ *on which* $\|z(t)\| \leq A$ *for some* $t_0 \in R_+$.

Based on this definition, one can formulate other notions relative to (4.8.2) whenever necessary.

One can now prove a result similar to Theorem 4.5.1,

Theorem 4.8.1. *Assume that*

(i) $V \in C(R_+ \times R^n, R_+^N)$, $V(t,x)$ *is locally Lipschitzian in* x *and for* $(t, x) \in R_+ \times S(\rho)$, $D^+V(t,x) \leq g(t, x, V(t,x))$;

(ii) $g \in C(R_+ \times R^n \times R_+^N, R^N)$, $g(t, 0, 0) = 0$ *and* $g(t, x, u)$ *is quasimonotone nondecreasing in* u;

(iii) $b(\|x\|) \leq \sum_{i=1}^{N} V_i(t,x) \leq a(\|x\|)$, $(t,x) \in R_+ \times S(\rho)$ *where* $a, b \in \kappa$.

Then the stability properties of the trivial solution of (4.8.2) *imply the corresponding stabilty properties of the trivial solution of* (4.8.1).

Proof. The proof is very much similar to the proof of Theorem 4.5.1. Hence we shall only indicate the proof of equistability.

Let $0 < \epsilon < \rho$ and $t_0 \in R_+$ be given. Assume that the trivial solution of (4.8.2) is equistable. Then given $b(\epsilon) > 0$ and $t_0 \in R_+$, there exist $\delta_1 = \delta_1(t_0, \epsilon)$, $\delta_2 = \delta_2(t_0, \epsilon) > 0$ such that

$$\|x_0\| < \delta_2 \quad \text{and} \quad \sum_{i=1}^{N} u_{0i} < \delta_1 \quad \text{imply} \quad \sum_{i=1}^{N} u_i(t, t_0, x_0, u_0) < b(\epsilon) \tag{4.8.4}$$

on any interval $t_0 \leq t \leq t_1$ on which $\|z(t)\| \leq \epsilon$. Chosse $u_0 = V(t_0, x_0)$ and $\delta^* = \delta^*(t_0, \epsilon) > 0$ such that $a(\delta^*) < \delta_1$. Let $\delta = \min(\delta^*, \delta_2)$ and $\|x_0\| < \delta$. Then we claim that $\|x(t)\| < \epsilon$ for $t \geq t_0$ for any solution $x(t) = x(x, t_0, x_0)$ of (4.8.1) with $\|x_0\| < \delta$ and $t_1 > t_0$ satisfying

$$\|x(t_1)\| = \epsilon \quad \text{and} \quad \|x(t)\| \leq \epsilon, \quad t_0 \leq t \leq t_1. \tag{4.8.5}$$

Hence with $x(t) = z(t)$, we get, by Theorem 4.3.3

$$V(t, x(t)) \leq r(t, t_0, x_0, u_0), \quad t_0 \leq t \leq t_1, \tag{4.8.6}$$

where $r(t, t_0, x_0, u_0)$ is the maximal solution of (4.8.3). Now the relations (4.8.4), (4.8.5) and (4.8.6) together with (iii) lead to

$$b(\epsilon) \leq \sum_{i=1}^{N} V_i(t_1, x(t_1)) \leq \sum_{i=1}^{N} r_i(t_1, t_0, x_0, u_0) < b(\epsilon),$$

because of the fact $\sum_{i=1}^{N} u_{0i} \leq a(\|x_0\|) < a(\delta) < \delta_1$ and $\|x_0\| < \delta_2$. This contradiction proves equistability and the proof is complete.

As a typical example of coupled comparison system, we consider the case

$$g(t, x, u) = Au + w(t, x) \tag{4.8.7}$$

and prove the following result.

Theorem 4.8.2. *Suppose that*

(i) the $N \times N$ matrix $A = (a_{ij})$ satisfies $a_{ij} \geq 0$, $i \neq j$ and has dominant diagonal property, that is, for each i

$$a_{ii} + \sum_{\substack{j=1 \\ i \neq j}}^{N} a_{ij} \leq -\gamma, \quad \gamma > 0;$$

(ii) $w \in C(R_+ \times R^n, R^N)$, $\|w_i(t, x)\| \leq \lambda_i(t)$ whenever $\|x\| \leq \epsilon$ where $\lambda_i \in C(R_+, R_+)$ and $\int_t^{t+1} \lambda_0(s) ds \to 0$ as $t \to \infty$ with $\lambda_0(t) = \sum_{i=1}^{N} \lambda_i(t)$.

Then the system (4.8.2) is eventually asymptotically stable.

Proof. Let $0 < \epsilon < \rho$ and suppose that on some interval $t_0 \leq t \leq t_1$, $\|z(t)\| \leq \epsilon$. Then (4.8.3) and (4.8.7) imply because of (ii),

$$u' \leq Au + \lambda(t), \quad t_0 \leq t \leq t_1.$$

Setting $v(t) = \sum_{i=1}^{N} u_i(t)$ and using (i) and (ii), we arrive at

$$v'(t) \leq -\gamma v(t) + \lambda_0(t), \quad t_0 \leq t \leq t_1, \quad v(t_0) = \sum_{i=1}^{N} u_{0i}. \tag{4.8.8}$$

It is therefore enough to consider the properties of $v(t)$. Observe that

$$\int_{t_0-1}^{t} p(s)\mathrm{d}s = \int_{t_0-1}^{t} \Big[\int_{s}^{s+1} \lambda_0(\sigma)\mathrm{d}\sigma\Big]\mathrm{d}s$$
$$\geq \int_{t_0}^{t} \Big[\int_{\sigma-1}^{\sigma} \lambda_0(\sigma)\mathrm{d}s\Big]\mathrm{d}\sigma = \int_{t_0}^{t} \lambda_0(\sigma)\mathrm{d}\sigma \quad \text{for} \quad t \geq t_0 \geq 1.$$

Also, for $\beta > 0$

$$\int_{t_0}^{t} e^{\beta s}\lambda_0(s)\mathrm{d}s \leq \int_{t_0-1}^{t} \Big[\int_{s}^{s+1} e^{\beta\sigma}\lambda_0(\sigma)\mathrm{d}\sigma\Big]\mathrm{d}s$$
$$\leq \int_{t_0-1}^{t} e^{\beta(s+1)}p(s)\mathrm{d}s,$$

whence,

$$e^{-\beta t}\int_{1}^{t} e^{\beta s}\lambda_0(s)\mathrm{d}s \leq e^{-\beta t}\int_{0}^{t} e^{\beta(s+1)}p(s)\mathrm{d}s.$$

Applying L'Hospital's rule on $\frac{1}{e^{\beta t}}\int_0^t e^{\beta(s+1)}p(s)\mathrm{d}s$, it is easy to verify that

$$\lim_{t\to\infty} e^{-\beta t}\int_{1}^{t} e^{\beta s}\lambda_0(s)\mathrm{d}s = 0 \quad \text{for all} \quad \beta > 0. \tag{4.8.9}$$

Let $b \in \kappa$ be given. Choose $0 < \delta < b(\epsilon)$ and set $\delta_1 = \delta_2 = \frac{\delta}{2}$. In view of (4.8.9) there exists a $\tau(\epsilon) \geq 1$ such that

$$\int_{1}^{\tau(\epsilon)} e^{-\gamma(t-s)}\lambda_0(s)\mathrm{d}s < \frac{\delta(\epsilon)}{2}. \tag{4.8.10}$$

Hence we let $t_0 \geq \tau(\epsilon)$, $\|x_0\| < \delta_2$ and $\sum_{i=1}^{N} u_{0i} < \delta_1$. Then we claim that $v(t) < b(\epsilon)$ for $t_0 \leq t \leq t_1$. If not, suppose that $v(t^*) \geq b(\epsilon)$ for some $t^* \in [t_0, t_1]$. It then follows from (4.8.8) and (4.8.10)

$$b(\epsilon) \leq v(t^*) \leq v(t_0)\, e^{-\gamma(t^*-t_0)} + \int_{t_0}^{t^*} e^{-\gamma(t^*-s)}\lambda_0(s)\mathrm{d}s$$
$$< \delta_1 + \frac{\delta}{2} = \delta \leq b(\epsilon),$$

which is a contradiction. Hence $v(t) < b(\epsilon), \quad t \geq t_0 \geq \tau(\epsilon)$. It therefore follows that

$$v(t) \leq v(t_0)\, e^{-\gamma(t-t_0)} + \int_{t_0}^{t} e^{-\gamma(t-s)}\lambda_0(s)\mathrm{d}s, \quad t \geq t_0 \geq \tau(\epsilon),$$

which yields in view of (4.8.9), that $\lim_{t\to\infty} v(t) = 0$. This proves eventual asymptotic stability of the coupled comparison system (4.8.2).

4.9 Large Scale Dynamic Systems

A large dynamic system composed of several subsystems cannot be expected to behave in a desired manner over long periods of operation since for some reason or other, the subsystems are disconnected and reconnected during the functioning of the system. Such on-off participations of subsystems represent changes in structure of the system which may destroy stability and cause the system to fail. To prevent the collapse, systems have to be built to have desired stability properties that can be preserved under structural perturbations. Let us consider a dynamic system described by the differential system

$$x' = f(t, x), \quad x(t_0) = x_0, \tag{4.9.1}$$

where $f \in C(R_+ \times R^n, R^n)$. Assume that (4.9.1) admits a decomposition of the form

$$x_i' = F_i(t, x_i) + R_i(t, x), \quad x_i(t_0) = x_{i0}, \tag{4.9.2}$$

where the vector x is decomposed into N subvectors x_i, namely,

$$x = (x_1, x_2, \ldots, x_N),$$

with each $x_i \in R^{n_i}$, $i = 1, 2, \ldots, N$ such that $n = \sum_{i=1}^{N} n_i$. The functions $F_i(t, x_i)$ in (4.9.2) represent isolated decoupled subsystems

$$x_i' = F_i(t, x_i), \quad x_i(t_0) = x_{i0}, \tag{4.9.3}$$

and $R_i(t, x)$ are the interconnections among the subsystems ($4.9.3_i$), which have the form

$$R_i(t, x) = R_i(t, e_{i1}x_1, e_{i2}x_2, \ldots, e_{iN}x_N), \tag{4.9.4}$$

where $e_{ij} \in C(R_+ \times R^n, [0, 1])$ are the elements of an $N \times N$ interconection matrix. Sometimes, we also assume that the interconnections have the form

$$R_i(t, x) = \sum_{j=1}^{N} e_{ij} R_{ij}(t, x_j), \tag{4.9.5}$$

where each R_{ij} satisfy the constraints

$$\|R_{ij}(t, x_j)\| \leq \beta_{ij} \|x_j\|. \tag{4.9.6}$$

If, in a system (4.9.1), all possible interconnections are present and its structure is described by a matrix $E = (e_{ij})$, then another choice of interconnection matrix $E_0 = (e_{ij}^0)$ represents a structural perturbation of the system (4.9.1).

If we know that susbsystems ($4.9.3_i$), have good qualitative properties such as uniform asymptotic stability or exponential asymptotic stability, then, in view of converse theorems, there exists a Lyapunov function which can then be used to discuss the stability properties of the composite system (4.9.1). Thus several Lyapunov functions result in a natural way when the given large scale system admits a decomposition of the form (4.9.2). Let us prove the following result in this direction.

Theorem 4.9.1. *Assume that*

(i) $V_i \in C(R_+ \times R^{n_i}, R_+)$ *for* $(t,x) \in R_+ \times S(\rho)$, $V_i(t,x_i)$ *is Lipschitzian in* x_i *with a constant* L_i *and*

$$D^+V_i(t,x_i)_{(4.9.3_i)} \le G_i(t, V_i(t,x_i)), \quad i = 1,2,\dots,N,$$

where $G_i \in C(R_+ \times R^{n_i}, R)$;

(ii) $b(\|x\|) \le V_0(t,x) \le a(\|x\|)$, $(t,x) \in R_+ \times S(\rho)$, *where* $a, b \in \kappa$ *and* $V_0(t,x) = \sum_{i=1}^{N} d_i V_i(t,x_i)$ *with* $d_i > 0$;

(iii) $b_i(\|x_i\|) \le V_i(t,x_i)$, $(t,x) \in R_+ \times S(\rho)$, $b_i \in \kappa$;

(iv) $\|R_i(t,x)\| \le H_i(t, \|x_1\|, \|x_2\|, \dots, \|x_N\|)$ *for all interconnections* e_{ij}, *where* $H_i \in C(R_+ \times R^N, R_+)$ *and* $H_i(t,u)$ *is nondecreasing in* u. *Then the stability properties of the trivial solution of*

$$u' = g(t,u) \quad u(t_0) = u_0 \ge 0,$$

where $g_i(t,u) = G_i(t,u_i) + H_i(t, b_1^{-1}(u_1), \dots, b_N^{-1}(u_N))$, b_i^{-1} *being the inverse function of* b_i, *imply the corresponding stability properties of the trivial solution of* (4.9.1).

Proof. For $(t,x) \in R_+ \times S(\rho)$, we compute $D^+V_i(t,x_i)_{(4.9.1)}$ so that in view of assumptions (i), (iii) and (iv), we arrive at

$$\begin{aligned} D^+V_i(t,x_i)_{(4.9.1)} &\le D^+V_i(t,x_i)_{(4.9.3_i)} + L_i\|R_i(t,x)\| \\ &\le G_i(t, V_i(t,x_i)) + L_i H_i(t, b_1^{-1}(V_1(t,x_1)), \dots, b_N^{-1}(V_N(t,x_N)) \\ &\equiv g_i(t, V_1(t,x_1), \dots, V_N(t,x_N)), \end{aligned}$$

for all interconnections e_{ij}. We note that $g(t,u)$ is quasimonotone nondecreasing because of assumptions on G_i and H_i. If we set that $V_i(t,x) \equiv V_i(t,x_i)$, then we have

$$D^+V(t,x)_{(4.9.1)} \le g(t, V(t,x)), \quad (t,x) \in R_+ \times S(\rho),$$

and $b(\|x\|) \le V_0(t,x) \le a(\|x\|)$. Consequently, we can apply Theorem 4.5.1 to obtain stability properties of the large scale system (4.9.1).

We can reduce Theorem 4.9.1 to Theorem 4.8.1 if we replace condition (iv) suitably. This we state in the following result.

Theorem 4.9.2. *Let the assumptions (i), (ii) and (iii) of Theorem 4.9.1 hold. Suppose further*

(iv)* $\|R_i(t, e_{i1}x_1, \dots, e_{iN}x_N)\| \le w_i(t, x_1, x_2, \dots, x_N)$ *for all interconnections* e_{ij} *where* $w_i \in C(R_+ \times R^n, R_+)$.

Then the stability properties of the trivial solution of the coupled comparison system (4.8.2) *implies the corresponding stability properties of the trivial solution of* (4.9.1) *where*

$$g_i(t,x,u) = G_i(t,u_i) + L_i w_i(t,x).$$

As an illustration, let us suppose that the trivial solution of the subsystem ($4.9.3_i$) is exponentially asymptotically stable. Then we know that there exists a Lyapunov function $V_i(t,x_i)$ for each i, satisfying the following conditions for $(t,x) \in R_+ \times S(\rho)$:

(i) $b_i|x_i| \le V_i(t,x_i) \le a_i|x_i|, \quad a_i, b_i > 0;$

(ii) $D^+V_i(t,x_i)_{(4.9.3_i)} \le -\alpha_{ii}V_i(t,x_i), \quad \alpha_{ii} > 0;$

and

(iii) $L_i > 0$ is the Lipschitz constant for $V_i(t,x_i)$.

It then follows that for some $d_i > 0$,

$$b(\|x\|) \le \sum_{i=1}^{N} d_i V_i(t,x_i) \le a(\|x\|),$$

where $a(u) = \sum_{i=1}^{N} d_i a_i u_i, \quad b(u) = \sum_{i=1}^{N} d_i b_i u_i$ and

$$\begin{aligned} D^+V_i(t,x_i)_{(4.9.1)} &\le -\alpha_{ii}V_i(t,x_i) + L_i \sum_{j=1}^{N} \beta_{ij}|x_j| \\ &\le -\alpha_{ii}V_i(t,x_i) + \frac{L_i}{b_i} \sum_{j=1}^{N} \beta_{ij}V_j(t,x_j). \end{aligned}$$

Thus, we have setting $V_i(t,x) = V_i(t,x_i)$,

$$D^+V(t,x) \le AV(t,x),$$

where $A = (a_{ij})$ is a $N \times N$ matrix given by

$$a_{ii} = -(\alpha_{ii} - \frac{L_i}{b_i}\beta_{ii}), \quad a_{ij} = \beta_{ij}, \quad i \ne j.$$

Clearly $g(t,u) = Au$ satisfies the quasimonotone property and if A is a stability matrix, then we derive the stability of (4.9.1) from Theorem 4.9.1.

4.10 A Technique in Perturbation Theory

In this section, we develop a new comparison theorem that connects the solutions of perturbed and unperturbed differential systems in a manner useful in the theory of perturbations. This comparison result blends, in a sense, the two approaches namely, the method of Lyapunov functions and the method of variation of parameters, and consequently provides a flexible mechanism to preserve the nature of perturbations. The results that are given in this section show that the usual comparison theorem in terms of a vector Lyapunov function is included as a special case and that perturbation theory could be studied in a more fruitful way.

Consider the two differential systems

$$y' = f(t, y), \quad y(t_0) = x_0, \tag{4.10.1}$$

and

$$x' = F(t, x), \quad x(t_0) = x_0, \tag{4.10.2}$$

where $f, F \in C(R_+ \times R^n, R^n)$. Relative to the system (4.10.1), let us assume that the following assumption (H) holds:

(H) the solutions $y(t, t_0, x_0)$ of (4.10.1) exist for all $t \geq t_0$, unique and continuous with respect to the initial data and $\|y(t, t_0, x_0)\|$ is locally Lipschitzian in x_0. For any $V \in C(R_+ \times R^n, R^N_+)$ and any fixed $t \in [0, \infty]$, we define

$$\begin{aligned} D\_V(s, y(t, s, x)) \equiv \liminf_{h \to 0^-} \frac{1}{h}[V(s+h, y(t, x+h, x+hF(s, x))) \\ - V(s, y(t, s, x))] \end{aligned} \tag{4.10.3}$$

for $t_0 < s \leq t$ and $x \in R^n$.

The following comparison result which relates the solutions of (4.10.2) to the solutions of (4.10.1) is an important tool in the subsequent discussion.

Theorem 4.10.1. *Assume that the assumption (H) holds. Suppose that*

(i) $V \in C(R_+ \times R^n, R^N_+)$, $V(s, x)$ is locally Lipschitzian in x and for $t_0 < s \leq t$, $x \in R^n$,

$$D\_V(s, y(t, s, x)) \leq g(t, V(s, y(t, s, x))); \tag{4.10.4}$$

(ii) $g \in C(R_+ \times R^N_+, R^N)$, $g(t, u)$ is quasimonotone nondecreasing in u and the maximal solution $r(t, t_0, u_0)$ of

$$u' = g(t, u), \quad u(t_0) = u_0 \geq 0 \tag{4.10.5}$$

exists for $t \geq t_0$.

Then, if $x(t) = x(t, t_0, x_0)$ is any solution of (4.10.2), we have

$$V(t, x(t, t_0, x_0)) \leq r(t, t_0, u_0), \quad t \geq t_0, \tag{4.10.6}$$

provided $V(t_0, y(t, t_0, x_0)) \leq u_0$.

Proof. Let $x(t) = x(t, t_0, x_0)$ be any solution of (4.10.2). Set

$$m(s) = V(s, y(t, s, x(s))), \quad t_0 \le s \le t$$

so that $m(t_0) = V(t_0, y(t, t_0, x_0))$. Then using the assumption (II) and (i), it is easy to obtain

$$D_{-}m(s) \le g(s, m(s)), \quad t_0 \le s \le t$$

which yields by Theorem 1.7.1 the estimate

$$m(s) \le r(s, t_0, u_0), \quad t_0 \le s \le t \tag{4.10.7}$$

provided $m(t_0) \le u_0$. Since $m(t) = V(t, y(t, t, x(t))) = V(t, x(t, t_0, x_0))$, the desired result (4.10.6) follows from (4.10.7) by setting $s = t$.

Taking $u_0 = V(t_0, y(t, t_0, x_0))$, the inequality (4.10.6) becomes

$$V(t, x(t, t_0, x_0)) \le r(t, t_0, V(t_0, y(t, t_0, x_0)), \quad t \ge t_0, \tag{4.10.8}$$

which shows the connection between the solutions of systems (4.10.1) and (4.10.2) in terms of the maximal solution of (4.10.5).

A number of remarks can be made:

(1) The trivial function $f(t, y) \equiv 0$ is admissible in Theorem 4.10.1 to yield the estimate (4.10.6) provided $V(t_0, x_0) \le u_0$. In this case $y(t, t_0, x_0) = x_0$ and the hypothesis (H) is trivially verified. Since $y(t, s, x) = x$, the definition (4.10.3) reduces to

$$D_{-}V(s, x) = \lim_{h \to 0^-} \inf \frac{1}{h}[V(s + h, x + hF(s, x)) - V(s, x)] \tag{4.10.9}$$

which is the usual definition of generalised derivative of the Lyapunov function relative to the system (4.10.2). Consequently, Theorem 4.10.1 reduces, in this special case, to Theorem 4.3.1.

(2) Suppose that $f(t, y) = A(t)y$ where $A(t)$ is an $n \times n$ continuous matrix. The solution $y(t, t_0, x_0)$ of (4.10.1) then satisfies $y(t, t_0, x_0) = \Phi(t, t_0)x_0$, where $\Phi(t, t_0)$ is the fundamental matrix solution of $y' = A(t)y$, with $\Phi(t_0, t_0) = I$ (identity matrix). The assumption (H) is clearly verified. Suppose also that $g(t, u) \equiv 0$. Then (4.10.6) yields

$$V(t, x(t, t_0, x_0)) \le V(t_0, \Phi(t, t_0)x_0), \quad t \ge t_0. \tag{4.10.10}$$

If, on the other hand, $g(t, u) = Bu$ where $B = (b_{ij})$ is an $N \times N$ matrix such that $b_{ij} \ge 0$ for $i \ne j$, we get a sharper estimate

$$V(t, x(t, t_0, x_0)) \le V(t_0, \Phi(t, t_0)x_0) \exp(B(t - t_0)), \quad t \ge t_0. \tag{4.10.11}$$

Clearly the relation (4.10.11) helps in improving the behavior of solutions of (4.10.2). relative to the behavior of solutions of (4.10.1). This is a great asset in perturbation theory and it can be seen by setting $F(t, x) = f(t, x) + R(t, x)$ where $R(t, x)$ is the perturbation term.

(3) Suppose that $f(t, y)$ is nonlinear, $f_y(t, y)$ exists and is continuous for $(t, y) \in R_+ \times R^n$. Then, it is well known that the solutions $y(t, t_0, x_0)$ are differentiable with respect to (t_0, x_0) and we have

$$\begin{cases} \dfrac{\partial y}{\partial t_0}(t, t_0, x_0) = -\Phi(t, t_0, x_0) f(t_0, x_0), \quad t \geq t_0, \\ \dfrac{\partial y}{\partial x_0}(t, t_0, x_0) = \Phi(t, t_0, x_0) \end{cases} \tag{4.10.12}$$

where $\Phi(t, t_0, x_0)$ is the matrix solution of the variational equation

$$z' = f_y(t, y(t, t_0, x_0))z.$$

If $V(s, x)$ is also assumed to be differentiable, then by (4.10.12), we have, for a fixed t,

$$\begin{aligned} D\_V(s, y(t, s, x)) &\equiv V_s(s, y(t, s, x)) \\ + V_x(s, y(t, s, x)) &\cdot \Phi(t, s, x) \cdot [F(s, x) - f(s, x)]. \end{aligned} \tag{4.10.13}$$

The relation (4.10.13) gives an intuitive feeling of the definition (4.10.3).

(4) When the solutions of (4.10.1) are known, a possible Lyapunov function for (4.10.2) is

$$W(s, x) = V(s, y(t, s, x)) \tag{4.10.14}$$

where $V(s, x)$ and $y(t, s, x)$ are as before.

As an application of Theorem 4.10.1, we shall consider some results on practical stability of the system (4.10.2).

Theorem 4.10.2. *Assume that (H) holds and (i) of Theorem 4.10.1 is verified. Suppose that $g \in C(R_+ \times R_+^N, R^N)$, $g(t, u)$ is quasimonotone nondecreasing in u and for $(t, x) \in R_+ \times S(A)$,*

$$b(\|x\|) \leq V_0(t, x) \leq a(\|x)\|, a, b \in \kappa, \tag{4.10.15}$$

where $V_0(t, x) = \sum_{i=1}^{N} V_i(t, x)$. Furthermore, suppose that $0 < \lambda < A$ are given and $a(\lambda) < b(A)$. If the unperturbed system (4.10.1) is (λ, λ) practically stable, then the practical stability properties of (4.10.1) imply the corresponding practical stability properties of the perturbed system (4.10.2).

Proof. Assume that (4.10.5) is strongly practically stable. Then, we have, given $(\lambda, A, B, T) > 0$ such that $\lambda < A, B < A$,

$$\sum_{i=1}^{N} u_i(t, t_0, u_0) < b(A), t \geq t_0 \text{ if } \sum_{i=1}^{N} u_{0i} < a(\lambda) \tag{4.10.16}$$

and

$$\sum_{i=1}^{N} u_{0i} < a(\lambda) \quad \text{implies} \quad \sum_{i=1}^{N} u_i(t, t_0, u_0) < b(B), t \geq t_0 + T. \tag{4.10.17}$$

Since (4.10.1) is (λ, λ) practically stable, we have

$$\|y(t, t_0, x_0)\| < \lambda, t \geq t_0, \text{ if } \|x_0\| < \lambda. \tag{4.10.18}$$

We claim that $\|x_0\| < \lambda$ also implies that $\|x(t, t_0, x_0)\| < A$, $t \geq t_0$, where $x(t, t_0, x_0)$ is any solution of (4.10.2). If this is not true, there would exist a solution $x(t, t_0, x_0)$ of (4.10.2) with $|x_0| < \lambda$ and a $t_1 > t_0$ such that $\|x(t_1, t_0, x_0)\| \leq A, t_0 \leq t \leq t_1$. Then by Theorem 4.10.1, we have

$$V(t, x(t, t_0, x_0)) \leq r(t, t_0, V(t_0, y(t, t_0, x_0))), t_0 \leq t \leq t_1.$$

Consequently, we get

$$\begin{aligned} b(A) \leq V_0(t_1, x(t_1, t_0, x_0)) &\leq \sum_{i=1}^{N} r_i(t_1, t_0, a(\|y(t_1, t_0, x_0)\|)) \\ &\leq \sum_{i=1}^{N} r_i(t_1, t_0, a(\lambda)) < b(A). \end{aligned}$$

This contradiction proves that

$$\|x_0\| < \lambda \text{ implies } \|x(t)\| < A, t \geq t_0.$$

To show strong practical stability, we see from the foregoing argument that we have

$$b(\|x(t, t_0, x_0)\|) \leq V_0(t, x(t, t_0, x_0)) \leq \sum_{i=1}^{N} r_i(t, t_0, V(t_0, y(t, t_0, x_0)))$$

for all $t \geq t_0$, if $\|x_0\| < \lambda$. From this it follows that

$$b(\|x(t, t_0, x_0)\|) \leq \sum_{i=1}^{N} r_i(t, t_0, a(\lambda)), t \geq t_0.$$

Now (4.10.17) yields the strong practical stability of the system (4.10.2) and the proof is complete.

Setting $F(t,x) = f(t,x) + R(t,x)$ in Theorem 4.10.2, we see that although the unperturbed system (4.10.2) is only practically stable, the perturbed system (4.10.2) is strongly practically stable, an improvement caused by the perturbing term.

Let us present a simple but illustrative example.

$$x' = e^{-t}x^2, x(t_0) = x_0, y' = -y, y(t_0) = y_0, \tag{4.10.19}$$

whose solutions are given by

$$x(t,t_0,x_0) = \frac{x_0}{1 + x_0(e^{-t} - e^{-t_0})}, y(t,t_0,y_0) = y_0 e^{-(t-t_0)}, t \geq t_0. \tag{4.10.20}$$

The fundamental matrix solutions of the corresponding variational equations are

$$\phi(t,t_0,x_0) = \frac{1}{[1 + x_0(e^{-t} - e^{-t_0})]^2}, \psi(t,t_0,y_0) = e^{-(t-t_0)}.$$

Consequently, choosing $V_1(t,x) = x^2, V_2(t,y) = y^2$, we see that

$$V_1'(t,x) = 2x(t,s,x)\phi(t,s,x)R(s,x,y),$$

$$V_2'(t,y) = 2y(t,s,y)\psi(t,s,y)L(s,x,y),$$

where R, L are perturbations so that the perturbed differential system is given by

$$\begin{cases} x' = e^{-t}x^2 + R(t,x,y), x(t_0) = x_0, \\ y' = -y + L(t,x,y), y(t_0) = y_0. \end{cases} \tag{4.10.21}$$

Let $R(t,x,y) = \frac{-x^2}{2}$ and $L(t,x,y) = \frac{-yx^2}{2}$. Then it is easy to compute

$$g_1(t,V_1,V_2) = -V_1^{\frac{3}{2}}, g_2(t,V_1,V_2) = -V_1V_2,$$

so that the comparison system reduces to

$$\begin{cases} u_1' = -u_1^{\frac{3}{2}}, u_1(t_0) = u_{10}, \\ u_2' = -u_1u_2, u_2(t_0) = u_{20}. \end{cases} \tag{4.10.22}$$

Choosing $u_{10} = V_1(t_0, y(t,t_0,x_0)), u_{20} = V_2(t_0, y(t,t_0,x_0))$ we find the solutions of (4.10.22) are given by

$$u_1(t,t_0,u_0) = \frac{4u_{10}}{[2 + u_{10}^{\frac{1}{2}}(t-t_0)]^2},$$

$$u_2(t,t_0,u_0) = u_{20}exp[-\frac{2u_{10}(t-t_0)}{2 + u_{10}^{\frac{1}{2}}(t-t_0)}], t \geq t_0.$$

Thus by Theorem 4.10.1, the estimates for the solutions $\overline{x}(t,t_0,x_0,y_0), \overline{y}(t,t_0,x_0,y_0)$ of (4.10.21), are of the form

$$\|\overline{x}(t,t_0,x_0,y_0)\|^2 \leq \frac{\|x_0\|^2}{[1 + x_0((e^{-t} - e^{-t_0}) + (\frac{t-t_0}{2}))]^2},$$

$$\|\overline{y}(t, t_0, x_0, y_0)\|^2 \leq \|y_0\|^2 e^{-2(t-t_0)} exp[\frac{\|x_0\|^2(t-t_0)}{[1+x_0(e^{-t}-e^{-t_0}+(\frac{t-t_0}{2}))][1+x_0(e^{-t}-e^{-t_0}}$$

for $t \geq t_0$, which show that all solutions $\overline{x}(t, t_0, x_0, y_0), \overline{y}(t, t_0, x_0, y_0) \to 0$ as $t \to \infty$, although from (4.10.20), it is clear that the solutions of the unperturbed system (4.10.19) do not enjoy this nice property. In fact, for $t_0 = 0$ and $x_0 = 1$, we get

$$x(t, t_0, x_0) = e^t, y(t, t_0, y_0) = e^{-t}, t \geq 0.$$

4.11 Quasisolutions

Recall that decomposition of (4.9.1) in the form (4.9.2) was crucial for the method of vector Lyapunov functions to be effective for structurally perturbed systems. Also, as we have seen, quasimonotone property of the comparison system is required when we utilize several Lyapunov functions and the corresponding theory of differential inequalities. Since the comparison systems with the desired stability property exist without satisfying quasimonotone property, this requirement of quasimonotone property is a drawback in applications. To circumvent this unpleasant fact, the notion of quasisolutions is helpful in certain situations. Consequently, we shall discuss quasisolutions in this section.

Let us consider the comparison system

$$u' = g(t, u), u(t_0) = u_0 \geqslant 0, \tag{4.11.1}$$

where $g \in C(R_+ \times R_+^N, R^N)$. To define quasisolutions of (4.11.1), we fix for each i, $1 \leqslant i \leqslant N$, two nonnegative integers p_i, q_i such that $p_i + q_i = N - 1$ and split $u \in R^N$ into

$$u = (u_i, [u]_{p_i}, [u]_{q_i}).$$

Then (4.11.1) becomes

$$u_i' = g_i(t, u_i, [u]_{p_i}, [u]_{q_i}), 1 \leqslant i \leqslant N.$$

Definition 4.11.1. *Let $a \in C(R_+, R_+^N)$. Then $w \in C^1(R_+, R_+^N)$ is said to be a quasisolution of* (4.11.1) *with respect to $a(t)$ if*

$$w_i' = g_i(t, w_i, [w]_{p_i}, [a(t)]_{q_i}), w(t_0) = u_0 \geqslant 0. \tag{4.11.2}$$

If $q_i = 0$ for all i, then quasolutions are just solutions of (4.11.1). If $p_i = 0$ for all i, then quasisolutions that result are most useful since they can be determined most easily. It is also clear from Definition 4.11.1 that one can define maximal and minimal quasisolutions of (4.11.1) relative to $a(t)$.

Definition 4.11.2. *The function $g(t, u)$ is said to possess a mixed quasimonotone property (mqmp for short) if for each $i, g_i(t, u_i, [u]_{p_i}, [u]_{q_i})$ is monotone nondecreasing in $[u]_{p_i}$ and monotone nonincreasing in $[u]_{q_i}$.*

We then have the following comparison result.

Theorem 4.11.1. *Let $a \in C(R_+, R_+^N)$, $g \in C(R_+ \times R_+^N, R^N)$ and $g(t,u)$ has mqmp. Let $m \in C(R_+, R_+^N)$, $m(t) \geqslant a(t)$ and*

$$D^+m(t) \leqslant g(t, m(t)), t \geqslant t_0.$$

Then $m(t_0) \leqslant u_0$ implies that $m(t) \leqslant r(t), t \geqslant t_0$, where $r(t) = r(t, t_0, u_0)$ is maximal quasisolution of (4.11.2) relative to $a(t)$ existing on $[t_0, \infty)$.

Proof. Since g(t,u) has mqmp and $m(t) \geqslant a(t), t \in R_+$, we have

$$g_i(t, m_i, [m]_{p_i}, [m]_{q_i}) \leqslant g_i(t, m_i, [m]_{p_i}, [a]_{q_i}),$$

which implies

$$D^+m_i(t) \leqslant g_i(t, m_i, [m]_{p_i}, [a]_{q_i}).$$

The stated result follows by Theorem 1.7.1 since

$$\widetilde{g}_i(t, u) = g_i(t, u_i, [u]_{p_i}, [a]_{q_i})$$

satisfies the conditions of Theorem 1.7.1

Example 4.11.1. *Consider $u' = Au$ where $a_{ii} < 0, a_{ij} \leq 0$ for $j < i$ and $a_{ij} \geq 0$ for $j > i$. Let $a(t) \equiv 0$. Then the quasisolutions are determined by $v' = \widetilde{A}v$ where $\widetilde{a}_{ii} = a_{ii}$, $\widetilde{a}_{ij} = 0$ for $j < i$ and $\widetilde{a}_{ij} = a_{ij}$ for $j > i$. The matrix $\widetilde{A}$ is stable since it has eigenvalues $a_{11}, a_{22}, ..., a_{NN} < 0$.*

Example 4.11.2. *Consider the system*

$$x_1' = -x_1 + x_2^{\frac{1}{2}}, \quad x_1(0) = 2,$$
$$x_2' = 1 - x_1^2 - x_2, \quad x_2(0) = 1.$$

For $a(t) = (1, 0)$, we get quasisolution $r(t) = (2e^{\frac{-t}{2}}, e^{-t})$.

We shall show that the idea of quasisolutions leads automatically to isolated subsystems and consequently the decomposition needed is always possible. Let us rewrite (4.9.1) in the form

$$x_i' = f_i(t, x_1, x_2, ..., x_N), x_i(t_0) = x_{i0},$$

where, as before, $x = (x_1, x_2, ..., x_N), x_i \in R^{n_i}$ and $n = \sum_{i=1}^{N} n_i$. Let $a \in C(R_+, R^N)$ and consider the subsystems

$$x_i' = F_i(t, x_i),$$

where $F_i(t, x_i) = f_i(t, a_1(t), ..., a_{i-1}(t), x_i, a_{i+1}(t), ..., a_N(t))$ so that the resulting large scale system (4.9.1) can be written as

$$x_i' = F_i(t, x_i) + R_i(t, x, a(t)),$$

where $R_i(t,x,a) = f_i(t,x) - f_i(t,a_1,...,a_{i-1},x_i,a_{i+1},...,a_N)$. Thus it is clear that with a judicial choice of the function a(t), the decomposition of (4.9.1) into the form (4.9.2) is always possible. This suggests that it would be nice to compute the error estimates between quasisolutions and the choice functions a(t), as well as, the solutions and quasisolutions, so that suitable choice of a(t) may ne made whenever possible. Let us obtain such error estimates for a simple situation.
Suppose that the function $f(t,x)$ of (4.9.1) satisfies a Lipschitz condition of the form

$$|f_i(t,x) - f_i(t,y)| \leq L \sum_{i=1}^{n} |x_i - y_i|.$$

Let $a \in C(R_+, R^n)$ and $y(t)$ be the quasisolution given by

$$y_i' = f_i(t, y_i, [y]_{p_i}, [a]_{q_i}), y(0) = x_0.$$

Then setting $v_i(t) = |y_i(t) - a_i(t)|$, we get the differential inequality

$$D^+ v_i(t) \leq L \sum_{j \in \{i,p_i\}} v_j(t) + \sigma_j(t), v_i(0) = |y_i(0) - a_i(0)|,$$

where $\sigma_i(t) = |f_i(t,a) - D^+ a_i(t)|$. Hence we have the estimate

$$v(t) \leq v(0)e^{At} + \int_0^t \sigma(s)e^{A(t-s)} \mathrm{d}s \equiv \delta(t), t \geq 0, \tag{4.11.3}$$

where the matrix A has for each i, q_i zeros in the ith row and the rest L's. We next let $m_i(t) = |x_i(t) - y_i(t)|$ where x(t) is the solution of (4.9.1). We note that $m(0) = 0$. Then we have

$$D^+ m_i(t) \leq L \sum_{j=1}^{n} m_j(t) + L \sum_{j \in \{p_i\}} v_j(t).$$

Consequently, using (4.11.3), we arrive at

$$D^+ m(t) \leq Bm(t) + \eta(t), m(0) = 0,$$

where $b_{ij} = L$ and $\eta_i(t) = L \sum_{j \in \{q_i\}} \delta_j(t)$. It therefore follows that

$$m(t) \leq \int_0^t \eta(s)e^{B(t-s)} \mathrm{d}s, t \geq 0. \tag{4.11.4}$$

The estimates (4.11.3) and (4.11.4) are the desired error estimates.

4.12 Cone-valued Lyapunov Functions.

As we have noted in Section 4.11, an unpleasant fact in the approach of several Lyapunov functions is the requirement of quasimonotone property of the comparison system. In Section 4.11, we have utilized the notion of quasisolutions to circumvent this drawback. In this section, we analyze the choice of cones used. We observe that this difficulty is due to the choice of the cone relative to the comparison system, namely, R^N_+, the cone of nonnegative elements of R^N and a possible answer lies in choosing a suitable cone other than R^N_+ to work in a given situation.

Using the comparison results Theorems 4.3.4 and 4.3.5, it is now easy to discuss the method of cone valued Lyapunov functions. We shall merely state two typical results.

Theorem 4.12.1. *Assume that*

(i) $V \in C(R_+ \times S(\rho), K)$, *$V(t,x)$ is locally Lipschitsian in x relative to the cone $K \subset R^N$ and for $(t,x) \in R_+ \times S(\rho)$,*

$$D^+V(t,x) \leq_K g(t, V(t,x));$$

(ii) $g \in C(R_+ \times K, R^N)$, *$g(t,0) \equiv 0$ and $g(t,u)$ is quasimonotone nondecreasing in u relative to cone K;*

(iii) $f(t,0) \equiv 0$ *and for some $\psi_0 \in K_0^*, \psi_0(V(t,x))$ is positive definite and decrescent for $(t,x) \in R_+ \times S(\rho)$, where $K_0 = K \backslash \{0\}$ and K_0^* is the adjoint of K_0.*

Then, the stability properties of the trivial solution $u = 0$ of

$$u' = g(t,u), u(t_0) = u_0, \tag{4.12.1}$$

imply the corresponding stability properties of the trivial solution $x = 0$ of (4.12.1)

The following version of Theorem 4.12.1 is in a more flexible setting so as to be useful in applications.

Theorem 4.12.2. *Assume that*

(i) P and Q are two cones in R^N such that $P \subset Q$.

(ii) $V \in C(R_+ \times S(\rho), Q)$, *$V(t,x)$ is locally Lipschitizian in x relative to P and*

$$D^+V(t,x) \leq_P g(t, V(t,x)), (t,x) \in R_+ \times S(\rho);$$

(iii) $g \in C(R_+ \times Q, R^N)$, *$g(t,0) \equiv 0$ and $g(t,u)$ is quasimonotone nondecreasing in u relative to P; and*

(iv) $f(t,0) \equiv 0$ *and for some* $\psi_0 \in Q_0^*, \psi_0(V(t,x))$ *is positive definite and decrescent for* $(t,x) \in R_+ \times S(\rho)$.

Then the stability properties of the trivial solution $u = 0$ *of* (4.12.1) *imply the corresponding stability properties of* $x = 0$ *of* (4.9.1).

If $K = R_+^N, \psi_0 = (1, 1, ..., 1)$, we obtain Theorem 4.5.1 from Theorem 4.12.1 since $\psi_0(V(t,x)) = \sum_{i=1}^{N} V_i(t,x) = V_0(t,x)$. One could also use other measures in place of $\psi_0(V(t,x))$. For example, let $\Phi \in C(K, R_+), \Phi(u)$ is nondecreasing in u relative to K. Then it is enough to suppose $\Phi(V(t,x))$ be positive definite and decrescent in Theorem 4.12.1. Moreover, if $P \subset Q = R_+^N$ in Theorem 4.12.2, the unpleasant fact concerning the quasimonotonicity of $g(t,u)$ mentioned earlier can be removed. This, of course, means that we have to choose an appropriate cone P which necessarily depends on the nature of $g(t,u)$. Let us demonstrate this by means of a simple example.

Example 4.12.1. *Consider the comparison system*

$$\begin{cases} u_1' = a_{11}u_1 + a_{12}u_2, u_1(t_0) = u_{10}, \\ u_2' = a_{21}u_1 + a_{22}u_2, u_2(t_0) = u_{20}. \end{cases} \tag{4.12.2}$$

Let $Q = R_+^2$. Suppose that we do not demand a_{12}, a_{21} to be nonnegative. Then, the function $g(t,u)$ violates the quasimonotone nondecreasing condition in $u = (u_1, u_2)$ relative to Q. Hence, the differential inequalities

$$\begin{cases} D^+V_1(t,x) \leq g_1(t, V_1(t,x), V_2(t,x)), \\ D^+V_2(t,x) \leq g_2(t, V_1(t,x), V_2(t,x)). \end{cases} \tag{4.12.3}$$

do not yield the componentwise estimates of $V(t, x(t))$ in terms of the maximal solution of (4.12.1).

Suppose now that there exist two numbers α, β such that $0 < \beta < \alpha$ and

$$\alpha^2 a_{21} + \alpha a_{22} \geq \alpha a_{11} + a_{12}, \tag{4.12.4}$$

$$\beta^2 a_{21} + \beta a_{22} \leq \beta a_{11} + a_{12}. \tag{4.12.5}$$

These conditions can hold with no restriction of nonnegativity of a_{21} and a_{12}. We shall now choose the cone $P \subset Q = R_+^2$ defined by

$$P = \{u \in R_+^2 : \beta u_2 \leq u_1 \leq \alpha u_2\}.$$

This cone has two boundaries $\alpha u_2 = u_1$ and $\beta u_2 = u_1$. On the boundary $\alpha u_2 = u_1$, we take $\psi = (-\frac{1}{\alpha}, 1)$ so that $< (-\frac{1}{\alpha}, 1), (u_1, \frac{u_1}{\alpha}) >= 0$ and

$$< (-\frac{1}{\alpha}, 1), (a_{11}u_1 + a_{12}\frac{u_1}{\alpha}, a_{21}u_1 + a_{22}\frac{u_1}{\alpha}) > \geq 0 \text{ for all } u \neq 0.$$

This reduces to the condition (4.12.4). Similarly, we can obtain (4.12.5). Thus, if the inequalities (4.12.3) are relative to P, we obtain the componentwise estimates on V as

$$V_i(t, x(t)) \leq r_i(t, t_0, V(t_0, x_0)), \tag{4.12.6}$$

by Theorem 1.7.3. We note that the estimate (4.12.6) is precisely the one we would have obtained if $a_{12}, a_{21} \geq 0$, by the standard method of vector Lyapunov functions. Since a_{12}, a_{21} need not be nonnegative in our example, the usefulness of conevalued Lyapunov function is clear.

In some situations, it may be difficult to choose all the components of the vector Lyapunov function to be nonnegative and to satisfy necessary requirements for the method of vector Lyapunov functions. This would dictate the use of a suitable measure for the vector Lyapunov function and partial stability properties for the comparison system. We shall therefore consider, in this section, such a possibility and develop suitable mechanism.

Let us consider the differential system

$$x' = f(t, x), x(t_0) = x_0, \tag{4.12.7}$$

where $f \in C(R_+ \times S(\rho), R^n)$ and prove the following result.

Theorem 4.12.3. *Assume that*

(i) for $1 \leq p < N, V \in C(R_+ \times S(\rho), R_+^p \times R^{N-p})$, $V(t,x)$ *is locally Lipschitzian in* x *and*

$$D^+V(t,x) \leq g(t, V(t,x)), (t,x) \in R_+ \times S(\rho),$$

where $g \in C(R_+ \times R_+^p \times R^{N-p}, R^N)$ *and* $g(t,u)$ *is quasimonotone nondecreasing in* u*;*

(ii) $b(\|x\|) \leq \sum_{i=1}^{p} V_i(t,x)$, $\sum_{i=1}^{p} V_i(t,x) + \sum_{i=p+1}^{N} \|V_i(t,x)\| \leq a(\|x\|)$ *for* $(t,x) \in R_+ \times S(\rho)$, *where* $a, b \in \kappa$.

Then p-partial stability properties of the trivial solution of

$$u' = g(t,u), u(t_0) = u_0, \tag{4.12.8}$$

imply the corresponding stability properties of the trivial solution of the system (4.12.7).

Proof. Let $0 < \varepsilon < \rho$ and $t_0 \in R_+$ be given and suppose that the trivial solution of (4.12.8) is p-partially equistable. Then given $b(\varepsilon) > 0$ and $t_0 \in R_+$, there exists a $\delta_1 = \delta_1(t_0, \varepsilon) > 0$ such that

$$\sum_{i=1}^{p} u_{0i} + \sum_{i=p+1}^{N} \|u_{0i}\| < \delta_1 \quad \text{implies} \quad \sum_{i=1}^{p} u_i(t, t_0, u_0) < b(\varepsilon), t \geq t_0, \tag{4.12.9}$$

where $u(t, t_0, u_0)$ is any solution of (4.12.8) with $u_{i0} \geq 0, i = 1, 2, ..., p$ and u_{i0} arbitrary for $i = p + 1, ..., N$. Choose $u_0 = V(t_0, x_0)$ and let $\delta = \delta(t_0, \varepsilon) > 0$ be selected such that $a(\delta) < \delta_1$. Suppose that $\|x_0\| < \delta$. Then we claim that $\|x(t)\| < \varepsilon, t \geq t_0$, where $x(t)$ is any solution of (4.12.7). If this is not true, then there would exist a solution x(t) of (4.12.7) with $\|x_0\| < \delta$ and a $t_1 > t_0$ satisfying

$$\|x(t_1)\| = \varepsilon \text{ and} \|x(t)\| \leq \varepsilon, t_0 \leq t \leq t_1. \tag{4.12.10}$$

Hence by Theorem 4.3.1, we get

$$V(t, x(t)) \leq r(t, t_0, u_0), t_0 \leq t \leq t_1, \tag{4.12.11}$$

where $r(t, t_0, u_0)$ is the maximal solution of (4.12.8). Since (ii) and (4.12.9) show, in view of the choice of u_0, that

$$\sum_{i=1}^{p} u_{0i} + \sum_{i=p+1}^{N} \|u_{0i}\| < \delta_1,$$

the relations (4.12.9), (4.12.10) and (4.12.11) yield

$$b(\varepsilon) \leq \sum_{i=1}^{p} V_i(t, x(t)) \leq \sum_{i=1}^{p} r_i(t, t_0, u_0) < b(\varepsilon).$$

This is a contradiction which proves equistability of (4.12.7)

We remark that if $p = N$, then Theorem 4.12.3 is precisely Theorem 4.5.1 As an illustration of the situation discussed in the Theorem 4.12.3, we shall consider higher derivatives of a single Lyapunov function. This can be done provided that $f(t, x)$ is smooth enough. hence let us suppose that $f \in C^{N-1}(R_+ \times S(\rho), R^n)$. Let $V \in C^{(N)}(R_+ \times S(\rho), R_+)$ and define

$$V'(t, x) = V_t(t, x) + V_x(t, x) f(t, x),$$
$$V''(t, x) = V_t'(t, x) + V_x'(t, x) f(t, x),$$

and, in general

$$V^{(N)}(t, x) = V_t^{(N-1)}(t, x) + V_x^{(N-1)}(t, x) f(t, x).$$

Assume that we have the estimate

$$V^{(N)}(t, x) \leq g_0(t, V(t, x), V^1(t, x), ..., V^{(N-1)}(t, x)), \tag{4.12.12}$$

for $(t, x) \in R_+ \times S(\rho)$, where $g_0 \in C(R_+^2 \times R^{N-2}, R)$. Then we can prove the following result.

Theorem 4.12.4. *Suppose that*

(i) f, V and g_0 are defined above and (4.12.12) holds;

(ii) $g_0(t, u_1, u_2, ..., u_N)$ *is nondecreasing in* $u_1, u_2, ..., u_{N-1}$, $b(\|x\|) \leq V(t,x)$ *and*

$$V(t,x) + |V'(t,x)| + ... + |V^{(N-1)}(t,x)| \leq a(\|x\|)$$

for $(t,x) \in R_+ \times S(\rho)$, *where* $a, b \in \kappa$.

Then the stability properties of the trivial solution of

$$u^{(N)} = g_0(t, u, u', u'', ..., u^{(N-1)}), u^i(t_0) = u_{0i}, i = 0, 1, ..., N-1, \quad (4.12.13)$$

imply the corresponding stability properties of the trivial solution of (4.12.7).

Proof. We wish to reduce this theorem to Theorem 4.12.3. We set

$$V_1(t,x) = V(t,x), V_2(t,x) = V'(t,x), ..., V_N(t,x) = V^{(N-1)}(t,x),$$

so that we have the vector Lyapunov function $V = (V_1, V_2, ..., V_N)$. Also (4.12.13) implies that

$$V'(t,x) \leq g(t, V(t,x)), (t,x) \in R_+ \times S(\rho),$$

where $g_i(t,u) = u_{i+1}, i = 1, 2, ..., N-1$,

$$g_N(t,u) = g_0(t, u_1, u_2, ..., u_N).$$

Clearly g(t,u) satisfies the quasimonotone property in u. If we let $p = 1$, it is easy to see that all the assumptions of Theorem 4.12.3 are satisfied and hence the conclusion follows.

If, in Theorem 4.12.3, the function g(t,u) does not satisfy quasimonotone condition, we can still obtain the same conclusion of the theorem in some situations. The next result deals with this special case.

Theorem 4.12.5. *Assume that the hupothesis of Theorem 4.12.3 hold except that g(t,u) is not assumed to be quasimonotone. Suppose that there exists a nonsingular* $N \times N$ *matrix* B *such that* $B^{-1} \geq 0$ *and* $G(t,u) = B^{-1}g(t, Bu)$ *is quasimonotone nondecreasing in* u. *Then the conclusion of Theorem 4.12.3 remains true.*

Proof. The transformation $w = B^{-1}u$ yields in view of assumptions on B,

$$w' \leq B^{-1}g(t,u) = B^{-1}g(t, Bw) \equiv G(t,w),$$

which implies that the stability properties of

$$w' = G(t,w), w(t_0) = w_0, \quad (4.12.14)$$

and of (4.12.8) are equivalent. Hence the proof.

Let us give an example to demonstrate Theorem 4.12.5. Suppose that in Theorem 4.12.4, $p = 1$, $N = 2$ and

$$g_0(t, u, u') = -2\beta u' - \kappa^2 u, \beta \geq \kappa,$$

so that $g_1(t, u_1, u_2) = u_2, g_2(t, u_1, u_2) = -2\beta u_2 - \kappa^2 u_1$. Choosing

$$B^{-1} = \begin{pmatrix} 1 & 0 \\ \beta & 1 \end{pmatrix}$$

we find that

$$G_1(t, w_1, w_2) = -\beta w_1 + w_2, G_2(t, w_1, w_2) = (\beta^2 - \kappa^2)w_1 - \beta w_2.$$

Hence, it is easy to see that the trivial solution of (4.12.14) is exponentially asymptotically stable, since the matrix

$$\begin{pmatrix} -\beta, & 1 \\ (\beta^2 - \kappa^2), & -\beta \end{pmatrix}$$

is a stability matrix and thus $G(t, w)$ satisfies quasimonotone property. Consequently the comparison system (4.12.12) is partially asymptotically stable with $p = 1$ and therefore Theorem 4.12.5 shows that the trivial solution of (4.12.7) is equiasymptotically stable.

4.13 New Directions.

In this section, we initiate a new approach to the method of vector Lyapunov functions which helps in distributing the burden between groups of components of the vector Lyapunov function and the comparison function. As a result, this approach contributes to the enrichment of the method of vector Lyapunov functions by including and improving earlier results and enhancing the applicability of the method.

For convenience, we split a vector $u \in R^N$ such that $u = (u_p, u_q)$ where $p + q = N < n, u_p, u_q$ denoting groups of components of u.

Definition 4.13.1. *Let $Q \in C(R_+^d, R_+)$ with $Q(0) = 0$ and $Q(u)$ nondecreasing in u. Then we say that $Q \in K(R_+^d, R_+)$.*

Definition 4.13.2. *Let $Q_1 \in K(R_+^p, R_+)$, $Q_2 \in K(R_+^q, R_+)$ and $u(t, t_0, u_0)$ be any solution of (4.12.7) existing for all $t \geq t_0$. Then the zero solution of (4.12.7) is said to be equiuniform stable if for given $\varepsilon_1 > 0, \varepsilon_2 > 0$ and $t_0 \in R_+$, there exist $\delta_1 = \delta_1(t_0, \varepsilon_1) > 0, \delta_2 = \delta_2(\varepsilon_2) > 0$ such that*

$$Q_1(u_{0p}) < \delta_1 \text{ implies } Q_1(u_p(t, t_0, u_0)) < \varepsilon_1, t \geq t_0$$

and

$$Q_2(u_{0q}) < \delta_2 \text{ implies } Q_2(u_q(t, t_0, u_0)) < \varepsilon_2, t \geq t_0.$$

We are now in a position to prove the following results.

Theorem 4.13.1. *Assume that*

(H_1) $V \in C(R_+ \times S(\rho), R_+^N)$, $V(t,x)$ is locally Lipschitzian in x, satisfying

$$b(\|x\|) \leq Q_2(V_q(t,x)) \leq a_0(\|x\|) + a_1(Q_1(V_p(t,x)))$$

for $(t,x) \in R_+ \times S(\rho) \bigcap S^c(\eta)$ for every $0 < \eta < \rho$ and

$$Q_1(V_p(t,0)) \equiv 0,$$

where $Q_1 \in K(R_+^p, R_+)$, $Q_2 \in K(R_+^q, R_+)$ and $b, a_0, a_1 \in \kappa$ with $p + q = N$;

(H_2) $g \in C(R_+ \times R_+^N, R^N)$, $g(t,u)$ is quasimonotone nondecreasing in u and satisfies

(i) $D^+V_p(t,x) \leq g_p(t, V_p(t,x), 0), (t,x) \in R_+ \times S(\rho)$, and

(ii) $D^+V_q(t,x) \leq g_q(t, V(t,x)), (t,x) \in R_+ \times S(\rho) \bigcap S^c(\eta)$,

for every $0 < \eta < \rho$;

(H_3) the zero solution of (4.12.8) is equi-uniform stable.

Then, the zero solution of (4.12.7) is equistable.

Proof. Let $0 < \varepsilon < \rho$ and $t_0 \in R_+$ be given. By hypothesis (H_3), given $\varepsilon_1 > 0, \varepsilon_2 > 0$ and $t_0 \in R_+$, there exist $\delta_{10} = \delta_{10}(t_0, \varepsilon_1) > 0$ and $\delta_{20} = \delta_{20}(\varepsilon_2) > 0$ such that

$$\begin{cases} Q_1(u_{0p}) < \delta_{10} \text{ implies } Q_1(u_p(t, t_0, u_0)) < \varepsilon_1, t \geq t_0 \text{ and} \\ Q_2(u_{0q}) < \delta_{20} \text{ implies } Q_2(u_q(t, t_0, u_0)) < \varepsilon_2, t \geq t_0. \end{cases} \tag{4.13.1}$$

Choose $V_p(t_0, x_0) = u_{0p}$ and $\delta_2 = \delta_2(\varepsilon) > 0$ such that

$$a_0(\delta_2) < \frac{1}{2}\delta_{20}. \tag{4.13.2}$$

Let $\varepsilon_2 = b(\varepsilon)$ and $\varepsilon_1 = a_1^{-1}(\frac{1}{2}\delta_{20})$.

Since $V(t,x)$ is continuous and $Q_1(V_p(t,0)) \equiv 0$, it follows that there exists a $\delta_1 = \delta_1(t_0, \varepsilon) > 0$ such that

$$Q_1(V_p(t_0, x_0)) < \delta_{10} \text{ and } \|x_0\| < \delta_1 \tag{4.13.3}$$

hold simultaneously.
Let $\delta = min(\delta_1, \delta_2)$. It is clear that δ depends on t_0 and ε. With this δ, we claim that the zero solution of (4.12.7) is equistable.
If this is not true, there exist $t_2 > t_1 > t_0$ and a solution $x(t) = x(t, t_0, x_0)$ of (4.12.7) with $\|x_0\| < \delta$ such that

$$\begin{cases} \|x(t_2)\| = \varepsilon < \rho, \ \|x(t_1)\| = \delta_2(\varepsilon) \text{ and} \\ x(t) \in S(\rho) \bigcap S^c(\eta) \text{ with } \eta = \delta_2(\varepsilon) > 0. \end{cases} \tag{4.13.4}$$

It then follows from (H_2) that

$$\begin{cases} D^+m_p(t) \leq g_p(t, m_p(t), 0), t_0 \leq t \leq t_2, \\ D^+m_q(t) \leq g_q(t, m(t)), t_1 \leq t \leq t_2, \end{cases} \tag{4.13.5}$$

where $m(t) = V(t, x(t))$. Hence by comparison Theorem 4.3.1, we have for $t_1 \leq t \leq t_2$,

$$m_p(t) \leq u_p(t, t_1, m(t_1)), m_q(t) \leq u_q(t, t_1, m(t_1)). \tag{4.13.6}$$

Let $u^*(t) = u(t, t_1, m(t_1)) \geq 0$ be the extension of $u(t)$ to the left of t_1 up to t_0 and let $u^*(t_0) = u_0^*$. Consider now the differential inequality which results from (4.13.5)

$$D^+m_p(t) \leq g_p(t, m_p(t), u_q^*(t)), u_p(t_0) = m_p(t_0),$$

which by comparison Theorem 4.3.1 yields

$$m_p(t) \leq u_p(t, t_0, u_0), t_0 \leq t \leq t_1, u_0 = (u_p(t_0), u_{0q}^*). \tag{4.13.7}$$

Then it is clear that $u(t) = (u_p(t, t_0, u_0), u_q^*(t, t_1, m(t_1)))$ is a solution of (4.13.2) on $[t_0, t_1]$. Using (4.13.4), (4.13.6) and (H_1), we obtain

$$b(\varepsilon) = b(\|x(t_2)\|) \leq Q_2(V_q(t_2, x(t_2))) \leq Q_2(u_q(t_2, t_1, m(t_1))). \tag{4.13.8}$$

But from (4.13.1) and (4.13.7), we get

$$Q_1(V_p(t_1, x(t_1))) \leq Q_1(u_p(t_1, t_0, u_0)) \leq a_1^{-1}(\frac{1}{2}\delta_{20}(t_0, \varepsilon)),$$

provided $Q_1(u_{0p}) < \delta_{10}$. From (H_1) (4.13.2) and (4.13.4), we now have

$$\begin{aligned} Q_2(V_q(t_1, x(t_1))) &\leq a_0(\|x(t_1)\|) + a_1(Q_1(V_p(t_1, x(t_1))) \\ &\leq a_0(\delta_2(\varepsilon)) + a_1(a_1^{-1}(\frac{1}{2}\delta_{20})) < \delta_{20} \end{aligned}$$

and therefore from (4.13.1), we get

$$Q_2(u_q(t_2, t_1, m(t_1))) < b(\varepsilon),$$

which contradicts (4.13.8). Hence the proof is complete.

Theorem 4.13.2. *Let the assumptions of Theorem 4.13.1 hold except that $(H_2)(i)$ is strengthened to*

$$D^+V_p(t, x) + W_p(t, x) \leq g_p(t, V_p(t, x), 0), (t, x) \in R_+ \times S(\rho), \tag{4.13.9}$$

where $W_p \in C(R_+ \times S(\rho), R_+)$, $g_i(t, u)$ is nondecreasing in u_i for $1 \leq i \leq p$ and for some fixed $p_0, 1 \leq p_0 \leq p$, $W_{p_0}(t, x) \geq b_0(\|x\|), b_0 \in \kappa$. Suppose further that

(H_3^*) *the trivial solution of* (4.13.2) *is uniformly stable;*

(H_4) *there exists a sufficiently small* $0 < \rho_0 < \rho$ *such that*

$$Q_1(V_p(t,x)) \leq a(\|x\|), (t,x) \in R_+ \times S(\rho_0) \text{ where } a \in \kappa.$$

Then the zero solution of (4.12.7) *is equiasymptotically stable.*

Proof. Since the assumptions of Theorem 4.13.1 hold, it follows that the zero solution of (4.12.7) is equistable. Let $\varepsilon = \rho$ so that we designate $\delta_0 = \delta_0(t_0, \rho)$. Then we have

$$\|x_0\| < \delta_0 \text{ implies } \|x(t,t_0,x_0)\| < \rho, t \geq t_0. \tag{4.13.10}$$

Now let $0 < \varepsilon < \rho_0$ and $t_0 \in R_+$ be given and let $\delta, \delta_1, \delta_2$ be the numbers corresponding to (t_0, ε) chosen as in the proof of Theorem 4.13.1. Note that all the δ's are now independent of t_0. Choose a $T = T(\varepsilon) > 0$ satisfying

$$Q_1((0,...,0,b_0(\delta(\varepsilon))T,0,...,0)) \geq a_1^{-1}(\frac{1}{2}\delta_{20}(\rho)). \tag{4.13.11}$$

We claim that there exists a $t^* \in [t_0, t_0 + T]$ such that

$$\|x(t^*, t_0, x_0)\| < \delta.$$

If this is false, suppose that $\|x(t,t_0,x_0)\| \geq \delta, t \in [t_0, t_0+T]$. Then setting

$$m_p(t) = V_p(t,x(t)) + \int_{t_0}^{t} W_p(s,x(s))ds, m_q(t) = V_q(t,x(t)) \tag{4.13.12}$$

where $x(t) = x(t,t_0,x_0)$, we get, in view of the fact $g(t,u)$ is quasimonotone nondecreasing in u and $g_i(t,u)$ is nondecreasing in u_i for $1 \leq i \leq p$, from (H_2) and (4.13.9),

$$D^+m(t) \leq g(t,m(t)), t \in [t_0, t_0+T].$$

By comparison Theorem 4.3.1, we then have, for $u_0 = V(t_0, x_0)$,

$$m_p(t) \leq u_p(t,t_0,u_0), m_q(t) \leq u_q(t,t_0,u_0), t \in [t_0, t_0+T]. \tag{4.13.13}$$

Since $W_{p_0}(t,x)$ is positive definite, we have $W_{p_0}(t,x(t)) \geq b_0(\delta(\varepsilon)), t \in [t_0, t_0+T]$ and therefore, it follows that

$$\int_{t_0}^{t_0+T} W_{p_0}(s,x(s))ds \geq b_0(\delta(\varepsilon))T. \tag{4.13.14}$$

Consequently, using the fact that $Q_1(u)$ is nondecreasing in u, we obtain from (4.13.12),(4.13.13) and (4.13.14), the relation

$$\begin{cases} Q_1((0,...,0,b_0(\delta(\varepsilon))T,0,...,0) \\ \leq Q_1(\int_{t_0}^{t_0+T} W_p(s,x(s))ds) \leq Q_1(V_p(t_0+T,x(t_0+T)))+ \\ (\int_{t_0}^{t_0+T} W_p(s,x(s))ds) \leq Q_1(u_p(t_0+T,t_0,u_0))). \end{cases} \tag{4.13.15}$$

Since $\|x_0\| < \delta_0(t_0, \rho)$, it follows that $Q_1(V_p(t_0, x_0)) < \delta_{10}(t_0, \rho)$, which implies, because of (4.13.1), the relation

$$Q_1(u_p(t_0 + T, t_0, V(t_0, x_0))) < a_1^{-1}(\frac{1}{2}\delta_{20}(\rho)).$$

Thus (4.13.15) yields

$$Q_1((0, ..., 0, b_0(\delta(\varepsilon))T, 0, ..., 0)) < a_1^{-1}(\frac{1}{2}\delta_{20}(\rho))$$

which contradicts the choice of T in (4.13.11). Therefore, it follows that there exists a $t^* \in [t_0, t_0 + T]$ such that

$$\|x(t^*, t_0, x_0)\| < \delta \leq \delta_2(\varepsilon). \tag{4.13.16}$$

We now claim that $\|x(t, t_0, x_0\| < \varepsilon, t \geq t_0 + T$. If this is not true, there exist $t_2 > t_1 > t^*$ such that

$$\begin{cases} \|x(t_2)\| = \varepsilon, \|x(t_1)\| = \delta_2(\varepsilon) \text{ and} \\ \delta_2 \leq \|x(t)\| \leq \varepsilon < \rho_0, t \in [t_1, t_2]. \end{cases} \tag{4.13.17}$$

We may now proceed as in the proof of Theorem 4.13.1 replacing t_0 by t^* to arrive at a contradiction. Hence it follows that $\|x(t)\| < \varepsilon, t \geq t_0 + T$ whenever $\|x_0\| < \delta_0(t_0, \rho)$. This proves the theorem completely.

Theorem 4.13.3. *Let the hypotheses of Theorem 4.13.1 and $(H_3^*), (H_4)$ of Theorem 4.13.2 hold except that $(H_2)(ii)$ is replaced by*

$$D^+V_q(t, x) + W_q(t, x) \leq g_q(t, V(t, x))$$

for $(t, x) \in R_+ \times S(\rho) \bigcap S^c(\eta), 0 < \eta < \rho$, where

$$W_q \in C(R_+ \times S(\rho) \bigcap S^c(\eta), R_+),$$

$g_i(t, u)$ is nondecreasing in u_i for $1 \leq i \leq q$ and for some fixed q_0, $1 \leq q_0 \leq q, W_{q_0}(t, x) \geq b_0(\|x\|), b_0 \in \kappa$. Then the zero solution of (4.7.1) is equiasymptotically stable.

The proof is similar to the proof of Theorem 4.13.2 with suitable modifications. We omit the details.

We shall next discuss several cases of Theorems 4.13.1, 4.13.2 and 4.13.3.

Theorem 4.13.4. *(a) In Theorem 4.13.1*

(i) if $q = 0$ so that $p = N$ and if $Q_1(V(t, x)) \geq b(\|x\|), b \in \kappa$ then the zero solution of (4.12.7) is equistable.

(ii) if $p = 0$, so that $q = N$, then the zero solution of (4.12.7) is uniformly stable.

(b) *In Theorem 4.13.3, if* $p = 0$, *then the zero solution of* (4.12.7) *is uniformly asymptotically stable.*

If $Q_1(u) = \sum_{i=1}^{N} u_i$, *then Theorem 4.13.4 a(i) reduces to Theorem 4.5.1.*

Theorem 4.13.5. *Assume that*

(i) $V_1 \in C(R_+ \times S(\rho), R+), V_1(t,x)$ *is locally Lipschitzian in* x, $V_1(t,0) \equiv 0$ *and*

$$D^+V_1(t,x) \leq g_1(t, V_1(t,x)), (t,x) \in R_+ \times S(\rho),$$

where $g_1 \in C(R_+^2, R), g_1(t,0) \equiv 0$;

(ii) *for every* $\eta > 0$, *there exists a family of Lyapunov functions* $V_{2\eta} \in C(R_+ \times S(\rho) \bigcap S^c(\eta), R_+), V_{2\eta}$ *locally Lipschitzian in* x *and for* $(t,x) \in R_+ \times S(\rho) \bigcap S^c(\eta)$,

$$b(\|x\|) \leq V_{2\eta}(t,x) \leq a(\|x\|), b, a \in \kappa$$

and

$$D^+V_1(t,x) + D^+V_{2\eta}(t,x) \leq g_2(t, V_1(t,x) + v_{2\eta}(t,x))$$

where $g_2 \in C(R_+^2, R), g_2(t,0) \equiv 0$;

(iii) *the null solution of* $u' = g_1(t,u), u(t_0) = u_0 \geq 0$ *is equistable and the null solution of* $v' = g_2(t,v), v(t_0) = v_0 \geq 0$ *is uniformly stable.*

Then the trivial solution of (4.12.7) *is equistable.*

Proof. In Theorem 4.13.1, let $p = 1 = q$, $V_2(t,x) = V_1(t,x) + V_{2\eta}(t,x)$, $Q_1(u_1) = u_1$, $Q_2(u_2) = u_2$, $g_1(t,u) = g_1(t,u_1,0)$, $g_2(t,u) = g_2(t,u_1,u_2)$. Then, the conclusion follows from Theorem 4.13.1.

Theorem 4.13.6. *In Theorem 4.13.2 if* $q = 0$ *so that* $p = N, (H_3^*), (H_4)$ *are deleted, and if for* $(t,x) \in R_+ \times S(\rho)$,

(a) $Q_1(V(t,x)) \geq b(\|x\|), b \in \kappa$,

(b) $W_{p_0}(t,x) \geq b_0(\|x\|), b_0 \in \kappa$ *and*

(c) $D^+W_{p_0}(t,x)$ *is bounded below or above,*

then the zero solution of (4.12.7) *is equiasymptotically stable.*

Proof. Since the assumptions of Theorem 4.13.4 hold, we have the equistability of the zero solution of (4.12.7). Take $\varepsilon = \rho$ and the set $\delta_0 = \delta_0(t_0,\rho)$, so that $\|x_0\| < \delta_0$ implies $\|x(t)\| < \rho, t \geq t_0$. In view of the positive definiteness of $W_{p_0}(t,x)$, it is enough to show that

$$\lim_{t\to\infty} W_{p_0}(t,x) = 0.$$

Since $\|x_0\| < \delta_0$, we also have

$$Q_1(V(t_0,x_0)) < \delta_1(t_0,\rho) \quad \text{implies} \quad Q_1(u(t,t_0,V(t_0,x_0))) < b(\rho). \tag{4.13.18}$$

We first prove that $\liminf_{t\to\infty} W_{p_0}(t, x(t)) = 0$. If this is not true, there exists a $\gamma > 0$ and sufficiently large T such that

$$W_{p_0}(t, x(t)) \geq \gamma \,\text{for}\, t \geq T.$$

As before, using (H_2), we obtain

$$V(t, x(t)) + \int_{t_0}^{t} W_{p_0}(s, x(s))\mathrm{d}s \leq u(t, t_0, V(t_0, x_0))$$

for $t \geq t_0$ and from the nondecreasing nature of $Q_1(u)$,

$$\begin{cases} Q_1((0, ..., 0, \int_T^t W_{p_0}(s, x(s))\mathrm{d}s, 0, ..., 0)) \\ \leq Q_1(V(t, x(t)) + \int_{t_0}^t W_{p_0}(s, x(s))ds) \\ \leq Q_1(u(t, t_0, V(t_0, x_0)), t \geq t_0. \end{cases} \tag{4.13.19}$$

From (4.13.18) and (4.13.19), we get

$$Q_1((0, ..., 0, \gamma(t - T), 0, ..., 0)) \leq Q_1(u(t, t_0, V(t_0, x_0))) < b(\rho).$$

Since $Q_1(u) \to \infty$ as $u \to \infty$, this leads to a contradiction as $t \to \infty$. Now, suppose $\limsup_{t\to\infty} W_{p_0}(t, x(t)) \neq 0$. Then, for any $\beta > 0$, there exist divergent sequences $\{t_k\}, \{t_k^*\}$ such that $t_k < t_k^* < t_{k+1}$, k=1,2,... and

$$\begin{cases} W_{p_0}(t_k, x(t_k)) = \frac{\beta}{2}, W_{p_0}(t_k^*, x(t_k^*)) = \beta, \\ and \\ \frac{\beta}{2} \leq W_{p_0}(t, x(t)) \leq \beta, t \in [t_k, t_k^*]. \end{cases} \tag{4.13.20}$$

From assumption (c), we have $D^+W_{p_0}(t, x) \leq M$. Then, as before, using (4.13.18) and (4.13.20), we arrive at

$$Q_1(0, ..., 0, \frac{\beta}{2}\sum_{i=1}^{n}(t_i^* - t_i), 0, ..., 0)$$

$$\leq Q_1(V(t_n^*, x(t_n^*)) + \sum_{i=1}^{n}\int_{t_i}^{t_i^*} W_{p_0}(s, x(s))\mathrm{d}s)$$

$$\leq Q_1(u(t, t_0, V(t_0, x_0))) < b(\rho).$$

As $n \to \infty$, this leads to a contradiction. This completes the proof. Consider the system

$$\begin{cases} x_1' = x_1 e^{-t} + x_2 sint - x_1 x_4^2, \\ x_2' = x_1 sint + x_2 e^{-t} - x_2 x_4^2, \\ x_3' = e^{-t} - x_2^2 x_3 \\ x_4' = (e^{-t} - 2)x_4 - x_1^2 x_4 \\ x_5' = -3x_5 - x_2^2 x_5. \end{cases} \tag{4.13.21}$$

It is not difficult to see that the system (4.13.21) satisfies stability notion with respect to the components x_1, x_2, is bounded relative to the component x_3 and is asymptotically stable with respect to the components x_4 and x_5. To deal with such a behavior of systems, it is clear using the norm as a candidate is not sensitive enough and therefore one needs to introduce a vector norm to investigate such situations. For this purpose, several Lyapunov functions are definitely suitable. Progress in this direction is minimal.

4.14 Lyapunov Functions in Cones and Stability Theory in Two Measures

Since it is now clear that employing cone-valued Lyapunov functions is beneficial in applications, we shall develop the method of cone-valued Lyapunov functions relative to stability theory in terms of two different measures so as to unify a variety of existing results and to provide greater flexibility in applications. We consider the differential system

$$x' = f(t,x), \quad x(t_0) = x_0, \quad t_0 \geq 0 \tag{4.14.1}$$

where $f \in C(R_+ \times R^n, R^n)$. Let $K \subseteq R^N$, be a cone, that is, K is closed, convex with $\lambda K \subset K$ for all $\lambda \geq 0$ and $K \bigcap (-K) = \theta$ with interior $K^0 \neq \varnothing$. For any $x, y \in R^N$, we let $x \leq y$ iff $y - x \in K$ and for any functions $u, v : R_+ \to R^N, u \leq v$ iff $u(t) \leq v(t)$ on R_+. Also, let $K^* = [\phi \in R^N : \phi(x) \geq 0$ for all $x \in K]$ and $K_0^* = K^* - [0]$.Let us now define the quasimonotone property.

Definition 4.14.1. *A function $F : R^N \to R^N$ is said to be quasimonotone nondecreasing relative to the cone $K \subseteq R^N$ if there exists a $\phi \in K_0^*$ such that*

$$x \leq y \text{ and } \phi(y - x) = 0 \text{ implies } \phi(F(y) - F(x)) \geq 0.$$

If $K = R_+^N$, then we have

$$x \leq y \text{ and } y_i = x_i \quad \text{for some} \quad 1 \leq i \leq N \text{ implies } F_i(y) - F_i(x) \geq 0,$$

which reduces to requiring nonnegative off-diagonal elements of A where $F(x) = Ax$ and A is an $N \times N$ matrix.

Let us recall the following classes of functions for future use.

$$\mathcal{K} = [a \in C(R_+, R_+) : a(0) = 0 \quad \text{and} \quad a(w) \quad \text{is increasing in w}],$$
$$\mathcal{CK} = [a \in C(R_+^2, R_+) : a(t,w) \in \mathcal{K} \text{ for each } t \in R_+]$$
$$\Gamma = [h \in C(R_+ \times R^n, R_+) : \ \inf h(t,x) = 0, (t,x) \in R_+ \times R^n],$$
$$\Gamma_0 = [h \in C(R_+ \times R^n, R_+) : \ \inf h(t,x) = 0 \quad \text{for each} \quad t \in R_+],$$
$$\Sigma = [Q \in C(K, R_+) : Q(0) = 0 \text{ and } Q(w) \quad \text{is increasing relative to K}].$$

Relative to the given system (4.14.1), we need to consider the comparison system

$$w' = g(t,w), w(t_0) = w_0 \geq 0, \tag{4.14.2}$$

where $g \in C(R_+ \times K, R^N)$.

We shall next define the stability concepts for the system (4.14.1) in terms of two measures $(h_0, h) \in \Gamma$. For this purpose, let $x(t) = x(t, t_0, x_0)$ be any solution of (4.14.1).

Definition 4.14.2. *The differential system* (4.14.1) *is said to be* (h_0, h)*-equistable, if given any* $\varepsilon > 0$ *and* $t_0 \in R_+$*, there exists a function* $\delta = \delta(t_0, \varepsilon) > 0$ *that is continuous in* t_0 *for each* ε *such that*

$$h_0(t_0, x_0) < \delta \quad \textit{implies} \quad h(t, x(t)) < \varepsilon, \quad t \geq t_0.$$

Based on this definition, one can formulate other notions of stability, boundedness and practical stability of the system (4.14.1)

We recall the different choices of the two measures (h_0, h) given in section 4.6 and the consequent generality of the definition 4.14.2. Moreover, (h_0, h)-stability concepts offer a unification of a variety of stability notions that exist in the literature.

Corresponding to definition 4.14.1, we require stability notions with respect to the comparison system (4.14.2)

Definition 4.14.3. *Let* $Q_0, Q \in \Sigma$*. Then we say that the comparison system* (4.14.2) *is* (Q_0, Q)*-equistable, if given an* $\varepsilon > 0$ *and* $t_0 \in R_+$*, there exists a positive function* $\delta = \delta(t_0, \varepsilon) > 0$ *that is continuous in* t_0 *for each* ε *such that*

$$Q_0(w_0) < \delta \quad \textit{implies} \quad Q(w(t)) < \varepsilon, \quad t \geq t_0,$$

where $w(t) = w(t, t_0, w_0)$ *is any solution of* (4.14.2)

Let us now introduce the cone-valued Lyapunov-like functions. The following comparison theorem plays an important role when we employ cone-valued Lyapunov-like functions.

Theorem 4.14.1. *Assume that:*

(i) $V \in C(R_+ \times R^n, K), V(t, x)$ *is locally Lipschitzian in* x *relative to* K *and for* $(t, x) \in R_+ \times R^n$,

$$D^+V(t, x) = lim_{h \to 0^+} sup \frac{1}{h}[V(t+h, x+hf(t, x)) - V(t, x)] \leq g(t, V(t, x));$$

(ii) $g \in C(R_+ \times K, R^N)$ *and* $g(t, w)$ *is quasimonotone nondecreasing in* w *relative to* K *for each* $t \in R_+$.

If $r(t) = r(t, t_0, w_0)$ *is the maximal solution of* (4.14.2) *relative to cone* K *and* $x(t) = x(t, t_0, x_0)$ *is any solution of* (4.14.1) *such that* $V(t_0, x_0) < w_0$*, then, on the common interval of existence, we have*

$$V(t, x(t)) \leq r(t).$$

Corollary 4.14.1. *In theorem 4.14.1, the function $g(t,w) \equiv 0$ is admissible to yield*

$$V(t,x(t)) \leq V(t_0,x_0)$$

for those $t \geq t_0$ for which $x(t)$ exists.

We shall develop results corresponding to Lyapunov theory in the present set up. We begin with the following result.

Theorem 4.14.2. *Assume that:*

(i) $h_0, h \in \Gamma$ and $h(t,x) \leq \Phi(h_0(t,x))$ if $h_0(t,x) < \rho_0$ for some $\rho_0 > 0$ where $\Phi \in \mathcal{K}$;

(ii) $Q_0, Q \in \Sigma$ and $Q(w) \leq \psi[Q_0(w)]$ if $Q_0(w) < \lambda$ for some $\lambda > 0$ where $\psi \in \mathcal{K}$;

(iii) $V \in C(R_+ \times R^n, K)$ and $V(t,x)$ is locally Lipschitzian in x with respect to K;

(iv) there exists a $\rho > 0$ such that $\Phi(\rho_0) < \rho$ satisfying

$$b(h(t,x)) \leq Q[V(t,x)] \quad if \quad h(t,x) < \rho,$$

$Q_0[V(t,x)] \leq a(t,h_0(t,x))$ if $h_0(t,x) < \rho_0$, where $b \in \mathcal{K}$ and $a \in C\mathcal{K}$;

(v) $D^+V(t,x) \leq 0$ for $(t,x) \in S(h,\rho) = [(t,x) : h(t,x) < \rho]$;

Then the differential system (4.14.1) is (h_0,h)-equistable. If in addition, a in (iv) is such that $a \in \mathcal{K}$, then (4.14.1) is (h_0,h) uniformly stable.

Proof. Let $0 < \varepsilon < min(\rho,\lambda)$ and $t_0 \in R_+$ be given. Let $w_0 = V(t_0,x_0)$. Choose $\delta = \delta(t_0,\varepsilon) < min(\rho_0,\lambda_0)$ with $a(t_0,\lambda_0) \leq \lambda$ and

$$\psi(a(t_0,\delta)) < b(\varepsilon). \tag{4.14.3}$$

Let $h_0(t_0,x_0) < \delta$ and note that

$$\begin{aligned} b(h(t_0,x_0) &\leq Q[V(t_0,x_0)] \leq \psi[Q_0(V(t_0,x_0))] \\ &\leq \psi[a(t_0,h_0(t_0,x_0))] \leq \psi(a(t_0,\delta)) < b(\varepsilon), \end{aligned}$$

which implies that

$$h(t_0,x_0) < \varepsilon. \tag{4.14.4}$$

We claim that with this δ, the system (4.14.1) is (h_0,h)-equistable. If not, because of (4.14.4), there exists a $t_1 > t_0$ and a solution $x(t) = x(t,t_0,x_0)$ of (4.14.1) with $h_0(t_0,x_0) < \delta$ such that

$$h(t_1,x(t_1)) = \varepsilon \quad \text{and} \quad h(t,x(t)) \leq \varepsilon, \quad t_0 \leq t \leq t_1. \tag{4.14.5}$$

Hence, (v) yields by corollary 4.14.1 the estimate

$$V(t,x(t)) \leq V(t_0,x_0), \quad t_0 \leq t \leq t_1. \tag{4.14.6}$$

Consequently, using (iv), (4.14.3), (4.14.5) and (4.14.6), we get

$$b(\varepsilon) = b(h(t_1, x(t_1))) \leq Q[V(t_1, x(t_1))] \leq Q[V(t_0, x_0)] \leq \psi[Q_0(V(t_0, x_0))]$$
$$\leq \psi[a(t_0, h_0(t_0, x_0))] < \psi[a(t_0, \delta)] < b(\varepsilon).$$

which is a contradiction. Hence, (h_0, h) equistability of (4.14.1) follows. If $a \in \mathcal{K}$ in (iv), then it is easy to see that λ_0 and consequently δ, are independent of t_0. As a result, we have (h_0, h)-uniform stability of the system (4.14.1). The proof is complete.

The next result provides criteria for (h_0, h)-uniform asymptotic stability.

Theorem 4.14.3. *Let the assumption* $(i) - (iv)$ *of Theorem 4.14.2 hold with* $a \in \mathcal{K}$. *Suppose further that*

$$D^+V(t, x) \leq -c(h_0(t, x)), \quad (t, x) \in S(h, \rho), \tag{4.14.7}$$

where $c \in C(R_+, K)$ *and* $c(t)$ *is increasing in* t *relative to* K, *with* $c(0) = 0$. *Then the system* (4.14.1) *is* (h_0, h)*-uniformly asymptotically stable.*

Proof. Since $c \in K$, it follows that $D^+V(t, x) \leq 0$ in $S(h, \rho)$ and this implies, by Theorem 4.14.1 that the system (4.14.1) is (h_0, h)-uniformly stable. Let $\varepsilon = \varepsilon_0 = min(\rho, \lambda)$ and designate by $\delta_0 = \delta(\varepsilon_0)$ so that we have

$$h_0(t_0, x_0) < \delta_0 \quad \text{implies} \quad h(t, x(t)) < \varepsilon_0, \quad t \geq t_0. \tag{4.14.8}$$

Now let $h_0(t_0, x_0) < \delta_0$ and for any $\varepsilon < \varepsilon_0$, choose a $T = T(\varepsilon) > 0$ such that

$$Q[c(\delta)T] \geq \psi(a(\delta_0)), \tag{4.14.9}$$

where $\delta = \delta(\varepsilon)$ corresponds to ε in (h_0, h)-uniform stability. To prove (h_0, h)-uniform asymptotic stability, it is sufficient to show that there exists a $t^* \in [t_0, t_0 + T]$ with

$$h_0(t^*, x(t^*)) < \delta$$

where $x(t)$ is any solution of (4.14.1) with $h_0(t_0, x_0) < \delta_0$. If this is not true, then we have

$$h_0(t, x(t)) \geq \delta, \quad t \in [t_0, t_0 + T]. \tag{4.14.10}$$

Setting $m(t) = V(t, x(t)) + \int_{t_0}^{t} c[h_0(s, x(s))]ds$ we obtain from (4.14.7), because of (4.14.8), the estimate

$$m(t) = V(t, x(t)) + \int_{t_0}^{t} c[h_0(s, x(s))]ds \leq m(t_0) = V(t_0, x_0) \tag{4.14.11}$$

for $t \in [t_0, t_0 + T]$. Hence, using the assumptions $(ii), (iv)$ and the relations (4.14.10), (4.14.11), it follows that

$$Q(c(\delta)T) = Q\Big[c(\delta)\int_{t_0}^{t_0+T} ds\Big] \leq Q\Big[\int_{t_0}^{t_0+T} c[h_0(s, x(s))]ds\Big] \leq Q[m(t_0 + T)]$$
$$\leq Q(V(t_0, x_0)) \leq \psi[Q_0(V(t_0, x_0))] \leq \psi[a(h_0(t_0, x_0))] < \psi(a(\delta_0)).$$

This is a contradiction to (4.14.9) and, therefore, we have (h_0, h)-uniform asymptotic stability of the system (4.14.1) proving the theorem.

Let us next discuss (h_0, h)-equiasymptotic stability which is in the spirit of Marachkov's theorem.

Theorem 4.14.4. *Let the assumptions (i) – (iv) of theorem 4.14.2 hold. Suppose further that*

$$D^+V(t,x) \leq -c[h(t,x)], \quad (t,x) \in S(h,\rho), \tag{4.14.12}$$

where c is the same function defined in theorem 4.14.3. Assume also that $h(t,x)$ is locally Lipschitzian in x and $D^+h(t,x)$ is bounded above or below in $S(h,\rho)$ and $Q(w) \to \infty$ as $\|w\| \to \infty$. Then the system (4.14.1) is (h_0,h)-equiasymptotically stable.

Proof. As in theorem 4.14.3, since $a \in \mathcal{CK}$, we arrive at

$$h_0(t_0, x_0) < \delta_0 = \delta_0(t_0, \varepsilon_0) \quad \text{implies} \quad h(t, x(t)) < \varepsilon_0, \quad t \geq t_0,$$

where $x(t) = x(t, t_0, x_0)$ is any solution of (4.14.1). To prove the claim of the theorem, it is enough to show that $h(t, x(t)) \to 0$ as $t \to \infty$. We shall first note that $\lim\limits_{t\to\infty} inf\, h(t, x(t)) = 0$. If not,there exists a $T > t_0$ and an $\eta > 0$ such that

$$h(t, x(t)) \geq \eta, \quad t \geq T.$$

Hence, (4.14.12) yields, as before,

$$Q[c(\eta)(t-T)] \leq Q\Big[\int_T^t c(h(s, x(s)))\mathrm{d}s\Big] \leq \psi(a(\delta_0)),$$

which, in view of the assumption on Q, leads to a contradiction.

Suppose that $\lim\limits_{t\to\infty} sup\, h(t, x(t)) \neq 0$. Then for any $\eta > 0$ there exist divergent sequences $\{t_n\}, \{t_n^*\}$ such that $t_i < t_i^* < t_{i+1}, i = 1, 2, ...,$ and

$$h(t_i, x(t_i)) = \frac{\eta}{2}, h(t_i^*, x(t_i^*)) = \eta \quad \text{and} \quad \frac{\eta}{2} < h(t, x(t)) < \eta, \quad t \in (t_i, t_i^*) \tag{4.14.13}$$

Indeed, one could also have, instead of (4.14.13)

$$h(t_i, x(t_i)) = \eta, h(t_i^*, x(t_i^*)) = \frac{\eta}{2} \quad \text{and} \quad \frac{\eta}{2} < h(t, x(t)) < \eta, \quad t \in (t_i, t_i^*) \tag{4.14.14}$$

Suppose that $D^+h(t,x) \leq M$. Then, using (4.14.13). we get

$$t_i^* - t_i \geq \frac{\eta}{4M} = \gamma > 0.$$

In view of (4.14.12), it then follows, for large n,

$$Q\Big[c(\frac{\eta}{2})\gamma n\Big] \leq \psi(a(\delta_0)).$$

Since $Q(w) \to \infty$ as $\|w\| \to \infty$, this leads to a contradiction as $n \to \infty$. Thus, $h(t, x(t)) \to 0$ as $t \to \infty$.

The case $D^+h(t,x)$ bounded below can be proved similarly using (4.14.14). Hence, the proof is complete.

Remark 4.14.1. *The assumption* (4.14.12) *of the theorem 4.14.4 can be generalized to* $D^+V(t,x) \leq -c[w(t,x)]$ *in* $S(h,\rho)$*, where* $w(t,x)$ *is h-positive definite and* w *satisfies similar conditions as* h*. Then the conclusion of theorem 4.14.4 remains true.*

Employing the comparison theorem 4.14.1 we shall now consider a general set of criteria for (h_0, h)-stability properties which unify several stability concepts in a single set up.

Theorem 4.14.5. *Assume that:*

(A_0) $h_0, h \in \Gamma$ *and* h_0 *is uniformly finer than* h*;*

(A_1) $V \in C(R_+ \times R^n, K)$ *and* $V(t,x)$ *is locally Lipschitzian in* x *relative to* K*;*

(A_2) $Q_0, Q \in \sum$ *and* Q_0 *is finer than* Q*;*

(A_3) $g \in C(R_+ \times K, R^N)$ *and for* $(t,x) \in S(h,\rho)$*,*

$$D^+V(t,x) \leq g(t, V(t,x)),$$

where $g(t,w)$ *is quasimonotone nondecreasing in* w *relative to* K *for each* $t \in R_+$*;*

(A_4) $b(h(t,x)) \leq Q(V(t,x))$ *if* $h(t,x) < \rho$ *and* $Q_0(V(t,x)) \leq a(h_0(t,x))$ *if* $h_0(t,x) < \rho_0$ *where* $a, b \in \mathcal{K}$.

Then the (Q_0, Q)*-stability properties of the comparison system* (4.14.2) *imply the corresponding* (h_0, h)*-stability properties of the system* (4.14.1)

Proof. We shall prove only (h_0, h)-equistability, since based on this proof, one can construct proofs of the other (h_0, h)-stability properties.

Since Q_0 is finer than Q, there exists a $\lambda > 0$ and a $\psi \in \mathcal{K}$ such that

$$Q(w) \leq \psi[Q_0(w)] \quad Q_0(w) < \lambda. \tag{4.14.15}$$

Let $0 < \varepsilon < min(\rho, \lambda)$ and $t_0 \in R_+$ be given. Suppose that the comparison system (4.14.2) is (Q_0, Q)-equistable. Then given $b(\varepsilon) > 0$ and $t_0 \in R_+$, there exists a $\delta_1 = \delta_1(t_0, \varepsilon) > 0$ with $\delta_1 < min(\lambda, \psi^{-1}(b(\varepsilon)))$ such that

$$Q_0(w_0) < \delta_1 \quad \text{implies} \quad Q(w(t)) < b(\varepsilon), \quad t \geq t_0, \tag{4.14.16}$$

where $w(t) = w(t, t_0, w_0)$ is any solution of the system (4.14.2). Also, h_0 is finer than h implies that there exists a $\phi \in \mathcal{K}$ such that

$$h(t,x) \leq \phi(h_0(t,x)) \quad h_0(t,x) < \rho_0 \tag{4.14.17}$$

with $\phi(\rho_0) < \rho$. Choose $w_0 = V(t_0, x_0)$ and $\delta < min(\rho_0, \lambda_0)$, where $a(\lambda_0) \leq \lambda$ such that

$$a(\delta) < \delta_1 \tag{4.14.18}$$

Now let $h_0(t_0, x_0) < \delta$ and note that

$$\begin{aligned} b(h(t_0, x_0)) &\leq Q(V(t_0, x_0)) < \psi[Q_0(V(t_0, x_0))] \leq \psi[a(h_0(t_0, x_0))] \\ &\leq \psi[a(\delta)] < \psi[\delta_1] < b(\varepsilon), \end{aligned}$$

so that

$$h(t_0, x_0) < \varepsilon. \tag{4.14.19}$$

We claim that with this δ, it follows that

$$h_0(t_0, x_0) < \delta \quad \text{implies} \quad h(t, x(t)) < \varepsilon, \quad t \geq t_0,$$

where $x(t) = x(t, t_0, x_0)$ is any solution of (4.14.1). If this is not true, because of (4.14.19), there exists a $t_1 > t_0$ and a solution $x(t) = x(t, t_0, x_0)$ of (4.14.1) with $h_0(t_0, x_0) < \delta$ such that

$$h(t_1, x(t_1)) = \varepsilon \quad \text{and} \quad h(t, t, x(t)) < \varepsilon, \quad t_0 \leq t < t_1, \tag{4.14.20}$$

which shows that $(t, x(t)) \in S(h, \rho)$ for $t_0 \leq t \leq t_1$. Hence, by Theorem 4.14.2 we get

$$V(t, x(t)) \leq r(t, t_0, w_0), \quad t_0 \leq t \leq t_1, \tag{4.14.21}$$

where $r(t, t_0, w_0)$ is the maximal solution of (4.14.2) relative to K. Since $Q(w)$ is nondecreasing in w with respect to K, we obtain using (A_4), (4.14.20) and (4.14.21),

$$b(\varepsilon) = b(h(t_1, x(t_1))) \leq Q[V(t_1, x(t_1))] \leq Q[r(t_1, t_0, w_0)]. \tag{4.14.22}$$

But

$$Q_0(w_0) = Q_0(V(t_0, x_0)) \leq a(h_0(t_0, x_0)) < a(\delta) < \delta_1,$$

and, hence, by (4.14.16)

$$Q(r(t_1, t_0, w_0)) < b(\varepsilon),$$

which contradicts (4.14.22). It, therefore, follows that that the system (4.14.1) is (h_0, h)-equistable. The proof is complete.

We have assumed in Theorem 4.14.5 stronger conditions on V, h_0, h, Q_0, Q only to unify several (h_0, h)-stability criteria in one theorem. This clearly imposes a certain burden on the comparison system (4.14.2). For example, in order to obtain only nonuniform (eventual) (h_0, h)-stability properties,it is enough to suppose that h_0, Q_0 are finer (asymptotically finer) than h, Q respectively, rather than uniformly finer. On the other hand, if we desire to prove only (h_0, h)-uniform stability properties, it is sufficient to assume conditions only in the tube $S(h, \rho) \cap S^c(h_0, \eta)$ for every $\eta > 0$. The next result addresses this aspect.

Theorem 4.14.6. *Assume that* $(A_0), (A_1)$ *and* (A_2) *of theorem 4.14.5 hold. Suppose further that* (A_3) *and* (A_4) *are replaced by*

(A_3^*) $g \in C[R_+ \times K, R^N], g(t, w)$ *is quasimonotone nondecreasing in* w *relative to* K *for each* $t \in R_+$ *and*

$$D^+V(t, x) \le g(t, V(t, x)) \quad \textit{for} \quad (t, x) \in S(h, \rho) \cap S^c(h_0, \eta)$$

for every $\eta > 0$ *;*

(A_4^*) $b(h(t, x)) \le Q(V(t, x))$ *for* $(t, x) \in S(h, \rho) \cap S^c(h_0, \eta)$ *and*

$$Q_0(V(t, x)) \le a(h_0(t, x)),$$

for $(t, x) \in S(h_0, \rho_0) \cap S^c(h_0, \eta)$ *for every* $\eta > 0$ *where* $a, b \in \mathcal{K}$.

Then the (Q_0, Q)*-uniform stability properties of the system* (4.14.2) *imply the corresponding* (h_0, h)*-uniform stability properties of the system* (4.14.1)

Proof. As before, we shall only consider (h_0, h)-uniform stability. We shall indicate necessary changes in the proof of theorem 4.14.5 to suit the present situation. We proceed as in theorem 4.14.5 and note that δ in (4.14.18) is independent of t_0 since δ_1 is. Instead of (4.14.20), we now have existence of $t_1 > t_2 > t_0$ satisfying

$$\begin{cases} h(t_1, x(t_1)) = \varepsilon, \quad h_0(t_2, x(t_2)) = \delta \quad \text{and for} \quad t_2 \le t \le t_1, \\ x(t) \in S(h, \rho) \cap S^c(h_0, \eta). \end{cases} \tag{4.14.23}$$

Consequently (4.14.21) reads as, taking $\eta = \delta$,

$$V(t, x(t)) \le r[t, t_2, V(t_2, x(t_2))], \quad t_2 \le t \le t_1, \tag{4.14.24}$$

where we consider solutions of (4.14.2) through $(t_2, V(t_2, x(t_2)))$.

The relation (4.14.22) reduces to

$$b(\varepsilon) = b(h(t_1, x(t_1))) \le Q(V(t_1, x(t_1))) \le Q[r(t, t_2, V(t_2, x(t_2)))] \tag{4.14.25}$$

and, therefore, we get

$$Q_0(w_0) = Q_0(V(t_2, x(t_2))) \le \psi[a(h_0(t_2, x(t_2)))] = \psi[a(\delta)] < \delta_1,$$

which leads to contradiction as before. Hence, the proof is complete.

Let us finally show by a simple example that the method of cone-valued Lyapunov functions yields results when the method of vector Lyapunov functions fails to do the same.

Example 4.14.1.

$$x' = -xy^2 e^{-t}\phi(t, x, y),$$
$$y' = \beta e^{-t} x^2 y \phi(t, x, y) + \frac{ye^{-t}}{2},$$

where $\phi(t, x, y) \ge 0$ *is a continuous function.*

Choose $V_1 = x^2, V_2 = e^{-t}y^2$ so that

$$D^+V_1 \leq 0, \quad D^+V_2 \leq -\beta D^+V_1, \quad \beta > 0.$$

Let $Q(\omega) = \omega_1, Q_0(\omega) = \omega_2 + (1+\beta)\omega_1, h = x^2$ and $h_0 = x^2 + y^2$. Then the method of vector Lyapunov functions does not provide (h_0, h)-equistability. On the other hand, if we choose the cone

$$K = [V = d_1 w_1 + d_2 w_2, w_i \geq 0, i = 1, 2]$$

where $d_1 = \begin{pmatrix} 1 \\ -\beta \end{pmatrix}$ $d_2 = \begin{pmatrix} 0 \\ 1 \end{pmatrix}$, we can see that relative to this K it follows that

$$D^+V(t,x) \leq 0 \quad \text{in} \quad S(h,\rho),$$

$$Q(V(t,x)) = V_1(t,x) \geq b(h(t,x)), \quad \text{if} \quad h(t,x) < \rho,$$

and

$$Q_0(V(t,x)) = V_2(t,x) + (1+\beta)V_1(t,x) \leq a(t, h_0(t,x))(1+\beta), \quad \text{if } h_0(t,x) < \rho_0.$$

Consequently, by theorem 4.14.2, we conclude (h_0, h)-equistability of the system (4.14.1)

4.15 Notes and Comments

Most of the results of Chapter 4 are taken from Lakshmikantham, Matrosov and Sivasundaram [1]. The results are also assembled from works of Lakshmikantham and Leela and Matunyuk [1,2], Lakshmikantham and Leela [1,4], Matrosov [1-6], Bellman [1], Voronov and Matrosov [1], Rama Mohan Rao and Sivasundaram [1], Siljak [1,2], Bailey [1], Ladde, Lakshmikantham and Leela [1], Deimling and Lakshmikantham [1,2], Lakshmikantham Leela and Oguztoreli [1], Lakshmikantham, Leela and Rama Mohan Rao [1], Deo [1], Ikeda and Sijak [1], Chandra, Lakshmikantham and Leela [1]. The contents of Section 4.14 are taken from Lakshmikantham and Papageorgiou [1]. See also Butz [1], Ladde, Lakshmikantham and Vatsala [1], Grujic, Martynyuk and Ribbens-Pavela [1], Karatueva and Matrosov [1], Deimling [1], Deimling and Lakshmikantham [1-3], Lakshmikantham [1-6], Matrosov [1-6], Lakshmikantam and Leela [1-7].

Chapter 5

Miscellaneous Topics

5.1 Introduction

This chapter is devoted to the introduction of several topics in the framework of differential equations in cones that need further investigation. We begin Section 5.2 with the application of the method of vector Lyapunov functions for the overlapping large scale systems, where because of the complexity of the system involved, we first expand the given system so as to make overlapping systems appear disjoint. In Section 5.3, the measure of nonconvexity is introduced with its properties and the concept is applied to get convexity of solution set of differential equations in a Banach space.

Section 5.4 deals with existence of solutions in weak topology, employing the weak measure of noncompactness. In section 5.5, some basic results for differential equations with delay in a Banach space are proved employing measure of noncompactness as well as Lyapunov-like functions. Existence of solutions of second order boundary value problems are discussed in Section 5.6 where suitable conditions in terms of measure of noncompactness are utilized. Section 5.7 deals with the extension of fractional differential equations to a Banach space providing necessary tools for the purpose. Using fractional derivative in the study of differential equations is a new area of investigation. Integro-differential equations in a Banach space are discussed in Section 5.8. Finally, in Section 5.9, we provide notes and comments.

5.2 Large-scale Systems with Overlapping Decompositions.

In this section, we show that the method of perturbing the vector Lyapunov functions in the expanded system can be directly employed to establish stability

results under weaker conditions for large-scale dynamic systems with overlapping decompositions. It is well known that the problems of size and complexity inherent to such systems are most efficiently addressed using system decomposition and decentralized control, and that the applicability of decentralized control can be greatly enlarged by overlapping decompositions. However, to apply standard methods to large systems decomposed in this manner it is necessary to first expand the original state space so as to make the overlapping subsystems appear disjoint. The stability laws established in the expanded system are then contracted for implementation in the original space. This two-step expansion-contraction process provides a natural set-up for the application of the method described above.
Following this approach we prove several nonuniform stability results for large scale dynamic systems.

Consider the dynamic systems

$$S_x : x' = f(t, x), \; x(t_0) = x_0, \; t_0 \geq 0 \tag{5.2.1}$$

and

$$S_y : y' = \tilde{f}(t, y), \; y(t_0) = y_0, \tag{5.2.2}$$

where $f \in C(R_+ \times R^n, R^n)$ and $\tilde{f} \in C(R_+ \times R^m, R^m), m > n$. We suppose that the functions f and $\tilde{f}$ are sufficiently smooth to guarantee existence and uniqueness of solutions of (5.2.1) and (5.2.2) for $t \geq 0$. Also assume that $f(t, 0) \equiv 0$ and $\tilde{f}(t, 0) \equiv 0$ so that systems (5.2.1) and (5.2.2) possess the trivial solutions. In addition, we assume that systems (5.2.1) admits a decomposition of the form

$$x_i' = F_i(t, x_i) + R_i(t, x), \; x_i(t_0) = x_{i0} \tag{5.2.3}$$

where the vector x is decomposed into N subvectors x_i, namely

$$x = (x_1, x_2, ..., x_N)$$

with each $x_i \in R^{n_i}$, $i = 1, 2, 3, ..., N$ such that $n = \sum_{i=1}^{N} n_i$. The functions $F_i(t, x_i)$ represent isolated decoupled subsystems

$$x_i' = F_i(t, x_i), \; x_i(t_0) = x_0 \tag{5.2.4}$$

and $R_i(t, x)$ are interconnections among the subsystems (5.2.4). Similarly, we assume that system (5.2.2) admits a decomposition of the form

$$y_i' = \tilde{F}_i(t, y_i) + \tilde{R}_i(t, y), \quad y_i(t_0) = y_{i0} \tag{5.2.5}$$

where the vector y is decomposed into N subvectors y_i, namely

$$y = (y_1, y_2, ..., y_N)$$

with each $y_i \in R^{m_i}, i = 1, 2, 3, ..., N$ such that $m = \sum_{i=1}^{N} m_i$. The functions $\tilde{F}_i(t, y_i)$ represent isolated decoupled subsystems

$$y_i' = \tilde{F}_i(t, y_i), \ y_i(t_0) = y_{i0} \tag{5.2.6}$$

and $\tilde{R}_i(t, y)$ are the interconnections among subsystems (5.2.6).

Our aim is to show that the stability properties of the large scale dynamic system represented by (5.2.1) can be established under weaker conditions by enlarging (5.2.1) into (5.2.2). For this purpose we need the following known results. Let us begin with the following definition.

Definition 5.2.1. *A system S_x is said to be included in a system S_y (written $S_x \subset S_y$) if there exists an ordered pair of matrices (U, V) such that $UV = I$, and if for any initial state (t_0, x_0) of $S_x, y_0 = Vx_0$ implies that*

$$x(t, t_0, x_0) = Uy(t, t_0, Vx_0), \quad \forall t \geq t_0. \tag{5.2.7}$$

Conditions for inclusion of S_x in S_y are given in the following theorem.

Theorem 5.2.1. *Assume that*

(i) there exists an ordered pair of matrices (U, V) such that $UV = I$, where V and U are $m \times n$ full row rank and $n \times m$ full column rank, respectively;

(ii) $F(t, y) = Vf(t, Uy) + m(t, y)$ where $m \in C(R_+ \times R^m, R^m)$ is a complementary function satisfying either

(a) $m(t, Vx) = 0$ for $(t, x) \in R_+ \times R^n$ or

(b) $Um(t, y) = 0$ for $(t, y) \in R_+ \times R^m$.

Then $S_x \subset S_y$.

Next we state some nonuniform stability results obtained under weaker conditions after expanding (5.2.1) into (5.2.2) as described in Definition 5.2.1.

For any function $V \in C(R_+ \times R^n, R_+^N)$, we define for $(t, x) \in R_+ \times R^n$

$$D^+V(t, x)_{(5.2.1)} = \lim_{h \to 0} sup \frac{1}{h}[V(t + h, x + hf(t, x)) - V(t, x)]$$

where $D^+V(t, x)_{(5.2.1)}$ means that $D^+V(t, x)$ is computed relative to (5.2.1). Let $S_n(\rho) = [x \in R^n : \|x\| < \rho]$ for some $\rho > 0$.

Theorem 5.2.2. *Assume that*

(i) there exists $V_1 \in C(R_+ \times S_n(\rho), R_+^N), V_1(t,0) \equiv 0,$ *and for* $(t,x) \in R_+ \times S_n(\rho)$

$$D^+V_1(t,x)_{(5.2.1)} \leq g_1(t, V_1(t,x))$$

where $g_1 \in C(R_+ \times R_+^N, R^N)$ *and* $g_1(t,u)$ *is quasimonotone nondecreasing in* u, *that is if* $u \leq v$ *and* $u_i = v_i$ *for some* $1 \leq i \leq N$, *then* $g_1(t,u) \leq g_1(t,v)$ *and* $g_1(t,0) \equiv 0$;

(ii) the conditions of theorem 5.2.1 hold;

(iii) there exists $V_2 \in C(R_+ \times S_m(\rho_0) \bigcap S_m^c(\eta), R_+^M)$, *for every* $0 < \eta < \rho_0$ *where* $\rho_0 = \frac{\rho}{|U|}$, *such that*

$$D^+V_2(t,y)_{(5.2.2)} \leq g_2(t, V_2(t,y)), \ (t,y) \in R_+ \times S_m(\rho_0) \cap S_m^c(\eta)$$

where $g_2 \in C(R_+ \times R_+^M, R^M)$, $g_2(t,u)$ *is quasimonotone nondecreasing in* u, $g_2(t,0) \equiv 0$, *and for* $(t,y) \in R_+ \times S_m(\rho_0) \cap S_m^c(\eta)$

$$b(\|y\|) \leq \sum_{i=1}^{M} V_{2i}(t,y) \leq a(\|y\|) + \sum_{i=1}^{N} V_{1i}(t, Uy)$$

where $a, b \in \mathcal{K} = \{\phi \in C([0,\rho_0), R_+) : \phi(u)$ *is strictly increasing and* $\phi(0) = 0\}$;

(iv) the trivial solution of $u' = g_1(t,u), u(t_0) = u_0$ *is equistable;*

(v) the trivial solution of $v' = g_2(t,v), v(t_0) = v_0$ *is uniformly stable.*

Then the trivial solution of (5.2.1) *is equistable.*

Corollary 5.2.1. *The conclusion of theorem 5.2.2 holds true if* $g_1(t,u) \equiv 0$ *and* $g_2(t,u) \equiv 0$.

First we state a variant of theorem 5.2.2 which is more useful in applications whenever g_1 and g_2 do not enjoy the required quasimonotone properties. We also generalize condition (iii) by substituting $a(t, \|x\|)$ for $a(\|x\|)$. This theorem will be useful in establishing nonuniform stability results for large scale dynamic systems.

Theorem 5.2.3. *Assume that*

(i) *there exists* $V_1 \in C(R_+ \times S_n(\rho), R_+^N), V_1(t,0) \equiv 0$ *and* $Q_1 \in C(R_+^N, R_+^{k_1}), k_1 \leq N, Q_1(0) \equiv 0$. *Let* $\mathcal{V}_1(t,x) \equiv Q_1(V_1(t,x))$ *and for* $(t,x) \in R_+ \times S_n(\rho)$

$$D^+\mathcal{V}_1(t,x)_{(5.2.1)} \leq G_1(t, \mathcal{V}_1(t,x))$$

where $G_1 \in C(R_+ \times R_+^{k_1}, R^{k_1})$, $G_1(t,u)$ *is quasimonotone nondecreasing in* u, *and* $G_1(t,0) \equiv 0$;

(ii) *the conditions of theorem 5.2.1 hold;*

(iii) *there exist* $V_2 \in C(R_+ \times S_m(\rho_0) \bigcap S_m^c(\eta), R_+^M)$ *for every* $0 < \eta < \rho_0$ *where* $\rho_0 = \frac{\rho}{|U|}$ *and* $Q_2 \in C(R_+^M, R_+^{k_2}), k_2 \leq M$, *such that if* $\mathcal{V}_2(t,y) \equiv Q_2(V_2(t,y))$ *and* $(t,y) \in R_+ \times S_m(\rho_0) \bigcap S_m^c(\eta)$ *then*

$$D^+\mathcal{V}_2(t,y)_{(5.2.2)} \leq G_2(t, \mathcal{V}_2(t,y))$$

where $G_2 \in C(R_+ \times R_+^{k_2}, R^{k_2})$, $G_2(t,u)$ *is quasimonotone nondecreasing in* u, $G_2(t,0) \equiv 0$ *and*

$$b(\|y\|) \leq \sum_{i=1}^{k_2} \mathcal{V}_{2i}(t, y_i) \leq a(t, \|y\|) + \sum_{i=1}^{k_1} \mathcal{V}_{1i}(t, Uy)$$

where $b \in \mathcal{K} = \{\phi \in C[[0,\rho_0), R_+] : \phi(u)$ *is strictly increasing and* $\phi(0) = 0\}$ *and* $a \in \mathcal{LK} = \{\gamma \in C[R_+ \times [0,\rho_0), R_+] : \gamma(t,u) \in \mathcal{K}\}$ *for each* t *and* $\gamma(t,u)$ *is strictly decreasing in* t *for each* u *and* $\lim_{t\to\infty} \gamma(t,u) = 0$ *for each* $u\}$;

(iv) *the trivial solution of* $u' = G_1(t,u), u(t_0) = u_0$ *is equistable;*

(v) *the trivial solution of* $v' = G_2(t,v), v(t_0) = v_0$ *is uniformly stable.*

Then the trivial solution of (5.2.1) *is equistable.*

The proof will be omitted.
Next we shall use theorem 5.2.3 to prove a nonuniform stability result under weaker conditions for large scale dynamic systems.

Theorem 5.2.4. *Assume that for each* $i = 1, 2, ..., N$

(i) *there exists* $V_{1i} \in C(R_+ \times S_n(\rho), R_+^{r_i}), V_{1i}(t,0) = 0$ *and* $Q_{1i} \in C(R_+^{r_i}, R_+^{k_i}), Q_{1i}(0) \equiv 0$ *such that, if* $Q_{1i}(V_{1i}(t,x_i)) \equiv \mathcal{V}_{1i}(t,x_i)$ *and* $0 \leq j \leq k_i, \mathcal{V}_{1ij}(t,u)$ *is* *locally Lipschitzian in u with Lipschitz constant L_{1ij}, and for $(t, x_i) \in R_+ \times S_n(\rho)$*

$$D^+\mathcal{V}_{1i}(t, x_i)_{(5.2.4)} \leq G_{1i}(t, \mathcal{V}_{1i}(t, x_i))$$

where $G_1 \in C(R_+ \times R_+^k, R^k)$, $k = \sum_{i=1}^{N} k_i$, $G_1(t, u)$ is quasimonotone nondecreasing in u, and $G_1(t, 0) \equiv 0$;

(ii) for $(t, x) \in R_+ \times S_n(\rho)$ and $1 \leq j \leq r_i$

$$L_{1ij}(t, x_i)\|R_i(t, x)\| \leq \lambda_{ij}(t)\mathcal{V}_{1ij}(t, x), \quad \lambda_{ij} \in L^1[t_0, +\infty);$$

(iii) the conditions of theorem 5.2.1 hold;

(iv) for every $0 < \eta < \rho_0$ where $\rho_0 = \frac{\rho}{|U|}$ there exists $V_{2i} \in C(R_+ \times S_m(\rho_0) \bigcap S_m^c(\eta),$ and $Q_{2i} \in C(R_+^{\tilde{r}_i}, R_+^{\tilde{k}_i})$ such that, if $Q_{2i}(V_{2i}(t, y_i)) \equiv \mathcal{V}_{2i}(t, y_i)$ and $0 \leq j \leq \tilde{k}_i$, $V_{2ij}(t, u)$ is locally Lipschitzian in u with Lipschitz constant L_{2ij}, and for $(t, y_i) \in R_+ \times S_m(\rho_0) \bigcap S_m^c(\eta)$

$$D^+\mathcal{V}_{2i}(t, y_i)_{(5.2.6)} \leq G_{2i}(t, \mathcal{V}_{2i}(t, y_i))$$

where $G_2 \in C(R_+ \times R_+^{\tilde{k}}, R^{\tilde{k}})$, $\tilde{k} = \sum_{i=1}^{N} \tilde{k}_i$, $G_2(t, u)$ is quasimonotone nondecreasing in u, and $G_2(t, 0) \equiv 0$ and

$$b(\|y\|) \leq \sum_{i=1}^{N} \sum_{j=1}^{\tilde{k}_i} \mathcal{V}_{2ij}(t, y_i) \leq \sum_{i=1}^{N} \sum_{j=1}^{k_i} \mathcal{V}_{1ij}(t, Uy_i) + a(t, \|y\|)$$

where $a \in \mathcal{LK}$ and $b \in \mathcal{K}$;

(v) for $(t, y) \in R_+ \times S_m(\rho_0) \bigcap S_m^c(\eta)$

$$L_{2ij}(t, y_i)\|\tilde{R}_i(t, y)\| \leq W_{ij}(t, \|y\|)$$

with $W_{ij}(t, u)$ nondecreasing in u;

(vi) the trivial solution of $u' = G_3(t, u), u(t_0) = u_0$ where $G_3(t, u) = \lambda(t)u + G_1(t, u)$ is equistable;

(vii) the trivial solution of $v' = G_4(t, v), v(t_0) = v_0$ where $G_4(t, v) = W(t, b^{-1}(v)) + G_2(t, v)$ is uniformly stable.

Then the trivial solution of (5.2.1) is equistable.

Proof. For $(t,x) \in R_+ \times S_n(\rho)$ and $1 \le j \le k_i$ we compute $D^+V_{1ij}(t,x_i)_{(5.2.3)}$. We also have

$$\mathcal{V}_{1ij}(t+h, x_i + hf_i(t,x)) - \mathcal{V}_{1ij}(t,x_i) = \mathcal{V}_{1ij}(t+h, x_i + hF_i(t,x_i)$$
$$+ hR_i(t,x)) - \mathcal{V}_{1ij}(t+h, x_i + hF_i(t,x_i)) + \mathcal{V}_{1ij}(t+h, x_i + hF_i(t,x_i)) - \mathcal{V}_{1ij}(t,x_i)$$
$$\le L_{1ij}(t,x_i)\|R_i(t,x)\| + \mathcal{V}_{1ij}(t+h, x_i + hF_i(t,x_i)) - \mathcal{V}_{1ij}(t,x_i).$$

Dividing by $h > 0$ and taking the limit as $h \to 0$ we have

$$D^+\mathcal{V}_{1ij}(t,x_i)_{(5.2.3)} \le L_{1ij}(t,x_i)\|R_i(t,x)\| + D^+\mathcal{V}_{1ij}(t,x_i)_{(5.2.4)}$$

which, in view of assumptions (i) and (ii), yields

$$D^+\mathcal{V}_{1ij}(t,x_i)_{(5.2.3)} \le \lambda_{ij}(t)\mathcal{V}_{1ij}(t,x_i) + G_{1ij}(t, \mathcal{V}_{1i}(t,x_i)) \equiv G_{3ij}(t, \mathcal{V}_{1i}(t,x_i)).$$

Hence,

$$D^+\mathcal{V}_1(t,x)_{(5.2.3)} \le G_3(t, \mathcal{V}_1(t,x))$$

where $G_3 \in C[R_+ \times R_+^k, R^k]$, $k = \sum_{i=1}^N k_i$, $G_3(t,0) = 0$ and $G_3(t,u)$ is quasimonotone nondecreasing in u.
Similarly, for $(t,y) \in R_+ \times S_m(\rho_0)$ and $1 \le j \le \tilde{r}_i$ we compute $D^+\mathcal{V}_2(t,y)_{(5.2.5)}$ so that in view of assumptions (iv) and (v) we arrive at

$$D^+\mathcal{V}_{2ij}(t,y_i)_{(5.2.5)} \le L_{2ij}(t,y_i)\|\tilde{R}_i(t,y)\| + D^+\mathcal{V}_{2ij}(t,y_i)_{(5.2.6)}$$
$$\le W_{ij}(t, \|y\|) + G_{2ij}(t, \mathcal{V}_{2i}(t,y_i)) = W_{ij}(t, b^{-1}(\mathcal{V}_{2i}(t,y_i))) + G_{2ij}(t, \mathcal{V}_{2i}(t,y_i))$$
$$\equiv G_{4ij}(t, \mathcal{V}_{2i}(t,y_i)).$$

Hence, $D^+\mathcal{V}_2(t,y) \le G_4(t, \mathcal{V}_2(t,y))$ where $G_4 \in C[R_+ \times R^{\tilde{k}_+}, R^{\tilde{k}}], \tilde{k} = \sum_{i=1}^N \tilde{k}_i$, and $G_4(t,u)$ is a quasimonotone nondecreasing in u. Then the conclusion of Theorem 5.2.4 follows directly from Theorem 5.2.3.

Example 5.2.1. Consider the following systems

$$S_x : \dot{x} = A(t,x)x \tag{5.2.8}$$

where $x = (x_1, x_2, x_3)^T$,

$$A \equiv \begin{bmatrix} a_{11} & a_{12} & a_{13} \\ a_{21} & a_{22} & a_{23} \\ g_{31} & a_{32} & a_{33} \end{bmatrix}$$

$$= \begin{bmatrix} -\frac{\phi_1}{2} + \frac{2}{3}x_1^2x_2^2\phi_2 & x_3 & -x_2 \\ -x_2x_1^3\phi_2 & -\frac{\phi_1}{2} & 0 \\ -x_2x_1^3\phi_2 & +x_1^4\phi_2 & -\frac{\phi_1}{2} - x_1^4\phi_2 \end{bmatrix},$$

$$\phi_1(t,x) = t + \frac{1}{3},$$

and

$$\phi_2(t,x) = \frac{e^{-\|x\|}}{1+\|x\|^4} \quad \text{where} \quad \|x\|^2 = \sum_{i=1}^{3} x_i^2.$$

Consider the decomposition of S_x into 2 subsystems $x = (X_1, X_2)^T$ with $X_1 = x_1$ and $X_2 = (x_2, x_3)^T$.

$$\begin{pmatrix} \dot{x_1} \\ \\ \cdots \\ \dot{x_2} \\ \\ \dot{x_3} \end{pmatrix} = \begin{bmatrix} a_{11} & \vdots & a_{12} & a_{13} \\ \cdots & \vdots & \cdots & \cdots \\ a_{21} & \vdots & a_{22} & a_{23} \\ a_{31} & \vdots & a_{32} & a_{33} \end{bmatrix} \begin{pmatrix} x_1 \\ \\ \cdots \\ x_2 \\ \\ x_3 \end{pmatrix}$$

for which we choose the following Lyapunov functions $V_{11}(t, X_1) = x_1^2$ and $V_{12}(t, X_2) = (x_2 - x_3)^2$. Their derivatives along the solutions of (5.2.8) are

$$\begin{aligned} \dot{V}_{11}(t, X_1) &= 2x_1\dot{x}_1 \\ &= 2(\frac{-\phi_1}{2} + \frac{2}{3}x_1^2x_2^2\phi_2)V_{11}(t, X_1) \end{aligned} \tag{5.2.9}$$

and

$$\begin{aligned} \dot{V}_{12}(t, X_2) &= 2(x_2 - x_3)(\dot{x}_2 - \dot{x}_3) \\ &= 2(x_2 - x_3)^2(-2\phi_2x_1^4 - t) \\ &= 2(-2\phi_2x_1^4 - t)\dot{V}_{12}(t, X_2). \end{aligned} \tag{5.2.10}$$

Now consider the transformation

$$y = Vx \quad x = Uy \tag{5.2.11}$$

where

$$V = \begin{bmatrix} 1 & 0 & 0 \\ 0 & 1 & 0 \\ 0 & 1 & 0 \\ 0 & 0 & 1 \end{bmatrix} \quad U = \begin{bmatrix} 1 & 0 & 0 & 0 \\ 0 & \frac{1}{2} & \frac{1}{2} & 0 \\ 0 & 0 & 0 & 1 \end{bmatrix} \quad UV = \begin{bmatrix} 1 & 0 & 0 \\ 0 & 1 & 0 \\ 0 & 0 & 1 \end{bmatrix}.$$

Under the transformations V, S_x becomes

$$\begin{pmatrix} \dot{x_1} \\ \dot{x_2} \\ \dot{x_2} \\ \dot{x_3} \end{pmatrix} = \begin{bmatrix} a_{11} & a_{12} & a_{13} \\ a_{21} & a_{22} & a_{23} \\ a_{21} & a_{22} & a_{23} \\ a_{31} & a_{32} & a_{33} \end{bmatrix} \begin{bmatrix} 1 & 0 & 0 & 0 \\ 0 & \frac{1}{2} & \frac{1}{2} & 0 \\ 0 & 0 & 0 & 1 \end{bmatrix} \begin{pmatrix} x_1 \\ x_2 \\ x_2 \\ x_3 \end{pmatrix} = \begin{bmatrix} a_{11} & \frac{a_{12}}{2} & \frac{a_{12}}{2} & a_{13} \\ a_{21} & \frac{a_{22}}{2} & \frac{a_{22}}{2} & a_{23} \\ a_{21} & \frac{a_{22}}{2} & \frac{a_{22}}{2} & a_{23} \\ a_{31} & \frac{a_{32}}{2} & \frac{a_{32}}{2} & a_{33} \end{bmatrix} \begin{pmatrix} x_1 \\ x_2 \\ x_2 \\ x_3 \end{pmatrix}$$

which can be written as

$$\begin{pmatrix} \dot{x_1} \\ \dot{x_2} \\ \dot{x_2} \\ \dot{x_3} \end{pmatrix} = \begin{bmatrix} a_{11} & a_{12} & 0 & a_{13} \\ a_{21} & a_{22} & 0 & a_{23} \\ a_{21} & 0 & a_{22} & a_{23} \\ a_{31} & 0 & a_{32} & a_{33} \end{bmatrix} \begin{pmatrix} x_1 \\ x_2 \\ x_2 \\ x_3 \end{pmatrix} + \begin{bmatrix} 0 & -\frac{a_{12}}{2} & \frac{a_{12}}{2} & 0 \\ 0 & -\frac{a_{22}}{2} & \frac{a_{22}}{2} & 0 \\ 0 & \frac{a_{22}}{2} & -\frac{a_{22}}{2} & 0 \\ 0 & \frac{a_{32}}{2} & -\frac{a_{32}}{2} & 0 \end{bmatrix} \begin{pmatrix} x_1 \\ x_2 \\ x_2 \\ x_3 \end{pmatrix}.$$

$$\textit{Let } m = \begin{bmatrix} 0 & \frac{a_{12}}{2} & -\frac{a_{12}}{2} & 0 \\ 0 & \frac{a_{22}}{2} & -\frac{a_{22}}{2} & 0 \\ 0 & -\frac{a_{22}}{2} & \frac{a_{22}}{2} & 0 \\ 0 & -\frac{a_{32}}{2} & \frac{a_{32}}{2} & 0 \end{bmatrix}.$$

It is easy to show that $mV = 0$. Hence, by theorem 5.2.1, the solutions of the following system

$$S_y : \dot{y} = \begin{pmatrix} \dot{y_1} \\ \dot{y_2} \\ \cdots \\ \dot{y_3} \\ \dot{y_4} \end{pmatrix} = \begin{bmatrix} a_{11} & a_{12} & \vdots & 0 & a_{13} \\ a_{21} & a_{22} & \vdots & 0 & a_{23} \\ \cdots & \cdots & \vdots & \cdots & \cdots \\ a_{21} & 0 & \vdots & a_{22} & a_{23} \\ a_{31} & 0 & \vdots & a_{32} & a_{33} \end{bmatrix} \begin{pmatrix} y_1 \\ y_2 \\ \cdots \\ y_3 \\ y_4 \end{pmatrix}, \tag{5.2.12}$$

where $y = (y_1, y_2, y_3, y_4) = (x_1, x_2, x_3, x_4)$, include those of (5.2.8). Now consider the decomposition of S_y into the 2 subsystems shown above by the dotted lines. We choose the following Lyapunov functions.

$$\begin{aligned} V_{21}(t,(y_1,y_2)) &= 3y_1^2 + 2y_2^2, \\ V_{22}(t,(y_3,y_4)) &= (y_3 - y_4)^2, \\ \text{and} \quad V_{23}(t,(y_3,y_4)) &= (y_3 + y_4)^2. \end{aligned}$$

Note that

$$2\|y\|^2 \leq \sum_{i=1}^{3} V_{2i}(t,y) \leq \sum_{i=1}^{2} V_{1i}(t,Uy) + 2\|y\|^2.$$

$$\sum_{i=1}^{3} V_{2i}(t, Vx) = 3x_1^2 + 4x_2^2 + 2x_3^2 \geq 2(x_1^2 + 2x_2^2 + x_3^2) = \|y\|^2$$

$$\begin{aligned}\sum_{i=1}^{3} V_{2i}(t, Vx) &= x_1^2 + (x_2 - x_3)^2 + 2x_1^2 + 2x_2^2 + (x_2 + x_3)^2 \\ &\leq \sum_{i=1}^{2} V_{1i}(t, x) + 2x_1^2 + 2x_2^2 + (x_2 + x_3)^2 + (x_2 - x_3)^2 \\ &\leq \sum_{i=1}^{2} V_{1i}(t, x) + 2(x_1^2 + 2x_2^2 + x_3^2) \\ &= \sum_{i=1}^{2} V_{1i}(t, x) + 2\|Vx\|^2.\end{aligned}$$

The derivatives of V_{21}, V_{22}, and V_{23} along the solutions of S_y can be expressed in terms of the $x_i^{'}s$ as follows:

$$\begin{aligned}\dot{V}_{21}(t, Vx) &= 6x_1\dot{x}_1 + 4x_2\dot{x}_2 = \phi_1 V_{21}(t, Vx) \\ \dot{V}_{22}(t, Vx) &= 2(x_2 - x_3)(\dot{x}_2 + \dot{x}_3) = 2(-\frac{\phi_1}{2} + \phi_2 x_1^4)V_{22}(t, Vx) \\ \dot{V}_{23}(t, Vx) &= 2(x_2 + x_3)(\dot{x}_2 + \dot{x}_3) = 2(-\frac{\phi_1}{2} + \phi_2 x_1^4)V_{23}(t, V_{23}(t, Vx).\end{aligned}$$

With $\phi_2 = \frac{e^{-\|x\|}}{1+\|x\|^4}$ and $\phi_1 = t + \frac{1}{3}$ we have

$$\begin{aligned}\dot{V}_{11} &= (-t - \frac{1}{3} + \frac{4}{3}\frac{x_2^2 x_1^2}{1 + \|x\|^4}e^{-\|x\|})V_{11} \\ &\leq (-t + 1)V_{11} \quad for \quad 0 \leq \|x\| \leq \rho.\end{aligned} \tag{5.2.13}$$

Similarly,

$$\begin{aligned}\dot{V}_{12} &= (-\phi_1 + 2\phi_2 x_1^4)V_{12} \\ &= (-t - \frac{1}{3} + 2\frac{x_1^4}{1 + \|x\|^2}e^{-\|x\|})V_{12} \\ &\leq (-t + \frac{5}{3})V_{12} \quad \text{for} \quad 0 \leq \|x\|.\end{aligned} \tag{5.2.14}$$

and,

$$\begin{aligned}\dot{V}_{21} &= (t - \frac{1}{3})V_{21} \\ \dot{V}_{22} &= (\phi_1 + 2\phi_2 x_1^4)V_{22} \\ &= (-t - \frac{1}{3} + \frac{2x_1^4}{1 + \|x\|^4}e^{-\|x\|})V_{22}.\end{aligned} \tag{5.2.15}$$

For $0 < \eta \leq \|Ux\| \leq \rho$ or $\|x\| \geq \frac{\eta}{\|U\|} \equiv \eta_0$,

$$\dot{V}_{22} \quad \leq (-t - \frac{1}{3} + 2e^{-\eta_0})V_{22}$$

with $e^{-\eta_0} < \frac{1}{6}$, i.e., $\eta_0 > ln6$, $2e^{-\eta_0} - \frac{1}{3} < 0$.

Let $c \equiv -2e^{-\eta_0} + \frac{1}{3}$, $c > 0$ then,

$$\dot{V}_{22} \quad \leq (-t - c)V_{22} \quad \text{for} \quad \|x\| > ln6. \tag{5.2.16}$$

The comparison problems corresponding to (5.2.13) -(5.2.16) are

I. a) $\dot{u_1} = (-t + 1)u_1 \qquad u_1(t_0) = u_{10}$

b) $\dot{u_2} = (-t + \frac{5}{3})u_2 \qquad u_2(t_0) = u_{20}$

II. a) $\dot{v_1} = (-t - \frac{1}{3})v_1 \qquad v_1(t_0) = v_{10}$

b) $\dot{v_2} = (-t - c)v_2 \qquad v_2(t_0) = v_{20}.$

Consider equation I_a. Its solution is

$$u_1(t) = u_{10}exp\{t(1 - \frac{t}{2}) - t_0(1 - \frac{t_0}{2})\}$$

The exponential reaches a maximum when

$$\frac{d}{dt}u_1(t) = 0 = (-t + 1)u_1(t),$$

i.e., when $t = 1$. Hence, if $t_0 < 1$

$$u_1(t) \leq u_{10}exp\{\frac{1}{2} - t_0 + \frac{t_0^2}{2}\} = u_{10}exp\frac{1}{2}(t_0 - 1)^2,$$

and

$$u_{10}exp\frac{1}{2}(t_0 - 1)^2 < \epsilon \Rightarrow u_{10} < \epsilon\, exp(-\frac{1}{2}(t_0 - 1)^2).$$

Let $\delta_1 = \epsilon\, exp(-\frac{1}{2}(t_0 - 1)^2)$. Then $u_{10} < \delta_1 \Rightarrow u_1(t) < \epsilon$. On the other hand, if $t_0 \geq 1$ then $\delta_1 = \epsilon$. Hence,

$$|u_{10}| < \delta(\epsilon, t_0) = \begin{cases} \epsilon & \text{if } t_0 \geq 1 \\ \epsilon\, exp-\frac{1}{2}(t_0 - 1)^2 & \text{if } t_0 < 1. \end{cases}$$

Therefore, the solution of I_a is stable.

A similar argument can be used to show that the trivial solution of Ib is also

stable.
Next we consider equation IIa which admits the solution

$$v_1(t) = v_{10}exp\{-(\frac{1}{2}(t^2 - t_0^2) + \frac{1}{3}(t - t_0))\}.$$

Since $t \geq t_0$ this is a function which is strictly decreasing. Therefore it reaches its maximum in $[t_0, +\infty)$ at $t = t_0$. Hence, $v_1(t) \leq v_{10}$ and $v_1(t) < \epsilon$ whenever $v_{10} < \delta(\epsilon) = \epsilon$. Consequently, the trivial solution of IIa is uniformly stable. Similarly, equation IIb is also uniformly stable. Hence, by theorem 5.2.4 the trivial solution of system S_x is stable.

We next prove a result on non uniform asymptotic stability in the same set-up.

Theorem 5.2.5. *Assume that conditions $(ii) - (vii)$ of Theorem 5.2.4 hold true and that condition (i) is replaced by*

(i) there exist $V_{1i} \in C(R_+ \times S_n(\rho), R_+^{r_i})$, $V_{1i}(t,0) = 0$, $Q_{1i} \in C(R_+^{r_i}, R_+^{k_i})$, $Q_{1i}(0) = 0$, such that, if $Q_{1i}(V_{1i}(t,x_i)) \equiv \mathcal{V}_{1i}(t,x_i)$ and $0 \leq j \leq k_i$, $\mathcal{V}_{1ij}(t,u)$ is locally Lipschitzian in u with Lipschitz constant L_{1ij}. Also assume that there exists $w \in C(R_+ \times S_n(\rho), R_+)$, $w(t,x)$ locally Lipschitzian in x, positive definite, $D^+w(t,x)$ bounded from above or below such that for $(t,x_i) \in R_+ \times S_n(\rho)$

$$D^+\mathcal{V}_{1ip}(t,x_i)_{(5.2.4)} \leq -w(t,x_i) \qquad 1 \leq i \leq N, \qquad 1 \leq p \leq \tilde{k}_i$$

$$D^+\mathcal{V}_{1iq}(t,x_i)_{(5.2.4)} \leq G_{1iq}(t, \mathcal{V}_{1i}(t,x_i)) \quad q \neq p \quad 1 \leq q \leq \tilde{k}_i$$

where $G_1 \in C(R_+ \times R_+^r, R_+^r)$, $G_1(t,u)$ is quasimonotone nondecreasing in u, $G_1(t,0) = 0$.

Then the trivial solution of (5.2.1) is asymptotically stable.

Proof. By Theorem 5.2.3 with $G_{1ip}(t,u) = 0$ it follows that system (5.2.1) is equistable. Hence, it is enough to prove that given $\epsilon > 0$, $t_0 \in R_+$ there exists a $\delta = \delta(t_0) > 0$ such that $\|x_0\| < \delta$ implies that $\|x(t,t_0,x_0)\| \to 0$ as $t \to \infty$. For $\epsilon = \lambda \leq \rho_0$ let $\delta = \delta(t_0, \lambda)$ be the δ of equistability and let $\|x_0\| < \delta$. Since $w(t,x)$ is positive definite, it is enough to prove that $\lim_{t\to\infty} w(t,x(t)) = 0$ for any solution $x(t) = x(t,t_0,x_0)$ of (5.2.1) with $\|x_0\| < \delta$.

First, we prove that $\lim_{t\to\infty} inf w(t,x(t)) = 0$. If not, there would exist $\beta > 0$ and $T = T(\beta) > 0$ such that

$$w(t,x(t)) \geq \beta \quad \text{for} \quad t \geq T. \tag{5.2.17}$$

Since $\lambda_{ij} \in L^1[t_0, +\infty)$ and $\lambda_{ij}(t) \geq 0$ for $t \in [t_0, +\infty) \exists N_{ij} > 0$ such that

$$\int_{t_0}^{t} \lambda_{ij}(s)\mathrm{d}s \leq \int_{t_0}^{\infty} \lambda_{ij}(s)\mathrm{d}s < N_{ij}. \tag{5.2.18}$$

Let $m(t) = \mathcal{V}_{1ip}(t, x(t)) + \int_{t_0}^{t} w(s, x(s))\mathrm{d}s - \int_{t_0}^{t} \lambda_{1p}(s)\mathrm{d}s$. Then

$$D^+ m(t) \leq D^+ \mathcal{V}_{1ip}(t, x(t)) + w(t, x(t)) - \lambda_{1p}(t) \leq 0$$

where the last inequality results from (i) and (ii). Hence $m(t) \leq m(t_0)$ or

$$\mathcal{V}_{1ip}(t, x(t)) \leq \mathcal{V}_{1ip}(t_0, x_0) - \int_{t_0}^{t} (w(s, s(x)))\mathrm{d}s + \int_{t_0}^{t} \lambda_{1p}(s)\mathrm{d}s$$

$$\leq \mathcal{V}_{1ip}(t_0, x_0) - \int_{T}^{t} w(s, x(s))\mathrm{d}s + N_{1p}$$

$$\leq \mathcal{V}_{1ip}(t_0, x_0) - \beta(t - T) + N_{1p}$$

where the last inequality results from (5.2.17). For t sufficiently large, $\mathcal{V}_{1ip}(t_0, x_0) - \beta(t-T) + N_{1p} < 0$ which contradicts the assumption that $\mathcal{V}_1(t, x(t)) \geq 0$. Consequently,

$$\lim_{t \to \infty} inf w(t, x(t)) = 0. \tag{5.2.19}$$

Next, we prove that $\lim_{t \to \infty} sup w(t, x(t)) = 0$. Suppose that $\lim_{t \to \infty} sup w(t, x(t)) \neq 0$. Then, there exist divergent sequences $\{t_n\}$ and $\{t_n^*\}$ and $\epsilon > 0$ such that

$$\begin{cases} t_i < t_i^* < t_{i+1} & i = 1, 2, 3, ... \\ w(t_i, x(t_i)) = \frac{\epsilon}{2} & \\ w(t_i^*, x(t_i^*)) = \epsilon & \\ \frac{\epsilon}{2} < w(t, x(t)) < \epsilon \quad \text{for} \quad t_i < t < t_i^*. \end{cases} \tag{5.2.20}$$

Suppose that $D^+ w(t, x(t)) \leq M$, a positive constant. Then it is easy to show that $t_i^* - t_i \geq \frac{\epsilon}{2M}$. Furthermore,

$$\mathcal{V}_{1ip}(t_n^*, x(t_n^*)) \leq \mathcal{V}_{1ip}(t_0, x_0) - \sum_{i=1}^{n} \int_{t_i}^{t_i^*} w(s, x(s))\mathrm{d}s + N_{1p}$$

$$\leq \mathcal{V}_{1ip}(t_0, x_0) - (\frac{\varepsilon}{2})(\frac{\varepsilon}{2M})n + N_{1p}$$

where the last inequality is obtained using (5.2.17), (5.2.18), and (5.2.20). The foregoing estimate yields a contradiction for sufficiently large n. Hence,

$\lim_{t\to\infty} sup w(t, x(t)) = 0$. The argument is similar if $D^+w(t, x(t))$ is bounded from below. This result together with (5.2.19) implies that $\lim_{t\to\infty} w(t, x(t)) = 0$. Therefore, $w(t, t_0, x_0) \to 0 \quad \text{as} \quad t \to \infty$.
The next result offers another set of conditions for asymptotic stability whose proof can be constructed with suitable modifications following the proof of theorem 5.2.5. The details will be omitted.

Theorem 5.2.6. *Assume that the conditions of theorem 5.2.4 hold except for condition (iii) which is replaced by*

(iii) for every $0 < \eta < \rho_0$ *where* $\rho_0 = \frac{\rho}{|U|}$ *there exists* $V_{2i} \in C(R_+ \times S_m(\rho_0) \bigcap S_m^c(\eta), R_+^{\tilde{r}_i})$, *and* $Q_{2i} \in C(R_+^{\tilde{r}_i}, R_+^{\tilde{k}_i})$,*such that, if* $Q_{2i}(V_{2i}(t, y_i)) \equiv \mathcal{V}_{2i}(t, y_i)$ *and* $0 \le j \le \tilde{k}_i$, $\mathcal{V}_{2ij}(t, u)$ *is locally Lipschitzian in* u *with Lipschitz constant* L_{2ij},*and for*

$$b(\|y\|) \le \sum_{i=1}^{N} \sum_{j=1}^{\tilde{k}_i} \mathcal{V}_{2ij}(t, y_i) \le a(t, \|y\|) + \sum_{i=1}^{N} \sum_{j=1}^{k_i} \mathcal{V}_{1ij}(t, Uy_i)$$

where $a \in \mathcal{LK}$ *and* $b \in \mathcal{K}$, *and assume there exists* $w \in C(R_+ \times S_m(\rho_0), R_+)$ *such that* $w(t, y)$ *is locally Lipschitzian in* y, $D^+w(t, y)$ *is bounded above or below and for* $(t, y) \in R_+ \times S_m(\rho_0) \bigcap S_m^c(\eta)$

$$c_1(\|y\|) \le w(t, y) \le c_2(\|y\|) \quad c_1, c_2 \in \mathcal{K}$$

$$D^+\mathcal{V}_{2ip}(t, y_i) \le -w(t, y) \quad 1 \le i \le N \quad 1 \le p \le \tilde{k}_i$$

$$D^+\mathcal{V}_{2ip}(t, y_i) \le G_{2iq}(t, y) \quad q \ne p \quad 1 \le q \le \tilde{k}_i$$

where $G_2 \in C(R_+ \times R_+^{\tilde{k}}, R^{\tilde{k}})$, $\quad \tilde{k} = \sum_{i=1}^{N} \tilde{k}_i$, $\quad G_2(t, u)$ *is quasimonotone nondecreasing in* u, $G_2(t, 0) \equiv 0$.
Then the trivial solution of (5.2.1) *is asymptotically stable.*

The foregoing stability theorems have been formulated for the trivial solution, $x = 0$, corresponding to the equilibrium. In cases where the systems (5.2.1) and (5.2.2) do not admit the zero solution, Lyapunov stability which implies the existence of a self invariant (*S.I.*) set, is ruled out. However, for sets that are S.I. in the asymptotic sense, Lyapunov theory can be extended to yield stability properties that closely resemble those of the S.I. An extension of Theorem 5.2.4 to asymptotically self-invariant (*A.S.I.*) sets follows. But first, let us state the following definitions.

Suppose $x(t) = x(t, t_0, x_0)$ is any solution of $x' = f(t, x), \quad x(t_0) = x_0$, where $f \in C(R_+ \times R^n, R^n)$.

Definition 5.2.2. *(i) A set S is said to be S.I. with respect to (5.2.1) if $x_0 \in S$ implies $x(t) \in S$ for $t \geq t_0$.*

(ii) A set G is said to be A.S.I. with respect to (5.2.1) if given any monotonic decreasing sequence $\{\epsilon_p\}$, $\epsilon \to 0$ as $p \to \infty$, such that $x_0 \in G, t_0 \geq t_p(\epsilon_p)$ implies $x(t) \in S(G, \epsilon_p) = \{x \in R^n : \|x(t) - x_0\| < \epsilon_p\}$

(iii) The A.S.I. set G of (5.2.1) is said to be equistable if for $\epsilon > 0, \exists t_1(\epsilon), t_1(\epsilon) \to \infty$ as $\epsilon \to 0$, and a $\delta = \delta(t_0, \epsilon), t_0 \geq t_1(\epsilon)$ which is continuous in t_0 for each ϵ such that $x(t) \in S(G, \epsilon), t \geq t_0 \geq t_1(\epsilon)$, provided $\|x_0\| < \delta$.

(iv) The A.S.I. set G of (5.2.1) is asymptotically stable if it is stable and $\lim_{t\to\infty} \|x(t)\| = 0$.

We now state the following theorem analogous to Theorem 5.2.4

Theorem 5.2.7. *Assume that conditions $(i)-(v)$ of Theorem 5.2.4 hold except for condition (ii) which is replaced by*

(ii) for $(t, x) \in R_+ \times S_n(\rho)$ and $i \leq j \leq r_i$

$$L_{iij}(t, x_i)\|R_i(t, x)\| \leq \lambda_{ij}(t) \quad \lambda_{ij} \in L^1.$$

Also assume that

(vi) the A.S.I. set $u = 0$ of $u' = G_3(t, u)$, $u(t_0) = u_0$, where $G_3(t, u) = \lambda(t) + G_1(t, u)$,is equistable,

(vii) the A.S.I. set $v = 0$ of $v' = G_4(t, v)$, $v(t_0) = v_0$ where $G_4(t, v) = W(t, b^{-1}(v)) + G_2(t, v)$ is uniformly stable.

Then the A.S.I. set $x = 0$ of (5.2.11) is equistable.

Proof Let $0 < \epsilon < \rho$ and $t_0 \in R_+$ be given.By uniform stability of (vii), given $b(\epsilon) > 0$, $\exists t_1(\epsilon)$, $t_1(\epsilon) \to \infty$ as $\epsilon \to 0$, and a $\delta_0 = \delta_0(\epsilon)$, $t_0 \geq t_1(\epsilon)$, such that

$$v(t, t_0, v_0) \leq b(\epsilon) \quad t \geq t_0 \geq t_1(\epsilon)$$

whenever $v_0 < \delta_0$, $v(t, t_0, v_0)$ is any solution of (vii). By the equistability of (v), given $\delta_0 > 0, \exists t_2(\varepsilon) > 0$ and $\delta_1 = \delta_1(t_0, \epsilon)$, $t_0 \geq t_2(\epsilon)$ such that

$$u(t, t_0, v_0) < \frac{\delta_0}{2} \quad \text{whenever} \quad u_0 < \delta_1 \quad \text{provided} \quad t_0 \geq t_2(\epsilon),$$

$u(t, t_0, u_0)$ being any solution of (vi). Let $u_0 = \mathcal{V}_1(t_0, x_0) = Q(V_1(t_0, x_0))$. Since $\mathcal{V}_1(t_0, x_0)$ is continuous and $\mathcal{V}_1(t, 0) \equiv 0$, there exists $\delta_2 = \delta_2(t, \epsilon) > 0$ such that

$$\|x_0\| < \delta_2 \quad \text{and} \quad \mathcal{V}_1(t_0, x_0) < \delta_1$$

hold simultaneously.
In addition given $a \in \mathcal{K}$, $\exists \delta_3(\epsilon) > 0$ such that

$$a(\delta_3) < \frac{\delta_0}{2}.$$

Now let $\delta = \delta(t_0, \epsilon) = min[\delta_2, \delta_3]$ and $t_3(\epsilon) = max[t_1(\epsilon), t_2(\epsilon)]$. We claim that $\|x_0\| < \delta \Rightarrow \|x(t, t_0, x_0)\| < \epsilon, \quad t \geq t_0 \geq t_3(\epsilon) \geq 0$.

The proof of this claim is identical to that given in Theorem 5.2.4 with suitable modifications. The details will be omitted.

Example 5.2.2. Consider the following system

$$S_x : \dot{x} \quad = \quad A(t, x)x \quad x(t_0) = x_0$$

where

$$A(t,x) = \begin{bmatrix} -t - \|x\| + 1 & x_3 & -x_2 \\ \frac{x_3^2(x_2 - x_3)x_1}{1 + x_1^2(x_2 - x_3)^2}\phi(t, \|x\|) & -t - \|x\| + 1 & 0 \\ 0 & 0 & -t - \|x\| + 1 \end{bmatrix}$$

and

$$\phi(t, \|x\|) = \frac{2.10^3 e^{-(t-1)}}{2.10^3 \|x\|^3 + 1} \quad t \geq t_0.$$

First we consider the decomposition of S_x into 2 subsystems $x = (X_1, X_2)^T$ where $X_1 = x_1$ and $X_2 = (x_2, x_3)^T$

$$\begin{pmatrix} \dot{x_1} \\ \cdots \\ \dot{x_2} \\ \dot{x_3} \end{pmatrix} = \left[\begin{array}{c:cc} -t - \|x\| + 1 & x_3 & -x_2 \\ \cdots & \cdots & \cdots \\ \frac{x_3^2(x_2 - x_3)x_1}{1 + x_1^2(x_2 - x_3)^2}\phi(t, \|x\|) & -t - \|x\| + 1 & 0 \\ 0 & 0 & -t - \|x\| + 1 \end{array}\right] \begin{pmatrix} x_1 \\ \cdots \\ x_2 \\ x_3 \end{pmatrix}$$

for which we choose the following Lyapunov functions

$$V_{11}(t, X_1) = x_1^2 \quad \text{and} \quad V_{12}(t, X_2) = (x_2 - x_3)^2 + 4 \cdot 10^3 e^{-(t-1)} x_3^2$$

Their derivatives with respect to t along the solutions of S_x are

$$\begin{aligned} \dot{V}_{11}(t, X_1) &= 2x_1 \dot{x}_1 \\ &= 2(-t - \|x\| + 1)V_{11}(t, X_1) \\ &\leq 2(-t + 1)V_{11}(t, X_1) \quad \text{for} \quad \|x\| \geq 0 \end{aligned} \tag{5.2.21}$$

$$\dot{V}_{12}(t, X_2) = 2(x_2 - x_3)(\dot{x}_2 - \dot{x}_3) - 4 \cdot 10^3 e^{-(t-1)} \cdot x_3^2 + 8 \cdot 10^3 e^{-(t-1)} x_3 \dot{x}_3$$

$$= \frac{2(x_1^2)(x_2 - x_3)^2}{1 + x_1^2(x_2 - x_3)^2} \phi + 2V_{12}(t, X_2)(-t - \|x\| + 1) - 4 \cdot 10^3 e^{-(t-1)} x_3^2$$

$$\leq 4 \cdot 10^3 \cdot \frac{e^{-(t-1)} x_3^2}{2 \cdot 10^3 \|x\|^3 + 1} + 2(-t - \|x\| + 1)V_{12}(t, X_2) - 4 \cdot 10^3 e^{-(t-1)} x_3^2$$

$$\leq 2(-t + 1)V_{12}(t, X_2) \quad \text{for} \|x\| \geq 0. \tag{5.2.22}$$

Now consider the transformations $y = Vx$ or $X = Uy$ where V and U are the matrices defined in example 5.2.1. We can show that the solutions of (5.2.1) are included in the solutions of the following system.(see example 5.2.1)

$$S_y : \dot{y} = \begin{pmatrix} y_1 \\ y_2 \\ \cdots \\ y_3 \\ y_4 \end{pmatrix} =$$

$$\begin{pmatrix} -t - \|x\| + 1 & x_3 & \vdots & 0 & -x_2 \\ \frac{(x_2 - x_3)x_1}{1 + x_1^2(x_2 - x_3)^2}\phi & -t - \|x\| + 1 & \vdots & 0 & 0 \\ \cdots & \cdots & \vdots & \cdots & \cdots \\ \frac{(x_2 - x_3)x_1}{1 + x_1^2(x_2 - x_3)^2}\phi & 0 & \vdots & -t - \|x\| + 1 & 0 \\ 0 & 0 & \vdots & 0 & -t - \|x\| + 1 \end{pmatrix} \begin{pmatrix} y_1 \\ y_2 \\ \cdots \\ y_3 \\ y_4 \end{pmatrix}$$

where $y = (y_1, y_2, y_3, y_4) = (x_1, x_2, x_2, x_3)$.

Now consider the decomposition of S_y into 2 subsystems as shown above by the dotted lines. We choose the following Lyapunov functions

$$\begin{aligned} V_{21}(t,(y_1,y_2)) &= 3y_1^2 + 2y_2^2 + 4\cdot 10^3 e^{-(t-1)} \\ V_{22}(t,(y_3,y_4)) &= (y_3 - y_4)^2 + 4\cdot 10^3 e^{-(t-1)} y_4^2 \\ V_{23}(t,(y_3,y_4)) &= (y_3 + y_4)^2 + 4\cdot 10^3 e^{-(t-1)} y_4^2 \end{aligned}$$

which satisfy the following inequalities

$$2\|y\|^2 \le \sum_{i=1}^{3} V_{2i}(t,y) \le \sum_{i=1}^{2} V_{1i}(t,U_y) + 2\|y\|^2 + 4.10^3 e^{-(t-1)}(2\|y\|^2 + 1).$$

Their derivatives with respect to time along solutions of S_y are (expressed in terms of the $x_i's$)

$$\dot V_{21}(t,Vx) = 6x_1\dot x_1 + 4x_2\dot x_2 - 4.10^3 e^{-(t-1)}$$

$$= 2(-t - \|x\| + 1)V_{21}(t,Vx) + \frac{4x_2x_1^2(x_2-x_3)}{1+x_1^2(x_2-x_3)^2}\cdot\frac{2\cdot 10^3 e^{-(t-1)}x_3^2}{2\cdot 10^3\|x\|^3+1} - 4\cdot 10^3 e^{-(t-1)}$$

$$\le 2(-t - \|x\| + 1)V_{21}(t,Vx) + 8\cdot 10^3\|x\|^4\cdot\frac{e^{-(t-1)}x_3^2}{2\cdot 10^3\|x\|^3+1} - 4\cdot 10^3 e^{-(t-1)}$$

$$\le -2tV_{21}(t,Vx) + (4\|x\|^3 - 4\cdot 10^3)\cdot e^{-(t-1)} \text{ for } t \ge t_0 \ge 1$$

$$\le -2tV_{21}(t,Vx) \; for \; 1 \le \|x\| \le 10. \qquad (5.2.23)$$

Similarly, we can show that

$$\begin{aligned} \dot V_{22}(t,Vx) &= 2(x_2-x_3)(\dot x_2 - \dot x_3) - 4\cdot 10^3 e^{-(t-1)}\cdot x_3^2 + 8\cdot 10^3 e^{-(t-1)} x_3\dot x_3 \\ &\le -2tV_{22}(t,Vx), \qquad 1 \le \|x\| \le 10, \quad t \ge t_0 \ge 1 \qquad (5.2.24) \end{aligned}$$

and,

$$\dot V_{23}(t,Vx) \le -2tV_{23}(t,Vx), \qquad 1 \le \|x\| \le 10, \quad t \ge t_0 \ge 1. \qquad (5.2.25)$$

The comparison problems corresponding to the equations (5.2.21) - (5.2.25) are

$$\text{for} \quad 0 \leq \|x\| \leq 10 \quad \begin{cases} I_a & \dot{u}_1 = 2(-t+1)u_1, & u_1(t_0) = u_{10} \\ I_b & \dot{u}_2 = 2(-t+1)u_2, & u_2(t_0) = u_{20} \end{cases}$$

$$\text{for} \quad 1 \leq \|x\| \leq 10,\ t \geq t_0 \geq 1 \begin{cases} II_a & \dot{v}_1 = -2tv_1, & v_1(t_0) = v_{10} \\ II_b & \dot{v}_2 = -2tv_2, & v_2(t_0) = v_{20} \\ II_c & \dot{v}_3 = -2tv_3, & v_3(t_0) = v_{30}. \end{cases}$$

For $t \geq t_0 \geq 1$ it is clear that the A.S.I. sets $\{u_1 = 0\}$ and $\{u_2 = 0\}$ are equistable (for a proof see example 5.2.1) whereas the A.S.I. sets $\{v_1 = 0\}$, $\{v_2 = 0\}$, $\{v_3 = 0\}$ are uniformly stable since $v_1(t) = v_{10}e^{-(t^2-t_0^2)} \leq v_{10} \Rightarrow v_1(t) < \varepsilon$ if $v_{10} < \delta = \varepsilon$. Hence, by theorem 5.2.7, the A.S.I. set $\{x = 0\}$ of (5.2.1) is equistable.

5.3 The Measure of Nonconvexity

As we have seen, the measure of noncompactness which was introduced by Kuratowski has now become an important tool in nonlinear analysis. Following Kuratowski, we shall introduce a measure of nonconvexity which has many properties in common with the measure of noncompactness and therefore we may now have convexity where we previously had compactness in the statement of some results.

Let E be a Banach space (with norm $\|\cdot\|$) and A is a subset in E. Denote by $co(A)$ the convex hull of A. We say that A is a α-measurable with measure $\alpha(A)$ if

$$\alpha(A) = \sup_{b \in co(A)} \inf_{a \in A} \|b - a\| < \infty. \tag{5.3.1}$$

Alternatively, if $H(X, Y)$ denotes the Hausdorff distance between two subsets X and Y,

$$\alpha(A) = H(A, co(A)). \tag{5.3.2}$$

Clearly, a bounded set is α-measurable.

From the definition, the following prperties of α can be derived in a straightforward manner.

$$\alpha(A) = 0 \text{ iff } \overline{A} \,(\text{the closure of } A) \text{ is convex}; \tag{5.3.3}$$

$$\alpha(\lambda A) = |\lambda|\alpha(A) \text{ for } \lambda \in \mathbb{R}^1 \ \big(\text{where } \lambda A = \{\lambda A \,|\, a \in A\}\big); \tag{5.3.4}$$
$$\alpha(A+B) \leq \alpha(A) + \alpha(B); \tag{5.3.5}$$
$$|\alpha(A) - \alpha(B)| \leq \alpha(A-B); \tag{5.3.6}$$
$$\alpha(\overline{A}) = \alpha(A); \tag{5.3.7}$$
$$\alpha(A) \leq \text{diam}(A) \quad (\text{the diameter of } A); \tag{5.3.8}$$
$$|\alpha(A) - \alpha(B)| \leq 2H(A,B). \tag{5.3.9}$$

Note that all of these properties are shared by the measure of noncompactness γ. Recall $\gamma(A) = \inf\{d > 0 \mid A$ can be covered by a finite number of sets of diameter $\leq d\}$. α is not monotone in the sense that $\alpha(A) \leq \alpha(B)$ if $A \subset B$. If it did, then every closed set would be convex which is not true. Unfortunately $\alpha(A)$ measures only the nonconvexity of $\overline{A}$ and not A itself if A is not closed.

As a consequence of (5.3.9) and a similar inequality for γ, $|\gamma(A) - \gamma(B)| \leq H(A,B)$, the measures α and γ are continuous with respect to the Hausdorff metric.

Proposition 5.3.1. *Let A_n be a sequence of subsets of E such that A_n approaches a subset A_∞ in the Hausdorff metric. Then,*

(i) if A_n are α-measurable,

$$\lim_{n\to\infty} \alpha(A_n) = \alpha(A_\infty); \tag{5.3.10}$$

(ii) if A_n are bounded,

$$\lim_{n\to\infty} \gamma(A_n) = \gamma(A_\infty). \tag{5.3.11}$$

Proposition 5.3.2. *(Kuratowski) Let (X,ρ) be a complete metric space and let $A_0 \supset A_1 \supset \ldots$ be a decreasing sequence of nonempty, closed subsets of E. Assume $\gamma(A_n) \to 0$. Then if we write $A_\infty = \cap_{n\geq 0} A_n$, A_∞ is a nonempty compact set and A_n approaches A_∞ in the Hausdorff metric.*

Proposition 5.3.3. *Let $A_0 \supset A_1 \supset \ldots$ be a decreasing sequence of closed bounded subsets of E. Let $A_\infty = \cap_{n\geq 0} A_n$. Then A_∞ is nonempty, convex and compact and A_n converges to A_∞ in the Hausdorff metric iff $\alpha(A_n) \to 0$ and $\gamma(A_n) \to 0$.*

Proof. Suppose $\gamma(A_n) \to 0$. It follows from Proposition 5.3.2 that A_n converges to the nonempty compact set A_∞ in the Hausdorff metric. If, in addition,

$\alpha(A_n) \to 0$ then in view of (5.3.10), $\alpha(A_\infty) = 0$. Since A is also closed, A_∞ is convex by (5.3.3).

Suppose $A_n \to A_\infty$ in the Hausdorff metric, and $\alpha(A_\infty) = \gamma(A_\infty) = 0$. Then, by (5.3.10) and (5.3.11), $\alpha(A_n) \to 0$ and $\gamma(A_n) \to 0$.

Proposition 5.3.4. *Let $A_0 \supset A_1 \supset \ldots$ be a decreasing sequence of closed, bounded subsets of E such that $\alpha(A_n) \to 0$ and $\gamma(A_n) \to 0$. Suppose T is a continuous map of $A_0 \to A_0$ such that*

$$Tx \in A_n, \text{ if } x \in A_n, \ n = 0, 1, \ldots . \tag{5.3.12}$$

Then there exists an $x \in A_\infty = \cap_{n\geq 0} A_n$ such that

$$Tx = x. \tag{5.3.13}$$

Proof. The result is a corollary of the Schauder principle since, from Proposition 5.3.3, A_∞ is nonempty, convex and compact and T maps A_∞ into itself.

Closely associated with the notion of measure of noncompactness is the concept of k-set-contraction. Let (X_1, d_1) and (X_2, d_2) be metric spaces and suppose $T : X_1 \to X_2$ is a continuous map. We say T is a k-set-contraction if given any bounded set A in X_1, $T(A)$ is bounded and $\gamma_2(T(A)) \leq k\gamma_1(A)$ where γ_i denotes the measure of noncompactness in X_i, $i = 1, 2$.

Proposition 5.3.5. *Let C be a closed, bounded, convex set and $T : C \to C$ a k-set-contraction, $k < 1$. Then T has a fixed point, i.e., a point x satisfying* (5.3.13).

The above generalization of the Schauder principle was further extended by introducing a comparison function ψ which has the following properties: (i) ψ maps a conical segment of regular cone in a partially ordered space into itself; (ii) ψ is monotone; (iii) ψ is upper semi-continuous from the right; (iv) $\psi(x) = x$ if $x = \theta$ (the zero of the space). Then Darbo's condition $\gamma(T(A)) \leq k\gamma(A)$, $k < 1$, is replaced by the weaker condition

$$\gamma(T(A)) \leq \psi(\gamma(A)). \tag{5.3.14}$$

Definition 5.3.1. *A function $\psi : [0, \infty) \to [0, \infty)$ is a comparison function if (i) $\psi(t) < t$ for $t > 0$, (ii) $\psi(0) = 0$, ψ is upper semi-continuous from the right.*

Proposition 5.3.6. *Let ψ be a comparison map and let $S_0, S_1, \cdots$ be a sequence of nonnegative real numbers such that $S_n \leq \psi(S_{n-1})$, $n = 1, 2, \cdots$. Then the sequence S_n converges to zero.*

Proof. Since, $S_n \leq \psi(S_{n-1}) \leq S_{n-1}$, the sequence S_n converges monotonically. Suppose $S_\infty = \lim S_n > 0$. Then $\psi S_\infty < S_\infty < S_n$, $n = 1, 2, \cdots$. But this contradicts the upper semi-continuity from the right.

Proposition 5.3.7. *Let $\psi : [0, a) \to [0, a)$ be nonincreasing, upper semicontinuous from the right, and $\psi(t) = t$ iff $t = 0$. Then ψ has an extension to $[0, \infty)$ which is a comparison function.*

Proof. Since the interval $[0, a]$ is a segment of the regular coneR_+, it follows that if $t \leq \psi(t)$ then $t \leq t_0$ where t_0 is the maximal solution of $\psi(t) = t$. By assumption $t_0 = 0$. Thus $t \leq \psi(t)$ iff $t = 0$. If we define $\phi(t) = \psi(a)$, $t \geq a$ then ϕ is a comparison function.

Definition 5.3.2. *Let $(X_1, \|\cdot\|_1)$ and $(X_2, \|\cdot\|_2)$ be Banach spaces and suppose $T : X_1 \to X_2$ is a continuous map. We say that T is a ψ-set-contraction with respect to convexity $\langle$ compactness$\rangle$ if given any α_1-measurable $\langle$ bounded$\rangle$ set A in X_1, $T(A)$ is α_2-measurable $\langle$ bounded$\rangle$ and*

$$\alpha_2(T(A)) \leq \psi(\alpha_1(A)) \tag{5.3.15}$$

$$\langle \gamma_2(T(A)) \leq \psi(\gamma_1(A)) \rangle \tag{5.3.16}$$

where $\alpha_i \langle \gamma_i \rangle$ denotes the measure of nonconvexity $\langle$ noncompactness$\rangle$ in X_i, $i = 1, 2$. We say that T is a ψ-contraction if $\|Tx - Ty\|_2 \leq \psi(\|x - y\|_1)$ for every $x, y \in X_1$.

The following result is a generalization of a similar result in regard to relating the notion of k-concentration, i.e., a ψ-contraction with $\psi(t) = kt$, to the notion of k-set-contraction.

Proposition 5.3.8. *Let $(X_1, \|\cdot\|_1)$ and $(X_2, \|\cdot\|_2)$ be Banach spaces. Let T be a ψ-contraction, then (i) T is a ψ-set-contraction with respect to compactness; (ii) $H(TA, TB) \leq \psi(H(A, B))$ whenever $H(A, B) < \infty$; (iii) if for every α-measurable set A, $co(TA) \subset \overline{T(\overline{co}(A))}$ (where $\overline{co}(X)$ denotes the convex closure of X), then T is ψ-set-contraction with respect to convexity.*

Proof. (i) Let A be a bounded set in X_1 and suppose $\gamma_1(A) = d$. Then given $\epsilon > 0$, we can write $A = \cup_{j=1}^m S_j$, $\text{diam}(S_j) \leq d + \epsilon$. Thus $T(A) = \cup_{j=1}^m T(S_j)$ and since T is a ψ-contraction, $\text{diam}(T(S_j)) \leq \psi(d + \epsilon)$. Let ϵ_i be a sequence of positive numbers converging to zero such that $\psi(d + \epsilon_i)$ converges and let $b = \lim \psi(d + \epsilon_i)$. Then by upper semi-continuity from the right, $b \leq \psi(d)$. Hence $\gamma_2(TA) \leq \psi(d)$. (ii) Let A and B be sets such that $H(A, B) = d < \infty$.

Let $b \in B$. Then $\inf\{\|Tb - Ta\|_2, a \in A\} \leq \inf\{\psi(\|b - a\|_1), a \in A\} \leq \psi d$ by the upper semi-continuity from the right of the function ψ. Similarly, $\inf\{\|Ta - Tb\|_2, b \in B\} \leq \psi d$. Thus $H(TA, TB) \leq d$. (iii) Let A be an α-measurable set in X_1. Then from (ii), $\alpha(TA) = H(TA, \overline{co}(TA)) \leq H(TA, \overline{T(\overline{co}(A))}) = H(TA, T(\overline{co}(A))) \leq \psi(H(A, \overline{co}A)) = \psi\alpha(A)$.

Proposition 5.3.9. *Let A be a closed subset of a Banach space and T a map from A onto itself. If T is set contractive with respect to convexity (compactness) then A is convex (compact). In particular, the set of fixed points of a set contractive with respect to convexity (compactness) map of a closed subset of a Banach space $\mathfrak{B}$ into $\mathfrak{B}$ is convex (compact).*

Proof. Set $m = \alpha(T(A)) = \alpha(A)$ $(m = \gamma(T(A)) = \gamma(A))$. Then $m \leq \psi(m)$. If $m > 0$ then $\psi(m) < m$. But this is impossible. Clearly $m = 0$.

Proposition 5.3.10. *Let C be a closed, bounded set and $T : C \to C$ a ψ_1-set-contraction with respect to convexity and a ψ_2-set-contraction with respect to compactness. The set of fixed points of T is nonempty, convex and compact.*

Proof. Let $C_0 = C$ and $C_{n+1} = \overline{T(C_n)}$. Then $C_{n+1} \subset C_n$. Let $s_n = \gamma(C_n)$, $t_n = \alpha(C_n)$, then it follows from Proposition 5.3.6 that $s_n \to 0$ and $t_n \to 0$. By Proposition 5.3.4, the set $F(T)$ of fixed points of T is nonempty and, by Proposition 5.3.9, it is also convex and compact.

Let E be a real Banach space and let $\|\cdot\|$ denote the norm in E. We let $B = \{x \in E \mid \|x\| \leq b\}$ denote the ball of radius b and let $R_0 = [t_0, t_0 + a] \times B$ where $t_0 \geq 0$, $a > t_0$.

Consider the differential equation

$$x' = f(t, x), \quad x(t_0) = x_0, \tag{5.3.17}$$

where $f \in C(R_0, E)$. There are several known results which guarantee the existence of solutions to (5.3.17). We mention only one regarding the convexity of the solution set. We need the following conditions:

(I) f is uniformly continuous in R_0.

Another is a compactness condition which is similar to the convexity condition (II) stated below.

For any subset $A \subset B$ and for small $h > 0$ set

$$A_h(f) = \{y \mid y = x + hf(t, x) : x \in A\}.$$

We introduce a (comparison) scalar differential equation

$$u' = g(t, u), \quad u(t_0) = 0, \tag{5.3.18}$$

where $g \in C([t_0, t_0 + a] \times R^+, R)$. Assume that $u \equiv 0$ is the unique solution of (5.3.18). Then the convexity condition on f is

(II) $$\liminf_{h\to 0^+}\{h^{-1}[\alpha(A_h(f))-\alpha(A)]\}\le g(t,\alpha(A))$$

for any subset $A\subset B$.

We also require the following condition on a set $A\subset B$:

(III) The set of solutions $x(t,x_0)$, $x_0\in A$ of (5.3.17) exists and is equicontinuous.

Proposition 5.3.11. *Let $A\subset B$ have convex closure and let conditions I, II and III be satisfied for* (5.3.17). *Then the set*

$$x(t,t_0,A)=\{x(t,t_0,x_0)\,|\,x_0\in A\}$$

has convex closure for $t\in[t_0,t_0+a]$.

Proof. Set $m(t)=\alpha(x(t,A))$ where α is the measure of nonconvexity and $x(t,A)=x(t,t_0,A)$. Our claim is then $m(t)=0$. Now $m(t+h)-m(t)=\alpha(x(t+h),A))-\alpha(x(t,A))=[\alpha(x(t+h),A)-\alpha(A_h(f))]+[\alpha(A_h(f))-\alpha(x(t,A))]$. If we know that

$$\liminf_{h\to 0^+} h^{-1}\big[\alpha(x(t+h,A))-\alpha(A_h(f))\big]\le 0 \tag{5.3.19}$$

then it follows from condition II that $D_+m(t)\le g(t,m(t))$ where D_+ denotes a Dini derivative. It follows further from the theory of differential inequalities that $m(t)\equiv 0$. Thus it remains to verify (5.3.19).

By properties of (5.3.4), (5.3.6) and (5.3.9)

$$\begin{aligned} h^{-1}\big[\alpha(x(t+h,A))-\alpha(A_h(f))\big] &\le \alpha\big[h^{-1}(x(t+h,A)-A_h(f))\big]\\ &\le 2\sup_{x_0\in A}\big|h^{-1}\big[(x(t+h,x_0)-x(t,x_0)\big]-f(t,x(t,x_0))\big|. \end{aligned}$$

Hence it suffices to show that

$$h^{-1}(x(t+h,x_0)-x(t,x_0))\to f(t,x(t,x_0))$$

uniformly in x_0. Now

$$\begin{aligned} &\|h^{-1}(x(t+h,x_0)-x(t,x_0))-f(t,x(t,x_0))\|\\ &\qquad\le h^{-1}\int_t^{t+h}\|f(t+s,x(t+s,x_0))-f(t,x(t,x_0))\|\mathrm{d}s. \end{aligned}$$

By the uniform continuity of f and by the equicontinuity of $x(t,x_0)$ this last expression can be made arbitrarily small, independent of x_0, by taking h sufficiently small. This concludes the argument.

Remark. Suppose further that A is compact and that the semi-group map $x_0 \to x(t, x_0)$ is continuous for each $t \in [t_0, t+a]$. Then $x(t, A)$ is closed and hence, by Proposition 5.3.11, also convex. In particular, if $y_0, z_0 \in A$, set $y = x(\bar{t}, y_0)$, $z = x(\bar{t}, z_0)$ for some $\bar{t} \in [t_0, t+a]$. Then we know that if w lies on the line segment connecting y and z there exists w_0 on the line segment connecting y_0 to z_0 such that $x(t, w_0) = w$, i.e., w is an attainable target.

5.4 Existence of Solution in Weak Topology

In this section, we wish to consider the Cauchy problem

$$x' = f(t, x), \quad x(t_0) = x_0 \tag{5.4.1}$$

where f satisfies the hypothesis:
(H_1) f is weakly continuous on R_0 and $\|f(t,x)\| \le M$ on R_0 where

$$R_0 = \{(t,x) : t_0 \le t \le t_0 + a, \quad \|x - x_0\| \le b\}.$$

We shall assume throughout this section that E_w (the Banach space E endowed with the weak topology) is weakly complete. Looking at hypothesis (H_1), it may seem more natural to assume boundedness of f in terms of each $\phi \in E^\star$. However, by Theorem 1.2.10, it is known that a subset of A of a normed space is bounded if and only if the set $\{\phi(x) : x \in A\}$ is bounded for each $\phi \in E^\star$.

As usual, the technique for proving the existence of solutions of (5.4.1) consists of three steps, namely (i) construction of a sequence of approximate solutions, (ii) convergence of the sequence of approximate solutions and (iii) showing that the limit function is a solution of the Cauchy problem. Statements concerning steps (i) and (iii) are given in the following lemmas.

Lemma 5.4.1. *Suppose (H_1) holds. Let $\{\varepsilon_n\}$ be a sequence such that $\varepsilon_n > 0$ and $\varepsilon_n \to 0$ as $n \to \infty$. Then there exists a sequence of approximate solutions $\{x_n(t)\}$ satisfying:*

(a) $x_n(t_0) = x_0$;

(b) $\{x_n(t)\}$ *is weakly equicontinuous and uniformly bounded on* $[t_0, t_0 + a]$;

(c) $x_n'(t) = f(t, x_n(t - \varepsilon_n))$, $t \in [t_0, t_0 + \alpha]$ *where* $\alpha = min(a, \frac{b}{M})$.

Proof. Let $\delta > 0$ be a suitably small constant and $x_0(t)$ a weakly differentiable function on $[t_0 - \delta, t_0]$ such that

$$x_0(t_0) = x_0, \; x_0'(t_0) = f(t_0, x_0), \quad \|x_0'(t)\| \le M \quad \text{and} \quad \|x_0 - x_0(t)\| \le b.$$

Notice that $x_0(t) = x_0 + (t - t_0) f(t_0, x_0)$ satisfies all the conditions with $\delta < \frac{b}{M}$. For $\varepsilon_n \leq \delta$, define $x_n(t)$ on $[t_0 - \delta, t_0 + \alpha]$ by

$$x_n(t) = \begin{cases} x_0(t), & t_0 - \delta \leq t \leq t_0, \\ x_0 + \int_{t_0}^{t} f(s, x_n(s - \varepsilon_n))ds, & t_0 \leq t \leq t_0 + \alpha. \end{cases} \tag{5.4.2}$$

This formula defines $x_n(t)$ on $[t_0, t_0 + \alpha_1]$, $\alpha_1 = min\{\alpha, \varepsilon_n\}$. For, by definition of $x_0(t)$, we have $\|x_n(s - \varepsilon_n) - x_0\| \leq b$ and thus, $f(s, x_n(s - \varepsilon_n))$ is well defined. Moreover, by (H_1), $f(s, x_n(s - \varepsilon_n))$ is weakly continuous on R_0 and hence weakly integrable on R_0. Also, by the Hahn-Banach theorem, there exists a $\phi \in E^\star$ such that $\|\phi\| = 1$ and $|\phi(x_n(t) - x_0)| = \|x_n(t) - x_0\|$. Then, for each $t \in [t_0, t_0 + \alpha]$, there exists a $\phi_t \in E^\star$ such that

$$\|x_n(t) - x_0\| = |\phi_t(x_n(t) - x_0)| = |\int_{t_0}^{t} \phi_t(f(s, x_n(s - \varepsilon_n)))ds|$$

$$\leq \int_{t_0}^{t} |\phi_t(f(s, x_n(s - \varepsilon_n)))|ds \leq M(t - t_0) \leq b.$$

This means that we can extend x_n up to $[t_0, t_0 + \alpha_2]$, where $\alpha_2 = min\{\alpha, 2\varepsilon_n\}$, and thus inductively on $[t_0, t_0 + \alpha]$. Observe that $\{x_n(t)\}$ is uniformly bounded because $\|x_n(t) - x_0\| \leq b$.

To see that $x_n(t)$ is equicontinuous; consider

$$x_n(t_1) - x_n(t_2) = \int_{t_2}^{t_1} f(s, x_n(s - \varepsilon_n))ds. \tag{5.4.3}$$

Again, by the Hahn-Banach theorem, there exists a $\phi \in E^\star$, $\|\phi\| = 1$ such that $|\phi(x_n(t_1) - x_n(t_2))| = \|x_n(t_1) - x_n(t_2)\|$. Applying this ϕ to $x_n(t_1) - x_n(t_2)$ in (5.4.3), we obtain $\|x_n(t_1) - x_n(t_2)\| \leq |t_1 - t_2|M$. We have thus shown that the sequence $\{x_n(t)\}$, as defined by (5.4.2), satisfies all the properties (a), (b) and (c). The lemma is proved.

Lemma 5.4.2. *Suppose* (H_1) *holds and that the sequence* $\{x_n(t)\}$ *obtained in lemma 5.4.1 converges weakly to* $x(t)$ *on* $[t_0, t_0 + \alpha]$. *Then* $x(t)$ *is a solution of* (5.4.1)

Proof. By the Ascoli-Arzela theorem, it is clear that $\{x_n(t)\}$ converges weakly uniformly to $x(t)$ and that $x(t)$ is weakly continuous. Let us now prove that

$f(t, x_n(t - \varepsilon_n))$ converges weakly uniformly to $f(t, x(t))$: Let $\widehat{\phi} \in E^\star$ be fixed and $\varepsilon > 0$. Since $f(t, x(t))$ is weakly continuous, for each $t \in [t_0, t_0 + \alpha]$ there exist $\delta_t > 0$ and an open set U_t in the weak topology containing $x(t)$ such that

$$|\widehat{\phi}(f(t, x(t)) - f(s, y))| < \frac{\varepsilon}{2},$$

whenever $|t - s| < \delta_t$ and $y \in U_t$. Without loss of generality, U_t can be considered as the basic open set.

$$U_t = \bigcap_{i=1}^{n_t} \{y : |\phi_{t_i}(x(t) - y)| < r_{t_i}\}.$$

Let $V_t = \bigcap_{i=1}^{n_t} \{y : |\phi_{t_i}(x(t) - y)| < \frac{r_t}{2}\}$, where $r_t = min\, r_{t_i}$. Since $[t_0, t_0 + \alpha]$ is compact and $x(t)$ is weakly continuous, there exists a finite subcover $V_1, V_2, ..., V_k$. Let

$$\phi = \{\phi \in E^\star : \phi \text{ is used in one of the } V_j's, j = 1, 2, ..., k\}.$$

Clearly ϕ is finite. Let $r = \min\{\frac{r_j}{2}, \quad j = 1, 2, ..., k\}$. By the equicontinuity of the family, for each $\phi \in \Phi$ there exists γ_ϕ such that

$$\phi(x_n(s) - x_n(t))| < \frac{r}{2} \quad \text{whenever} \quad |s - t| < \gamma_\phi.$$

Let $\gamma = \min_{\Phi} \gamma_\phi$. Choose N_1 so that $\varepsilon_n < \gamma$ for $n > N_1$. By the uniform convergence of the sequence $\{x_n(t)\}$, for each $\phi \in \Phi$ there exists N_ϕ such that $|\phi(x_n(t) - x_n(s))| < \frac{r}{2}$ whenever $n > N_\phi$. Let $N = \max_{\Phi}\{N_1, N_\phi\}$.

Now, let $n > N$ and $t \in [t_0, t_0 + \alpha]$. Then

$$x(t) \in V_{t_j} = \bigcap_{i=1}^{n_j} \{y : |\phi_i(x(t_j) - y)| < \frac{r_j}{2}\} \subset U_{t_j}.$$

Thus, $|\widehat{\phi}(f(t, x(t))) - f(t, x(t_j))| < \frac{\varepsilon}{2}$. For all $\phi \in \Phi$,

$$|\phi(x_n(t - \varepsilon_n) - x(t))| \leq |\phi(x_n(t - \varepsilon_n) - x_n(t))| + |\phi(x_n(t) - x(t))|$$

$$\leq \frac{r}{2} + \frac{r}{2} = r.$$

In particular, for each $i = 1, 2, ..., n_j$,

$$|\phi_i(x_n(t) - x(t_j))| \leq |\phi_i(x_n(t - \varepsilon_n) - x(t))| + |\phi_i(x(t) - x(t_j))|$$

$$\leq r + \frac{r_j}{2} \leq r_j.$$

Hence $x_n(t - \varepsilon_n) \in U_{t_j}$ and $|\phi(f(t, x_n(t - \varepsilon_n)) - f(t, x(t_j)))| < \frac{\varepsilon}{2}$.Now

$$\widehat{\phi}(f(t, x_n(t - \varepsilon_n)) - f(t, x(t)))| \leq |\widehat{\phi}(f(t, x_n(t - \varepsilon_n)) - f(t, x(t_j)))| + |\widehat{\phi}(f(t, x(t_j)) - f(t, x(t))| \leq \frac{\varepsilon}{2} + \frac{\varepsilon}{2}.$$

Recalling that $\phi(x_n(t)) = \phi(x_0 + \int_{t_0}^{t} f(s, x_n(s - \varepsilon_n))ds)$ and taking the limit of the two sides as $n \to \infty$, we obtain the desired result.

We shall now focus our attention on showing the convergence of the sequence of approximate solutions. As in the case of existence of strong solutions, we utilize weak compactness type conditions and weak dissipative type conditions to achieve this goal. Note that if we assume E is reflexive, we do not need any additional assumptions for proving an existence result because $\{x_n(t)\}$ is weakly compact.

(a) Weak Compactness Type Condition

Here we shall employ the measure of weak noncompactness $\beta^\star$ discussed in Section 1.5, to impose conditions on f. Specifically, let us first prove the following result.

Theorem 5.4.1. *Let (H_1) hold. Suppose further that $\beta^\star(f(I \times A)) \leq g(\beta^\star(A))$, where $I = [t_0, t_0 + \alpha]$, $A \subset R_0$ and $g \in C(R_+, R_+)$. Assume that $u(t) \equiv 0$ is the unique solution of $u' = g(u)$, $u(t_0) = 0$ on $[t_0, t_0 + \alpha]$. Then there exists a solution $x(t)$ for the problem (5.4.1) on $[t_0, t_0 + \alpha]$, where $\alpha = min(a, \frac{b}{M})$.*

Proof. Let $\{x_n(t)\}$ be the sequence of approximate solutions of (5.4.1) constructed in Lemma 5.4.1. In view of the Ascoli-Arzela Theorem (Theorem 1.2.6 and Lemma 5.4.2, it is sufficient to show $\beta^\star(\{x_n(t)\}_{n=1}^{\infty}) = 0$. Define $m(t) = \beta^\star(\{x_n(t)\}_{n=1}^{\infty})$ and note that $m(t_0) = 0$. The continuity of $m(t)$ is clear from the equicontinuity of the sequence $\{x_n(t)\}$ and property (vi) of $\beta^\star$ in Theorem 1.6.1. Now

$$D^+ m(t) = \lim_{\tau \to 0^+} \sup_{h \in [0,\tau]} \frac{m(t+h) - m(t)}{h} \leq \lim_{\tau \to 0^+} \sup_{h \in [0,\tau]} \frac{\beta^\star(\{x_n(t+h) - x_n(t)\})}{h},$$

by property (vi) of $\beta^\star$.

Using Lemma 1.4.4 and property (v) of Theorem 1.6.1, we get

$$\begin{aligned}
D^+m(t) &\leq \lim_{\tau\to 0^+} \sup_{h\in[0,\tau]} \beta^\star\Big(\bigcup_{n=1}^{\infty} \overline{co}\{x_n'(s) : s\in[t,t+h]\}\Big) \\
&\leq \lim_{\tau\to 0^+} \beta^\star\Big(\overline{co}\bigcup_{n=1}^{\infty}\{f(s,x_n(s-\varepsilon_n)) : s\in[t,t+\tau]\}\Big) \\
&= \lim_{\tau\to 0^+} \beta^\star\Big(\bigcup_{n=1}^{\infty}\{f(s,x_n(s-\varepsilon_n)) : s\in[t,t+\tau]\}\Big) \\
&\leq \lim_{\tau\to 0^+} \beta^\star\Big(\bigcup_{n=1}^{\infty}\{f(I\times\{x_n(x-\varepsilon_n)\}) : s\in[t,t+\tau]\}\Big) \\
&\leq \lim_{\tau\to 0^+} g\Big(\beta^\star\Big(\bigcup_{n=1}^{\infty}\{x_n(s-\varepsilon_n) : s\in[t,t+\tau]\}\Big)\Big) \\
&\leq g\Big(\lim_{\tau\to 0^+} \beta^\star\Big(\bigcup_{n=1}^{\infty}\{x_n(s-\varepsilon_n) : s\in[t,t+\tau]\}\Big)\Big) \\
&= g\big(\beta^\star(\{x_n(t-\varepsilon_n)\}_{n=1}^{\infty})\big).
\end{aligned}$$

Given $\varepsilon > 0$, by the weak equicontinuity of the family $\{x_n(t)\}$, for ε_n sufficiently small or, equivalently, n sufficiently large $\|(x_n(t-\varepsilon_n) - x_n(t))\| < \varepsilon$.
By property (ix) of Theorem 1.6.1 we conclude that $\beta^\star(\{x_n(t-\varepsilon_n)-x_n(t)\}) < \varepsilon$. But ε being arbitrary, we have $\beta^\star(\{x_n(t-\varepsilon_n) - x_n(t)\}) = 0$.

Since

$$\{x_n(t-\varepsilon_n)\} = \{x_n(t-\varepsilon_n) - x_n(t) + x_n(t)\} \subset \{x_n(t-\varepsilon_n) + x_n(t)\} + \{x_n(t)\},$$

it follows that

$$\beta^\star(\{x_n(t-\varepsilon_n)\} \leq \beta^\star(\{x_n(t-\varepsilon_n) - x_n(t)\}) + \beta^\star(\{x_n(t)\}) = 0 + \beta^\star(\{x_n(t)\}).$$

Similarly

$$\beta^\star(\{x_n(t)\}) \leq \beta^\star(\{x_n(t-\varepsilon_n)\})$$

and therefore

$$\beta^\star(\{x_n(t-\varepsilon_n)\}) = \beta^\star(\{x_n(t)\}).$$

Consequently, we obtain

$$D^+m(t) \le g(m(t)), \quad t \in [t_0, t_0+\alpha].$$

This implies, by Theorem 1.6.1, that $m(t) \le r(t)$ where $r(t)$ is the maximal solution of $u' = g(u), \quad u(t_0) = 0$. Since by assumption $r(t) \equiv 0$, we have $m(t) \equiv 0$. The proof of the theorem is thus complete.

(b) **Weak Dissipative Type Condition**

Here we shall utilize a weak dissipative type condition on f to prove existence and uniqueness of solutions of the problem (5.4.1).

Theorem 5.4.2. *Suppose that:*

(i) f is weakly weakly uniformly continuous on R_0 and $\|f(t,x)\| \le M$ on R_0;

(ii) for each $\phi \in E^\star$ with $\|\phi\| = 1$,

$$\lim_{h\to 0^+} sup \frac{|\phi[x-y+h(f(t,x)-f(t,y))]| - |\phi(x-y)|}{h} \le g(t, |\phi(x-y)|)$$

where $g \in C((t_0, t_0+a) \times [0, 2b], R_+)$;

(iii) $u(t_0) \equiv 0$ is the only solution $u' = g(t,u),\ u(t_0) = 0$ on $[t_0, t_0+a]$.

Then there exists a unique solution $x(t)$ of (5.4.1) on $[t_0, t_0+a]$ where $\alpha = min\{a, \frac{b}{M}\}$.

Proof. As before, let $\{x_n(t)\}$ be the family of approximate solutions for (5.4.1) constructed as in Lemma 5.4.1. We want to show that $\{x_n(t)\}$ is weakly Cauchy, i.e., given $\phi, \widehat{\varepsilon}$ we need to find N such that $n, m > N$ implies $|\phi(x_n(t) - x_m(t))| < \widehat{\varepsilon}$.

For ε sufficiently small, the maximal solutions $r(t,\varepsilon)$ of $u' = g(t,u)+\varepsilon$, $u(t_0) = \varepsilon$ converge to zero as $\varepsilon \to 0$, by virtue of Lemma 1.7.1. Thus, for ε sufficiently small, $\varepsilon < \varepsilon_1, \quad r(t,\varepsilon) < \widehat{\varepsilon}$. By the weak weak uniform continuity of f, $|\phi_i(f(t,x) - f(s,y))| < \frac{\varepsilon_1}{2}$ whenever $|t-s| < \delta$ and $|\phi_i(x-y)| < \delta$, $i = 1, 2, ..., p$. By the weak equicontinuity of $\{x_n(t)\}, \quad |\phi_i(x_n(t) - x_n(s))| < \delta$ whenever $|t-s| < \gamma_i$. Let $\gamma = \min\limits_{i=1,2,\ldots,p} \{\gamma_i\}$. Choose N such that $\varepsilon_n < \gamma$ for $n > N$.

Let $n, m > N$ and define $m(t) = |\phi(x_n(t) - x_m(t))|$. Then

$$D^+m(t) \leq$$
$$\limsup_{h\to 0^+} \frac{|\phi(x_n(t)-x_m(t)+h(f(t,x_n(t))-f(t,x_m(t)))|-|\phi(x_n(t)-x_m(t))|}{h}$$
$$+\limsup_{h\to 0^+} \frac{|\phi(x_n(t+h)-x_m(t+h))-\phi(x_n(t)-x_m(t)+h(f(t,x_n(t)-f(t,x_m(t)))|}{h}.$$

This implies that

$$D^+m(t) \leq g(m(t)) + \limsup_{h\to 0^+} |\frac{\phi(x_n(t+h)-x_n(t)-hf(t,x_n(t)))}{h}|$$
$$+\limsup_{h\to 0^+} |\frac{\phi(x_n(t+h)-x_m(t)-hf(t,x_m(t)))}{h}|.$$

But

$$\limsup_{h\to 0^+} |\frac{\phi(x_n(t+h)-x_n(t)-hf(t,x_n(t)))}{h}|$$
$$\leq \limsup_{h\to 0^+} |\frac{\phi(x_n(t+h)-x_n(t)-hf(t,x_n(t-\varepsilon_n))))}{h}|$$
$$+\limsup_{h\to 0^+} |\phi(f(t,x_n(t-\varepsilon_n))-f(t,x_n(t)))| \leq 0 \ + \ \frac{\varepsilon_1}{2}.$$

Similarly

$$\limsup_{h\to 0^+} \frac{|\phi(x_m(t+h)-x_m(t)-hf(t,x_m(t)))|}{h} \leq \frac{\varepsilon_1}{2}.$$

Hence we have

$$D^+m(t) \leq g(t,m(t)) \ + \ \varepsilon_1, \quad m(t_0)=0.$$

Thus, by Theorem 1.7.1, $m(t) \leq r(t,\varepsilon_1) < \widehat{\varepsilon}$. We have now shown that $\{\phi x_n(t)\}$ is Cauchy for each ϕ. Since the space is weakly complete, $\{x_n(t)\}$ converges weakly to a function $x(t)$. Now, lemma 5.4.2 assures that $x(t)$ is a solution of (5.4.1). To see that the solution is unique; suppose that $x(t)$ and $y(t)$ are both solutions of (5.4.1). Let $p(t) = |\phi(x(t)-y(t))|$. Then

$$\limsup_{h\to 0^+} \frac{p(t+h)-p(t)}{h}$$
$$\leq \limsup_{h\to 0^+} \frac{|\phi(x(t)-y(t)+h(f(t,x(t))-f(t,y(t))))|-|\phi(x(t)-y(t))|}{h}$$
$$+ \ \limsup_{h\to 0^+} \frac{|\phi(x(t+h)-x(t)-hf(t,x(t))|}{h}$$
$$+ \ \limsup_{h\to 0^+} \frac{|\phi(y(t+h)-y(t)-hf(t,y(t)))|}{h}.$$

Thus

$$D^+p(t) \leq g(t, p(t)), \quad t \in [t_0, t_0 + \alpha], \quad p(t_0) = 0$$

By Theorem 1.7.1 and (iii), we have $0 \leq m(t) \leq r(t) \equiv 0$. We conclude that $m(t) \equiv 0$ and therefore $\phi x = \phi y$ for each $\phi \in E^\star$ which implies $x(t) \equiv y(t)$, thus completing the proof.

Remark 5.4.1. *In the dissipative condition assumed in Theorem 5.4.2 we could replace g by g_ϕ for each ϕ provided $|g_\phi(t, u)| \leq M$ for $t \in [t_0, t_0 + \alpha]$ and $0 \leq u \leq 2b$.*

5.5 Equations with Delay

Let E be a Banach space and let $\mathcal{C} = C([-\tau, 0], E)$, where $\tau > 0$ is a real number, be the space of continuous functions defined on the interval $[-\tau, 0]$ and with values in E. If, $\phi \in \mathcal{C}$, let us define

$$\|\phi\|_0 = \sup_{\theta \in [-\tau, 0]} \|\phi(\theta)\|,$$

where $\| \cdot \|$ denotes the norm in E. Let F be a closed subset of E and consider the subsets

$$\mathcal{C}_F = \{\phi \in \mathcal{C} : \phi(0) \in F\},$$

and

$$\hat{\mathcal{C}}_F = \{\phi \in \mathcal{C} : \phi(0) \in F \text{ and } \phi(\theta) \in \overline{co}F, \text{for every } \theta \in [-\tau, 0]\}.$$

Note that the sets $\mathcal{C}_F$ and $\hat{\mathcal{C}}_F$ are closed subsets of $\mathcal{C}$.
Let $a > 0$, $t_0 \in R_+$ and $\phi_0 \in \mathcal{C}_F$. Let us consider the function:
$y \in C((t_0 - \tau, t_0 + a), E)$ defined as follows:

$$y(t) = \begin{cases} \phi_0(t - t_0), & \text{if } t_0 - \tau \leq t \leq t_0 \\ \phi_0(0), & \text{if } t_0 \leq t \leq t_0 + a. \end{cases}$$

For $b > 0$ and $t \in [t_0, t_0 + a]$, define the set $\mathcal{C}_F^t(b)$ by

$$\mathcal{C}_F^t(b) = \mathcal{C}_F \cap \{\phi \in \mathcal{C} : \|\phi - y_b\|_0 \leq b\}$$

where y_t denotes the element of $\mathcal{C}$ defined by $y_t(\theta) = y(t + \theta)$, $-\tau \leq \theta \leq 0$, for each t.

If $f \in C(R_+ \times \mathcal{C}_F, E)$, it is possible to prove that there exists a $b > 0$ such that the function f is bounded on the set

$$\mathcal{C}_0(b) = \bigcup_{t \,\in\, [t_0, t_0 + a]} (\{t\} \times \mathcal{C}_F^t(b)).$$

The set $\mathcal{C}_0(b)$ is closed. In fact, let $\{(t_n, \phi_n\}$ be a sequence of elements of $\mathcal{C}_0(b)$ such that $t_n \to \hat{t}$ and $\phi_n \to \hat{\phi}$. Obviously $\hat{t} \in [t_0, t_0 + a]$ and $\hat{\phi} \in \mathcal{C}_F$. Also, we have

$$\|\hat{\phi} - y_{\hat{t}}\|_0 \le \|\hat{\phi} - \phi_n\|_0 + \|\phi_n - y_{t_n}\|_0 + \|y_{t_n} - y_{\hat{t}}\|_0,$$

and for arbitrary $\epsilon > 0$ and n sufficiently large,

$$\|\hat{\phi} - y_{\hat{t}}\|_0 \le 2\epsilon + b.$$

Since ϵ is arbitrary, this shows that $(\hat{t}, \hat{\phi}) \in \mathcal{C}_0(b)$.
We wish to establish the existence criteria for solutions of the Cauchy problem

$$x' = f(t, x_t), \; x(t_0) = \phi_0. \tag{5.5.1}$$

For purposes of reference, let us list the following hypotheses :

(A_1) $f \in C([t_0, t_0 + a) \times C_F, E)$, $t_0 \in R_+$, $\phi_0 \in C_F$, a, b and M are such that $\|f(t, \phi)\| \leqslant M - 1$ $(M \geqslant 1)$ on $\mathcal{C}_0(b)$;

(A_2) $\liminf_{h \to 0^+} \frac{1}{h} d(\phi(0) + hf(t, \phi).F) = 0$, for each $(t, \phi) \in [t_0, t_0 + a] \times C_F$, where $d(x, F) = \inf\{\|x - y\| : y \in F\}$;

(A_3) for $t \in [t_0, t_0 + a]$ and $\phi^t \subset C_F^t(b)$,

$$\liminf_{h \to 0^+} \frac{1}{h}[\alpha(\phi^t(0)) - \alpha(\{(\phi(0) - hf(t, \phi) : \phi \in \phi^t\})] \le g(t, \alpha(\phi^t(0)))$$

whenever $\alpha(\phi^t(\theta)) \le \alpha(\phi^t(0))$ for every $\theta \in [-\tau, 0]$;

(A_4) $g \in C([t_0, t_0 + a] \times [0, 2b], R_+)$, $g(t, 0 \equiv 0$ and $u(t) \equiv 0$ is the unique solution of $u' = g(t, u)$, $u(t_0 = 0$.

We shall now prove two auxiliary results which are needed to prove the main existence results. One is concerning the construction of an ϵ approximate solution of the Cauchy problem (5.5.1). The second result establishes that the limit of a uniformly convergent sequence of ϵ approximate solutions is indeed a solution of the Cauchy problem (5.5.1).

Lemma 5.5.1. *Suppose that (A_1) and (A_2) hold. Let $\gamma = min(a, \frac{b}{M})$. Then for each $\epsilon \in (0, 1]$ the Cauchy problem (5.5.1) has an ϵ- approximate solution, that is, there exists a function x from $[t_0 - \tau, t_0 + \gamma]$ into E which satisfies the following properties:*

(i) there exists a sequence $\{\sigma_i\}_{i=0}^{\infty}$ in $(t_0, t_0 + \gamma]$ such that $\sigma_0 = t_0$, $\sigma_i < \sigma_{i+1}$ if $\sigma_i < t_0 + \gamma$, $\sigma_{i+1} - \sigma_i \le \epsilon$ and $\lim_{i \to \infty} \sigma_i = t_0 + \gamma$;

(ii) $x(t) = \phi_0(t - t_0$ for $t \in (t_0 - \tau, t_0]$ and $\|x(t) - x(s)\| \le M|t - s|$ for $(t, s) \in [t_0, t_0 + \gamma)$;

(*iii*) *for each* $i > 0$ $\sigma_i, x_{\sigma_i} \in \mathcal{C}_0(b)$ *and* $x(t)$ *is linear on each of the intervals* $[\sigma_i, \sigma_{i+1}]$;

(*iv*) *if* $t \in (\sigma_i, \sigma_{i+1})$ *and* $\sigma_i < t_0 + \gamma$, *then* $\|x'(t) - f(\sigma_i, x_{\sigma_i})\| \leq \epsilon$.

Proof. We proceed to prove the lemma by induction on i. Let us assume that we have defined $\sigma_0 = t_0, ...\sigma i, \sigma_i < t_0 + \gamma,\ i \geq 1$ and the function $x(t)$ on $[t_0 - \tau, \sigma_i]$ such that properties $(i) - (iv)$ hold on $[t_0 - \tau, \sigma_i]$. We shall show that it is possible to define σ_{i+1} and the function $x(t)$ on $[\sigma_i, \sigma_{i+1}]$ such that $(i) - (iv)$ hold on $[t_0 - \tau, \sigma_{i+1}]$. Choose $\delta_i \in (0, \epsilon]$ satisfying

((1) $\sigma_i + \delta_i \leq t_0 + \gamma$;

(2) $d(x(\sigma_i)) + \delta_i f(\sigma_i, x_{\sigma_i}), F) \leq \frac{\epsilon}{2}\delta_i$;

(3) δ_i is the largest number such that (1) and (2) hold.

Set $\sigma_{i+1} = \sigma_i + \delta_i$ and choose $x(\sigma_{i+1}) \in F$ such that

$$\|x(\sigma_i) + \delta_i f(\sigma_i, x_{\sigma_i}) - x(\sigma_{i+1})\| \leqslant \epsilon\delta_i,$$

which is possible in view of (2). We then define

$$x(t) = \frac{x(\sigma_{i+1}) - x(\sigma_i)}{\sigma_{i+1} - \sigma_i}(t - \sigma_i) + x(\sigma_i)),\ t \in [\sigma_i, \sigma_{i+1}].$$

It is easy to check that $x(t)$ satisfies properties (ii) and (iv) on $[t_0 - \tau, \sigma_{i+1}]$. To verify that $x(t)$ satisfies property (iii), we have to prove that

$$x_{\sigma_{i+1}} \in \mathcal{C}_F \text{ and } \|x_{\sigma_{i+1}} - y_{\sigma_{i+1}}\|_0 \leqslant b.$$

Since $x(\sigma_{i+1}) \in F$, it is clear that $x_{\sigma_{i+1}} \in \mathcal{C}_F$. Also, since

$$\|x_{\sigma_{i+1}} - y_{\sigma_{i+1}}\| = \sup_{\theta \in [-\tau, 0]} \|x(\sigma_{i+1} + \theta) - y(\sigma_{i+1} + \theta)\|,$$

let us consider the value $\|x(\sigma_{i+1} + \theta) - y(\sigma_{i+1} + \theta)\|$ for

(*i*) $\sigma_{i+1} \geqslant t_0 + \tau$

(*ii*) $\sigma_{i+1} < t_0 + \tau$.

In the first case, $\sigma_{i+1} + \theta \geq t_0$ for every $\theta \in [-\tau, 0]$ and

$$\begin{aligned}\|x(\sigma_{i+1} + \theta) - y(\sigma_{i+1} + \theta)\| &= \|x(\sigma_{i+1} + \theta) - \phi_0(0)\| \\ &= \|x(\sigma_{i+1} + \theta) - x(t_0)\| \leq M|\sigma_{i+1} - t_0| \ \leq b\end{aligned}$$

using the definition of $y(\cdot)$ and property (ii), which is satisfied by $x(t)$ on $[t_0 - \tau, \sigma_{i+1}]$. In case (ii),

$$\|x(\sigma_{i+1}+\theta)-y(\sigma_{i+1}+\theta)\| = \begin{cases} \|x(\sigma_{i+1}+\theta)-x(t_0)\| \leqslant M|\sigma_{i+1}-t_0| \leqslant b \\ \quad \text{if } \sigma_{i+1}+\theta \geqslant t_0 \\ \|\phi_0(\sigma_{i+1}+\theta-t_0)-\phi_0(\sigma_{i+1}+\theta-t_0)\| = 0 \\ \quad \text{if } \sigma_{i+1}+\theta < t_0. \end{cases}$$

That, in either case, $\|x_{\sigma_{i+1}} - y_{\sigma_{i+1}}\|_0 \le b$ and $(\sigma_{i+1}, x_{\sigma_{i+1}}) \in \mathcal{C}_0(b)$, proving that property (iii) is verified.

To complete the proof, it only remains to show that $\lim_{i\to\infty} \sigma_i = t_0 + \gamma$. Let us assume, for contradiction, that for $i > 1$, $\sigma_i < t_0 + \gamma$ and $\lim_{i\to\infty} \sigma_i = \bar{\sigma} < t_0 + \gamma$. We shall first prove that the sequence $\{x_{\sigma_i}\}$ converges and for this, we need to consider the two cases

$((i)$ $\bar{\sigma} > t_0 + \tau$

$((ii)$ $\bar{\sigma} \leqslant t_0 + \tau$.

In case (i), by choosing l, k large enough, we can have $\sigma_l, \sigma_k \geqslant t_0 + \tau$. Then

$$\|x_{\sigma_l} - x_{\sigma_k}\|_0 = \sup_{\theta\in[-\tau,0]} \|x(\sigma_l+\theta) - x(\sigma_k+\theta)\| \leqslant M|\sigma_l - \sigma_k|,$$

which shows that $\{x_{\sigma_i}\}$ is a Cauchy sequence.

In case (ii), let us choose $\sigma_l < \sigma_k < \bar{\sigma}$. Then we have

$$\|x_{\sigma_k} - x_{\sigma_l}\|_0 = \sup_{\theta\in[-\tau,0]} \|x(\sigma_k+\theta) - x(\sigma_l+\theta)\|.$$

But

$$\|x(\sigma_k+\theta)-x(\sigma_l+\theta)\| = \begin{cases} \|\phi_0(\sigma_k+\theta-t_0)-\phi_0(\sigma_l+\theta-t_0)\|, \text{ if } \sigma_k+\theta < t_0, \\ \|x(\sigma_k+\theta)-\phi_0(\sigma_l+\theta-t_0)\|, \text{ if } \sigma_l+\theta \leqslant t_0 \leqslant \sigma_k+\theta, \\ \|x(\sigma_k+\theta)-x(\sigma_l+\theta)\| \text{ if } t_0 \leqslant \sigma_l+\theta. \end{cases}$$

Since ϕ_0 is uniformly continuous on $[-\tau, 0]$, we obtain

$$\|x_{\sigma_k} - x_{\sigma_l}\|_0 \to 0, \ k, l \to \infty.$$

Thus, $\bar{\phi} = \lim_{i\to\infty} x_{\sigma_i}$ exists and $\bar{\phi} \in \mathcal{C}_F$ since $\mathcal{C}_F$ is closed.

We shall now show that there exists a $\bar{\delta} \in (0, \epsilon]$ and an index i_0 such that for $i \geqslant i_0$,

(i) $\sigma_i + \bar{\delta} > t_0 + \gamma$;

(ii) $d(x(\sigma_i) + \bar{\delta} f(\sigma_i, x_{\sigma_i}), F) \leqslant \frac{\epsilon}{2}\bar{\delta}$.

In fact, for any $h > 0$,

$$d(x(\sigma_i) + hf(\sigma_i, x_{\sigma_i}), F) \leqslant \|x(\sigma_i) - \bar{\phi}(0)\| + h\|f(\sigma_i, x_{\sigma_i}) - f(\bar{\sigma}, \bar{\phi})\| + d(\bar{\phi}(0) + hf(\bar{\sigma}, \bar{\phi}), F).$$

Let $\bar{\delta} \leqslant t_0 + \gamma - \bar{\sigma}$, $\bar{\delta} > 0$ be such that

$$d(\bar{\sigma}(0) + \bar{\delta} f(\bar{\sigma}, \bar{\phi}), F) \leqslant \frac{\epsilon}{4}\bar{\delta},$$

which is possible in view of (A_2). Because of the convergence of $\{x_{\sigma_i}\}$ and $\{\sigma_i\}$ to $\bar{\phi}$ and $\bar{\sigma}$ respectively, and the continuity of f at $(\bar{\sigma}, \bar{\phi})$, we can deduce that there exists an index i_0 such that for every $i \geqslant i_0$,

$$d(x(\sigma_i) + \bar{\delta} f(\bar{\sigma}, x_{\sigma_i}), F) \leqslant \frac{\epsilon}{2}\bar{\delta}.$$

As δ_i is chosen to be the largest number such that (1) and (2) are satisfied, we have $\delta_i > \bar{\delta}$. But this is absurd since $\delta_i \to 0$ as $i \to \infty$. The lemma is proved.

Remark 5.5.1. *Lemma 5.5.1 remains valid when, in* (A_1) *and* (A_2)*, the set* $\mathcal{C}_F$ *is replaced by* $\hat{\mathcal{C}}_F$.

We now prove that if a sequence of ϵ-approximate solutions like the one we have constructed in Lemma 5.5.1 converges, then it converges to the solution of the Cauchy problem (5.5.1).

Lemma 5.5.2. *Let* (A_1) *and* (A_2) *hold. Let* $\{\epsilon_n\} \subset (0, 1)$ *be a nonincreasing sequence such that* $\lim_{n\to\infty} \epsilon_n = 0$. *Let* $\{x_n(t)\}$ *be the sequence of* ϵ_n*- approximate solutions of the Cauchy problem (5.5.1) which exists by Lemma 5.5.1. If* $\{x_n(t)\}$ *converges uniformly on* $[t_0 - \tau, t_0 + \gamma]$*, then* $x(t) = \lim_{n\to\infty} x_n(t)$ *is a solution of (5.5.1).*

Proof. We have $x_{t_0} = \phi_0$. Further $x(t)$ is continuous on $[t_0 - \tau, t_0 + \gamma]$ and thus, the function: $t \to x_t$, $t \in [t_0, t_0 + \gamma]$ is continuous. Let us set $\tau_n = \sigma^n_{i_n}$, if $t \in [\sigma^n_{i_n}, \sigma_{i_n+1}]$. we then have

$$\lim_{n\to\infty} \|x_t - x_{n,\tau_n(t)}\| = 0 \text{ uniformly},$$

because $x_{n,t} \to x_t$ uniformly, the function $t \to x_t$, $t \in [t_0, t_0 + \gamma]$, is continuous, and $|\tau_n(t) - t| \to 0$ as $n \to \infty$ uniformly. We also get that $x_t \in \mathcal{C}_F$ since

$x_{n,\tau_n(t)} \in \mathcal{C}_F$ and $\mathcal{C}$ is closed.

In order to complete the proof, we need to show that for $t \in [t_0, t_0 + \gamma]$,

$$\|x(t) - \phi_0(0) - \int_{t_0}^{t} f(s, x_s)ds\| = 0.$$

Let $t \in [t_0, t_0 + \gamma]$. Then for each positive integer n, we have

$$\begin{aligned} &\|x(t) - \phi_0(0) - \int_{t_0}^{t} f(s, x_s)ds\| \\ \leqslant\ &\|x(t) - x_n(t)\| + \|x_n(t) - \phi_0(0) - \int_{t_0}^{t} f(\tau_n(s), x_{n,\tau_n(s)})ds\| \\ &+ \int_{t_0}^{t} \|f(\tau_n(s), x_{n,\tau_n(s)}) - f(s, x_s)\|ds. \end{aligned}$$

Now, for any $\eta > 0$. we can find a positive integer $N(\eta)$ such that for any $n \geq N(\eta)$

(1) $\|x(t) - x_n(t)\| < \eta$;

(2) $\|x_n(t) - \phi_0(0) - \int_{t_0}^{t} f(\tau_n(s), x_{n,\tau_n(s)})ds\| \leqslant \|x_n(t) - \phi_0(0) - \int_{t_0}^{t} x_n'(s)ds\|$ $+ \int_{t_0}^{t} \|x_n'(s) - f(\tau_n(s), x_{n,\tau_n(s)})\|ds \leqslant \epsilon_N(t - t_0) \leqslant \epsilon_N\gamma < \eta$;

(3) $\int_{t_0}^{t} \|f(\tau_n(s), x_{n,\tau_n(s)}) - f(s, x_s)\|ds \leq \eta$, as (s, x_s) varies over a compact subset of $[t_0, t_0 + \gamma] \times \mathcal{C}_F$.

Hence for every $t \in [t_0, t_0 + \gamma]$, we have

$$\|x(t) - \phi_0(0) - \int_{t_0}^{t} f(s, x_s)ds\| < 3\eta,$$

for every $\eta > 0$. The proof is therefore complete.
We are now in a position to prove the main results concerning the local existence of a solution of the Cauchy problem.
(a) Let us first consider the case when f satisfies compactness type conditions.

Theorem 5.5.1. *Let $(A_1), (A_2), (A_3)$ and (A_4) hold. Further assume that f is uniformly continuous on $[t_0, t_0 + a] \times \mathcal{C}_F$. Then the Cauchy problem (5.5.1) has a solution $x(t)$ existing on $[t_0 - \tau, t_0 + \gamma]$ such that $x(t) \in F$, $t \in [t_0, t_0 + \gamma]$ where $\gamma = min(a, \frac{b}{M})$.*

Proof. Let $\{\epsilon_n\}_{n=1}^{\infty} \subset (0, 1)$ be a decreasing sequence such that $\lim_{n\to\infty} \epsilon_n = 0$. Let $\{x_n(t)\}$ be a sequence of ϵ_n- approximate solution of (5.5.1), which exists on

$[t_0-\tau, t_0+\gamma]$ by Lemma 5.5.1. By virtue of Lemma 5.5.2 the theorem is proved if we show that there exists a subsequence of $\{x_n(t)\}$ which converges uniformly on $[t_0-\tau, t_0+\gamma]$. However, in order to apply the Ascoli-Arzela theorem to get a uniformly convergent subsequence of $\{x_n(t)\}$, it is enough to prove that, for each $t \in [t_0, t_0+\gamma]$ the set $\{x_n(t)\}$ is a relatively compact subset of E, that is, $\alpha(\{x_n(t)\}) = 0$, since we already know by Lemma 5.5.1 that $\{x_n(t)\}$ is an equi-continuous and uniformly bounded sequence.
We define

$$p(t) = \alpha(\{x_n(t)\}), \; t \in [t_0-\tau, t_0+\gamma]$$

and observe that

(1) $p(t) = \alpha(\{x_n(t) : n \geq k\} = \alpha(\{x_n(\tau_n(t) : n \geq k\}$ for every positive integer k;

(2) $\alpha(\{x_n(t) - hf(\tau_n(t), x_{n,\tau_n(t)})\}) = \alpha(\{x_n(\tau_n(t)) - hf(\tau_n(t), x_{n,\tau_n(t)})\})$

Let $t \in [t_0, t_0+\gamma]$. If $t \in (\sigma_{i_n}, \sigma_{i_n+1})$ for each n, we have using (1).(2) and the property (vii) of α,

$$\begin{aligned}
\frac{p(t)-p(t-h)}{h} &= \frac{1}{h}[\alpha(\{x_n(\tau_n(t))\}) - \alpha(\{x_n(t-h)\})] \\
&\leqslant \frac{1}{h}[\alpha(\{x_n(\tau_n(t))\}) - \alpha(\{x_n(t) - hf(\tau_n(t), x_{n,\tau_n(t)})\})] \\
&\quad + \frac{1}{h}[\alpha(\{x_n(t) - x_n(t-h) - hf(\tau_n(t), x_{n,\tau_n(t)})\})] \\
&= \frac{1}{h}[\alpha(\{x_n(\tau_n(t))\}) - \alpha(\{x_n(\tau_n(t)) - hf(\tau_n(t), x_{n,\tau_n(t)})\})] \\
&\quad + \frac{1}{h}[\alpha(\{x_n(t) - x_n(t-h) - hf(\tau_n(t), x_{n,\tau_n(t)})\})].
\end{aligned}$$

But, in view of the property (iv) of the ϵ_n-approximate solution,

$$\begin{aligned}
&\|x_n(t) - x_n(t-h) - hf(\tau_n(t), x_{n,\tau_n(t)})\| \\
&\qquad \leqslant \int_{t-h}^{t} \|x_n'(s) - f(\tau_n(s), x_{n,\tau_n(s)})\| ds \\
&\quad + \int_{t-h}^{t} \|f(\tau_n(s), x_{n,\tau_n(s)}) - f(\tau_n(t), x_{n,\tau_n(t)})\| ds \\
&= \epsilon_n(h) + \int_{t-h}^{t} \|f(\tau_n(s), x_{n,\tau_n(s)}) - f(\tau_n(t), x_{n,\tau_n(t)})\| ds.
\end{aligned}$$

The uniform continuity of f yields that for every $\eta > 0$, there exists $\delta = \delta(\eta) > 0$ such that $\|f(\tau_n(s), x_{n,\tau_n(s)}) - f(\tau_n(t), x_{n,\tau_n(t)})\| < \eta$ whenever $|\tau_n(s) - \tau_n(t)| < \delta$ and $\|x_{n,\tau_n(s)} - x_{n,\tau_n(t)}\|_0 < \delta$. However, $|\tau_n(s) - \tau_n(t)| \leqslant |t-s| + \epsilon_n \leqslant h + \epsilon_n$ and

$$\|x_n(\tau_n(s)+\theta)-x_n(\tau_n(t)+\theta)\| \leqslant \begin{cases} M|\tau_n(s)-\tau_n(t)| \leqslant M(h+\epsilon_n), \\ \qquad \text{if } t_0 \leqslant \tau_n(s)+\theta, \\ \|\phi_0(\tau_n(s)+\theta-t_0)-\phi_0(\tau_n(t)+\theta-t_0)\|, \\ \qquad \text{if } \tau_n(t)+\theta \leqslant t_0, \\ \|\phi(\tau_n(s)+\theta-t_0)-\phi_0(0)\| \\ \qquad +\|x_n(t_0)-x_n(\tau_n(t)+\theta)\| \\ \qquad \text{if } t_0-\tau_n(t)<\theta<t_0-\tau_n(s). \end{cases}$$

As ϕ_0 is uniformly continuous on $[-\tau, 0]$, there exists a $\bar{\delta}$ such that $\|\phi_n(\tau_n(s) - \phi_n(\tau_n(t)\|_0 < \frac{\delta}{2}$, for all $\tau_n(s)$ and $\tau_n(t)$ in $[t_0, t_0+\gamma]$ (with $|\tau_n(s)-\tau_n(t)| \leq h+\epsilon_n < \bar{\delta}$). Then we get

$$\|x_n(\tau_n(s)+\theta)-x_n(\tau_n(t)+\theta)\| \leqslant \begin{cases} M(h+\epsilon_n), \ \theta \geqslant t_0-\tau_n(s) \\ \frac{\delta}{2}, \quad \theta \leqslant t_0-\tau_n(t) \\ \frac{\delta}{2}+M(h+\epsilon_n), \ t_0-\tau_n(t) \leqslant \theta \leqslant t_0-\tau_n(s). \end{cases}$$

Let us now choose $m = m(\eta)$ large enough and $\hat{h} = \hat{h}(\eta)$ small enough so that for $n \geqslant m$, and $h \leqslant \hat{h}$, we have $\epsilon_n < \eta$, $h+\epsilon_n < \bar{\delta}$ and $M(h+\epsilon_n) \leqslant \frac{\delta}{2}$. Then, for $\eta > 0$, $n \geqslant m(\eta)$ and $h \leqslant \hat{h}(\eta)$, we obtain,

$$\frac{1}{h}\|x_n(t)-x_n(t-h)-hf(\tau_n(t),x_{n,\tau_n(t)})\| \leqslant \epsilon_n+\eta < 2\eta$$

and consequently, by property $(viii)$ of the measure α, we deduce that

$$\begin{aligned} &\frac{1}{h}\alpha(\{x_n(t)-x_n(t-h)-hf(\tau_n(t),x_{n,\tau_n(t)})\}) \\ &= \alpha\Big(\Big\{\frac{x_n(t)-x_n(t-h)}{h} - f(\tau_n(t),x_{n,\tau_n(t)}) : n \geqslant m(\eta)\Big\}\Big) \leqslant 2(2\eta) \end{aligned}$$

and

$$\liminf_{h\to 0^+} \frac{1}{h}\alpha(\{x_n(t)-x_n(t-h)-hf(\tau_n(t),x_{n,\tau_n(t)})\}) \leqslant 4\eta.$$

Since this is true for every $\eta > 0$, we get

$$\liminf_{h\to 0^+} \frac{1}{h}\alpha(\{x_n(t)-x_n(t-h)-hf(\tau_n(t),x_{n,\tau_n(t)})\}) = 0.$$

Thus, we have

$$D_-p(t) \leqslant \liminf_{h\to 0^+} \frac{1}{h}[\alpha(\{x_n(\tau_n(t))\}) - \alpha(\{x_n(\tau_n(t)) - hf(\tau_n(t), x_{n,\tau_n(t)})\})].$$

Let us now consider the subset of $\mathcal{C}_F$ given by

$$\phi_N^t = \{x_{n,\tau_n(t)} : n \geqslant N, N \text{a positive integer}\}$$

and show that for N large enough, $\phi_N^t \subset \mathcal{C}_F^t(b)$. In fact, since

$$\|x_n(\tau_n(t)+\theta) - y(t+\theta)\| = \begin{cases} \|x_n(\tau_n(t)+\theta) - x_n(t_0)| \leqslant M\gamma \leqslant b, \\ \qquad \text{if } \tau_n(t)+\theta \geqslant t_0, \\ \|\phi_0(\tau_n(t)+\theta-t_0) - \phi_0(t+\theta-t_0)\|, \\ \qquad \text{if } t+\theta \leqslant t_0, \\ \|\phi(\tau_n(t)+\theta-t_0) - \phi_0(0)\| \\ \qquad \text{if } \tau_n(t)+\theta \leqslant t_0 \leqslant t_0+\theta. \end{cases}$$

If we choose N sufficiently large, we can have, for $n \geqslant N$, $|t - \tau_n(t)| \leq \epsilon_n \leqslant \epsilon_N < \delta_{\phi_0}(b)$ and, therefore $\|x_{n,\tau_n(t)} - y_t\|_0 \leq b$ on $[t_0 - \tau, t_0 + \gamma]$. Further, if $\alpha(\phi_N^t(\theta) \leqslant \alpha(\phi_N^t(0)$ for every $\theta \in [-\tau, 0]$, that is, $p(t+\theta) \leqslant p(t)$ for every $\theta \in [-\tau, 0]$, then by hypothesis (A_3) we get the differential inequality

$$D_-p(t) \leqslant g(t, p(t)).$$

Therefore for every $t \in [t_0, t_0+\gamma]\backslash S$, where S is a countable set, for which $\|p_t\|_0 \leqslant p(t)$ we have

$$D_-p(t) \leqslant g(t, p(t)).$$

Also $\|p_t\|_0 = \sup_\theta p(t_0+\theta) = \sup_\theta \alpha(\{x_n(t_0+\theta)\}) = \sup_\theta(\{\phi_0(\theta)\}) = 0$. Therefore, by Theorem 1.7.5,

$$p(t) \leqslant r(t, t_0, 0),$$

where $r(t, t_0, 0)$ is the maximal solution of $u' = g(t, u), u(t_0) = 0$, which, by hypothesis (A_4), yields that

$$p(t) = \alpha(\{x_n(t)\}) \equiv 0,$$

and the proof is complete.

Remark 5.5.2. *Theorem 5.5.1 is still valid if we require that (A_3) be satisfied for only those $\phi^t \subset C_F^t$ which are sequences of elements of C_F^t. In such a situation, the condition*

$$\alpha(\{\{f(t,\phi)\} : \phi \in \phi^t\}) \leqslant g(t, \alpha(\phi^t(0)))$$

is stronger than the compactness type condition in (A_3).

It is possible to give an existence theorem under a rather general compactness-type condition given in terms of Lyapunov-like function.
Let us consider the mapping $V : \cup_{t\in[t_0-\tau,t_0+a]}\{t\} \times \Omega^t \to R^t$, where, for each $t \in [t_0 - \tau, t_0 + a]$, $\Omega^t = \{A : A \subset B[y(t), b]\}$, $B[y(t), b]$ being the ball of radius b with center at $y(t)$. Let us suppose V has the following properties:

(i) $V(t, A)$ is continuous in t and α continuous in A;

(i) $V(t, A) \equiv 0$ iff $\bar{A}$ is compact;

(i) $|V(t, A) - V(t, B)| \leqslant L|\alpha(A) - \alpha(B)|$, $t \in [t_0, t_0 + a]$, $A, B \in \Omega^t$.

An existence result under general compactness-like condition in terms of the Lyapunov-like function V is the following.

Theorem 5.5.2. *Let the hypotheses of Theorem 5.5.1 hold with the assumption*

(A_3) *replaced by*

(A_3^*) *there exists a function V satisfying $(i), (ii)$ and (iii) and the condition*

$$\begin{cases} D_-V(t,\phi^t) & \equiv \liminf\limits_{h\to 0^+} \dfrac{1}{h}[V(t,\phi^t(0)) - V(t-h, \{\phi(0) - hf(t,\phi) : \phi \in \phi^t\})] \\ \leqslant g(t, V(t, \phi^t(0))),\ t \geqslant t_0 \end{cases} \tag{5.5.2}$$

whenever $\phi^t \in C_F^t$ is such that

$$V(t+\theta, \phi^t(\theta)) \leqslant V(t, \phi^t(0)),\ \ \theta \in [-\tau, 0].$$

Then the conclusion of Theorem 5.5.1 remains valid.

Proof. We shall briefly sketch the proof since details follow on similar lines to that of the proof of Theorem 5.5.1.
Set $p(t) = V(t, \{x_n(t)\})$, $t \in [t_0 - \tau, t_0 + \gamma]$. It can be shown, as in Theorem 5.5.1, that

$$D_-p(t) \leqslant g(t, p(t))$$

for all $t \in [t_0, t_0 + \gamma]\backslash S$ (S is a countable subset of $[t_0, t_0 + a]$ such that

$$\|p_t\|_0 \leqslant p(t).$$

Also,

$$\|p_t\|_0 = \sup_{\theta\in[-\tau,0]} V(t_0+\theta, \{x_n(t_0+\theta)\}) = \sup_{\theta\in[-\tau,0]} V(t_0+\theta, \{\phi_0(t_0+\theta)\}) = 0$$

an consequently, applying Theorem 1.7.5, we have

$$p(t) = V(t, \{x_n(t)\}) \equiv 0,$$

which yields that the set $\{x_n(t)\}$ is relatively compact.

Remark 5.5.3. *If we take $V(t, A) \equiv \alpha(A)$, $A \in \Omega^t$, the inequality (5.5.2)reduces to*

$$\liminf_{h\to 0^+} \frac{1}{h}[\alpha(\phi^t(0)) - \alpha(\{\phi(0) - hf(t,\phi) : \phi \in \phi^t\})] \leqslant g(t, \alpha(\phi^t(0)))$$

whenever $\alpha(\phi^t(\theta)) \leqslant \alpha(\phi^t(0))$, $\theta \in [-\tau, 0]$, which is the same as (A_3).

Remark 5.5.4. *Observe that if we assume the following less stringent boundary condition*

$$\liminf_{h\to 0^+} \frac{1}{h} d(\phi(0) + hf(t,\phi), F) = 0 \tag{5.5.3}$$

for every $\phi \in C$ such that $\phi(s) \in F$ for every s $[-\tau, 0]$, then Theorems 5.5.1 and 5.5.2 are not true. We offer the following counterexample.

Let the closed set F be $(-\infty, 0] \cup \{\frac{1}{n} : n \geq 1\}$. consider the function $T : R \to R$ be defined by

$$T(x) = \begin{cases} -x \text{ if } x < 0, \\ 0 \text{ if } x \geq 0, \end{cases}$$

and the function $f : C[[-1, 0], R] \to R$ defined by

$$f(\phi) = T(\phi(-1)).$$

It is easy to show that f satisfies condition (5.5.3). Let $\phi \in C[[-1, 0], R]$ be such that $\phi(s) \in F$ for every $s \in [-1, 0]$. The possible cases are $(i)\phi(0) > 0$ and $\phi(0) \leq 0$. In the first case, $\phi(s)$ is a continuous function on $[-1, 0]$ and $\phi(s) = \frac{1}{n}$, $s \in [-1, 0]$, for some integer n. Then we have

$$\frac{1}{h} d(\phi(0) + hf(\phi), F) = \frac{1}{h} d(\frac{1}{n}, F) = 0, \quad \text{for every} \quad h > 0.$$

In the second case, we could have either $\phi(0) \leq 0$, $\phi(-1) = 0$ or $\phi(0) \leq 0$, $\phi(-1) < 0$. Then

$$\frac{1}{h} d(\phi(0) + hf(\phi), F) = \begin{cases} \frac{1}{h} d(\phi(0), F) = 0, \\ \frac{1}{h} d(\phi(0) - h\phi(-1)), F), \end{cases}$$

and $\liminf_{h\to 0^+} \frac{1}{h} d(\phi(0) - h\phi(-1)), F) = 0$. Thus, condition (5.5.3) is verified. But, if we consider $\phi_0 \in C[[-1, 0], R]$ defined by $\phi_0(s) = s$, $s \in [-1, 0]$, the Cauchy problem

$$x'(t) = f(x_t), \ x_{t_0} = \phi_0, \tag{5.5.4}$$

does not have a solution in the closed set F, since we have, for a solution $x(t)$ of (5.5.4)

$$x'(t_0) = f(\phi_0) = T(-1) = 1 > 0 \quad \text{and} \quad x(t_0) = 0.$$

To see that the function f does not satisfy our boundary condition (A_2), consider a continuous function $\phi \in \mathcal{C}_F$ such that $\phi(0) = 1$, $\phi(-1) = -1$. We then have

$$\liminf_{h \to 0^+} \frac{1}{h} d(\phi(0) + hf(\phi), F) = \liminf_{h \to 0^+} \frac{1}{h} d(1 + h, F) = \liminf_{h \to 0^+} \frac{1}{h} \frac{1 + h - 1}{h} = 1.$$

We now consider the existence and uniqueness criteria for the Cauchy problem (5.5.1) in terms of dissipative conditions.

Theorem 5.5.3. *Suppose that the set F is convex and (A_1) and (A_2) hold. Assume that f satisfies the following dissipative type condition:*

(A_5) for $t \in [t_0, t_0 + a]$

$$\|f(t, \phi) - f(t, \psi)\| \leqslant g(t, \|\phi(0) - \psi(0))\|), \tag{5.5.5}$$

whenever $\phi, \psi \in \mathcal{C}_F^t(b)$ satisfy the relation $\|\phi(\theta) - \psi(\theta)\| \leqslant \|\phi(0) - \psi(0)\|$, for every $\theta \in [-\tau, 0]$, and g is the function satisfying (A_4). Then the problem (5.5.1) has a unique solution $x(t)$ on $[t_0 - \tau, t_0 + \gamma]$ such that $x(t) \in F$ for $t \in [t_0, t_0 + \gamma]$.

Remark 5.5.5. *The conclusion of Theorem 5.5.3 does not preclude the possibility of the existence of a solution of (5.5.1) which does not remain in F for all $t \in [t_0, t_0 + \gamma]$.*

Proof. Let n and m be positive integers and let $x_n(t)$ $x_m(t)$ be the ϵ_n ϵ_m approximate solutions of (5.5.1) assured by Lemma 5.5.1. Since F is convex, we have $x_n(t), x_m(t) \in F$ for every $t \in [t_0, t_0 + \gamma]$. Let us define

$$m(t) = \|x_n(t) - x_m(t)\|, \ t \in [t_0 - \tau, t_0 + \gamma].$$

Let $t \in (\sigma_i^n, \sigma_{i+1}^n) \cap (\sigma_j^n, \sigma_{j+1}^n)$, we then have

$$\begin{aligned} D^+ m(t) &\leqslant \|x_n'(t) - x_m'(t))\| \\ &\leqslant \|x_n'(t) - f(\sigma_i^n, x_{n,\sigma_i^n})\| + \|x_m'(t) - f(\sigma_j^m, x_{m,\sigma_j^m})\| \\ &\quad + \|f(\sigma_i^n, x_{n,\sigma_i^n} - f(t, x_{n,t})\| + \|f(\sigma_j^m, x_{m,\sigma_j^m} - f(t, x_{m,t})\| \\ &\quad + \|f(t, x_{n,t}) - f(t, x_{m,t})\|. \end{aligned}$$

Since the approximate solutions constructed in Lemma 5.5.1 also have the property that for $(t, \phi) \in [\sigma_i^n, \sigma_{i+1}^n] \times \mathcal{C}_F$, $\|\phi - x_{n,\sigma_i^n}\|_0 \leqslant 2M\delta_i^n$ and $|t - \sigma_i^n| \leqslant \delta_i^n$ imply that

$$\|f(t, \phi) - f(\sigma_i^n, x_{n,\sigma_i^n})\| \leqslant \epsilon_n,$$

we obtain, by using (iv) of Lemma 5.5.1, the inequality

$$D^+m(t) \leqslant 2(\epsilon_n + \epsilon_m) + \|f(t, x_{n,t}) - f(t, x_{m,t})\|.$$

If t is such that $\|x_{n,t}(\theta) - \|x_{m,t}(\theta)\| \leqslant \|x_n(t) - x_m(t)\|$ then, by (A_5), the above inequality can be written as

$$D^+m(t) \leqslant 2(\epsilon_n + \epsilon_m) + g(t, m(t)) \tag{5.5.6}$$

Thus $m(t) : [t_0 - \tau, t_0 + \gamma] \to R^+$ is a continuous function satisfying the differential inequality (5.5.6) for every $t \in [t_0, t_0 + \gamma] \backslash S$, where S is a countable set, for which $m_t(\theta) \leqslant m(t)$. Note also that $m(t_0) \equiv 0$. Therefore, by Theorem 1.7.5 we get

$$m(t) \leqslant r_{n,m}(t, t_0, 0), \;\; t \in [t_0, t_0 + \gamma],$$

where $r_{n,m}(t, t_0, 0)$ is the maximal solution of

$$u' = g(t, u) + 2(\epsilon_n + \epsilon_m), \quad u(t_0 = 0. \tag{5.5.7}$$

Now, by Lemma 1.7.1, we have $r_{n,m}(t, t_0, 0) = r(t, t_0, 0)$ as $n, m \to \infty$, uniformly on $[t_0, t_0 + \gamma]$, where $r(t, t_0, 0)$ is the maximal solution of

$$u' = g(t, u), \;\; u(t_0) = 0$$

and by hypotheses (A_4), $r(t, t_0, 0) \equiv 0$. Hence it follows that $\|x_n(t) - x_m(t)\| \to 0$ as $n, m \to \infty$, uniformly on $[t_0, t_0 + \gamma]$, which means that $\{x_n(t)\}$ converges uniformly on $[t_0 - \tau, t_0 + \gamma]$. Thus by Lemma 5.5.2, $x(t) = \lim_{n \to \infty} x_n(t)$ is a solution of (5.5.1) such that $x(t) \in F$ for every $t \in [t_0, t_0 + \gamma]$.

If $x(t), \hat{x}(t) : [t_0 - \tau, t_0 + \gamma] \to E$ are two solutions of (5.5.1) such that $x(t), \hat{x}(t) \in F \cap B[\phi_0(0), b]$, $t \in [t_0, t_0 + \gamma]$, by setting $m(t) = \|x(t) - \hat{x}(t)\|$, we obtain

$$D^+m(t) \leqslant \|f(t, x_t) - f(t, \hat{x}_t)\| \leqslant g(t, m(t))$$

whenever $t > t_0$ is such that $m_t(\theta) \leqslant m(t)$. Since $m_{t_0} \equiv 0$, Theorem 1.7.5 and hypothesis (A_4) together imply that $m(t) \equiv 0$, and the uniqueness of solutions of (5.5.1) is established. The proof is complete.

Corollary 5.5.1. *If in Theorem 5.5.1 the relation* (5.5.5) *is satisfied for* $\phi, \psi \in B(y_t, b]$ *such that* $\|\phi(\theta) - \psi(\theta)\| \leqslant \|\phi(0) - \psi(0)\|$, *for every* $\theta \in [-\tau, 0]$, *then there exists a unique solution* $x(t)$ *of (5.5.1) and the set* F *is positively invariant with respect to the equation* $x' = f(t, x_t)$.

Remark 5.5.6. *In view of Remark 5.5.1 we can replace the set* C_F *in the assumption of Theorem 5.5.3 and Corollary 5.5.1 by the set* $\hat{C}_F$ *which coincides with the set* $\{\phi \in C : \phi(\theta) = F$ *for every* $\theta \in [-\tau, 0]$, *since* F *is convex. Then we obtain existence and uniqueness criteria for the Cauchy problem (5.5.1) with* $\phi_0 = \hat{C}_F$.

In Theorem 5.5.3 we have assumed that F is convex. If we suppose that the set F is not convex, we can still prove an existence and uniqueness result in terms of dissipative type of conditions, and it becomes necessary to employ functional differential inequality theory.

Theorem 5.5.4. *Suppose that* (A_1) *and* (A_2) *hold. Assume that* f *satisfies the following dissipative type condition:*
(A_6) *for* $t \in [t_0, t_0 + a]$, $\phi, \psi C_F^t$,

$$\|f(t,\phi) - f(t,\psi)\| \leqslant g(t, \|\phi(0) - \psi(0)\|, \|\phi(\cdot) - \psi(\cdot)\|)$$

where $g \in C(R_+ \times R_+ \times C^+, R)$ *is such that* $g(t,0,0) \equiv 0$, $g(t,u,u_t)$ *is nondecreasing in* u *and* u_t *for each* (t,u_t) *and* (t,u) *respectively and* $u(t) \equiv 0$ *is the unique solution of*

$$u'(t) = g(t,u,u_t), \quad u_{t_0} = 0.$$

Then the Cauchy problem (5.5.1) has a unique solution $x(t)$ *on* $[t_0 - \tau, t_0 + \gamma]$ *such that* $x(t) \in F$, $t \in [t_0, t_0 + \gamma]$.

Proof. Let n and m be positive integers and let

$$m(t) = \|x_n(t) - x_m(t)\|, t \in [t_0 - \tau, t_0 + \gamma],$$

where $x_n(t), x_m(t)$ are ϵ_n, ϵ_m-approximate solutions of (5.5.1). If $t \in (\sigma_i^n, \sigma_{i+1}^n) \cap (\sigma_j^m, \sigma_{j+1}^{m+1})$, we get

$$\begin{aligned}
D^+ m(t) &\leqslant \|x'_n(t) - x'_m(t))\| \\
&\leqslant \|f(t, x_{n,\sigma_i^n}) - f(t, x_{m,\sigma_j^n})\| + \|x'_n(t) - f(\sigma_i^n, x_{n,\sigma_i^n})\| \\
&\quad + \|x'_m(t) - f(\sigma_i^m, x_{m,\sigma_i^m})\| + \|f(t, x_{n,\sigma_i^n}) - f(\sigma_i^n, x_{n,\sigma_i^n})\| \\
&\quad + \|f(t, x_{m,\sigma_j^m}) - f(\sigma_j^m, x_{m,\sigma_j^m})\| \\
&\leqslant g(t, \|x_n(\sigma_i^n) - x_m(\sigma_j^m)\|, \|x_{n,\sigma_i^{n(\cdot)}} - x_{m,\sigma_j^{m(\cdot)}}\|) + 2(\epsilon_n + \epsilon_m),
\end{aligned}$$

because of (5.5.7), the continuity of f and the property (iv) of Lemma 5.5.1. Now, using (ii) of Lemma 5.5.1, we have

$$\begin{aligned}
\|x_n(\sigma_i^n) - x_m(\sigma_j^m)\| &\leqslant \|x_n(t) - x_m(t)\| + \|x_n(t) - x_n(\sigma_i^n)\| \\
&\quad + \|x_m(t) - x_m(\sigma_j^m)\| \leqslant m(t) + M(\epsilon_n + \epsilon_m).
\end{aligned}$$

Also, for each $\theta \in [-\tau, 0]$, using (ii) of Lemma 5.5.1, we obtain

$$\|x_{n,\sigma_i^{n(\theta)}} - x_{m,\sigma_j^{m(\theta)}}\| \leqslant m(t+\theta) + 2M(\epsilon_n + \epsilon_m).$$

Hence by the increasing character of $g(t,u,u_t)$ in u and u_t, we can now write

$$D^+m(t) \leqslant g(t, m(t) + 2\beta_{m,n} m_t(\cdot) + 2\beta_{m,n}) + \eta_{m,n}$$

where $\beta_{m,n} = M(\epsilon_n + \epsilon_m)$ and $\eta_{m,n} = 2(\epsilon_n + \epsilon_m)$. Setting $v(t) = m(t) + 2\beta_{m,n}$, the above differential inequality reduces to

$$D^+v(t) \leqslant g(t, v(t), v_t) + \eta_{m,n}.$$

Also, $v_{t_0}(\theta) = m_{t_0}(\theta) + 2\beta_{m,n} = 2\beta_{m,n}$. Hence, by applying Theorem 1.7.6, we obtain

$$v(t) \leqslant r_{m,n}(t, t_0, 2\beta_{m,n}),\ t \in [t_0, t_0 + \gamma],$$

where $r_{m,n}(t, t_0, 2\beta_{m,n})$ is the maximal solution of

$$u' = g(t, u, u_t) + \eta_{m,n},\ u_{t_0} = 2\beta_{m,n}.$$

But, as a consequence of applying Lemma 1.7.1 we have

$$r_{m,n}(t, t_0, 2\beta_{m,n}) \to r(t, t_0, 0) \quad \text{as} \quad m, n \to \infty$$

uniformly on $[t_0, t_0 + \gamma]$, where $r(t, t_0, 0)$ is the maximal solution of (5.5.7). Hence

$$\|x_n(t) - x_m(t)\| = m(t) < v(t) + r(t, t_0, 0)$$

as $m, n \to \infty$, which in view of (A_6) yields that

$$\lim_{m,n\to\infty} \|x_n(t) - x_m(t)\| = 0 \quad \text{uniformly on} \quad [t_0, t_0 + \gamma],$$

and by Lemma 5.5.2, $x(t) = \lim_{n\to\infty} x_n(t)$ is a solution of (5.5.1) such that $x(t) \in F,\ t \in [t_0, t_0 + \gamma]$.
Let $x(t), \hat{x}(t)$ be two solutions of (5.5.1) such that $x(t), \hat{x}(t) \in F \cap B[\phi_0(0), b]$. By setting

$$m(t) = \|x(t) - \hat{x}(t)\|, \quad t \in [t_0 - \tau, t_0 + \gamma],$$

we have, by (5.5.7), the inequality

$$\begin{aligned} D^+m(t) &\leqslant \|x'(t) - \hat{x}'(t)\| = \|f(t, x_t) - f(t, \hat{x}_t)\| \\ &\leqslant g(t, m(t), m_t),\ t \in [t_0, t_0 + \gamma]. \end{aligned}$$

Since $m_{t_0} \equiv 0$, by applying Theorem 1.7.6 we get

$$m(t) \leqslant r(t, t_0, 0)$$

where $r(t, t_0, 0)$ is the maximal solution of (5.5.7). By (A_6), we can now conclude that $m(t) \equiv 0$ and this proves the uniqueness of solutions of (5.5.1) in F.

Remark 5.5.7. *As in Corollary 5.5.1, if we assume that the hypothesis of Theorem 5.5.4 are verified with the condition (5.5.7) being satisfied for $\phi, \psi \in B(y_t, b]$, it is possible to conclude that there exists a unique solution of (5.5.1) and that the set F is positively invariant with respect to the equation $x' = f(t, x_t)$. Moreover, Theorem 5.5.4 remains valid if the set C_F is replaced by $\hat{C}_F$, in (A_1), (A_2) and (A_6).*

5.6 Boundary Value Problems

This section is concerned with the existence of solutions of boundary value problems (BVP, for short) for nonlinear second order ordinary differential equations of the type

$$x'' = H(t, x, x'), \quad 0 < t < 1, \tag{5.6.1}$$

$$ax(0) - bx'(0) = x_0 \quad \text{and} \quad ax(1) + dx'(1) = x_1, \tag{5.6.2}$$

where $a, b, c, d \geq 0$, $ad + bc > 0$ and $x \in E$, where E is a real Banach space.

In case $E = R^n$, existence is proved by first obtaining a priori bounds for $\|x(t)\|$, $\|x'(t)\|$ of a solution of (5.6.1) and (5.6.2) and then employing a theorem of Scorze-Dragoni. The methods involve assuming inequalities in terms of the second derivative of Lyapunov-like functions relative to H, using comparison theorems for scalar second order equations and utilizing Leray Schauder's alternative or, equivalently, the modified function approach.

Here we wish to extend this fruitful method to the case when E is an arbitrary Banach space. Let us denote the interval $[0, 1]$ by J. The following fixed point theorem is due to Darbo.

Theorem 5.6.1. *Let S be a closed, bounded and convex subset of a Banach space E. If $T \in C(S, S)$ is such that $\alpha(TA) \leq \beta\alpha(A)$, with $\beta < 1$, for each bounded subset of S, then T has a fixed point.*

Let $G(t, s)$ be the Green's function associated with the scalar BVP

$$y'' = h(t), \tag{5.6.3}$$

$$ay(0) - by'(0) = 0 \quad \text{and} \quad cy(1) + dy'(1) = 0. \tag{5.6.4}$$

Also,let ψ be the unique function satisfying

$$\psi'' = 0, \tag{5.6.5}$$

$$a\psi(0) - b\psi'(0) = x_0 \quad \text{and} \quad c\psi(1) + d\psi'(1) = x_1. \tag{5.6.6}$$

We let

$$\begin{cases} \max[1, \sup\limits_{J\times J} |G(t,s)|] = P, & \sup\limits_{J\times J} |G_t(t,s)| = R, \\ \sup\limits_{J} \|\psi(t)\| \le Q \quad \text{and} & \sup\limits_{J} \|\psi^{'}(t)\| \le S. \end{cases} \tag{5.6.7}$$

One easily verifies that

$$\begin{cases} R = 1, \quad Q = \dfrac{a+b+c+d}{ad+bc}\max[\|x_0\|, \|x_1\|], \\ \text{and} \quad S = \dfrac{a+c}{ad+bc}\max[\|x_0\|, \|x_1\|]. \end{cases} \tag{5.6.8}$$

Clearly P is a function of only a, b, c and d.

We are now in a position to prove the following result concerning existence in the large.

Theorem 5.6.2. *Assume that*

(i) $H \in C(J \times E \times E, E)$ *and for all bounded subsets* A, B *in* E,

$$\alpha(H(J \times A \times B)) \le k \max[\alpha(A), \alpha(B)];$$

(ii) $\|H(t,x,y)\| \le L$, *for* $(t,x,y) \in J \times E \times E$.

Then there exists a solution $x \in C^2[J,E]$ *of the boundary value problem* (5.6.1) *and* (5.6.2) *provided* $k < \dfrac{1}{2P}$, *where* $P = \max[1; \sup\limits_{J\times J} |G(t,s)|]$.

Proof. Let $G(t,s)$ be the Green's function associated with the scalar BVP (5.6.3) and (5.6.4) and let $\psi(t)$ be the unique function satisfying (5.6.5) and (5.6.6).

Define $T : E_0 \to E_0$, where $E_0 = C^1(J,E)$, by

$$(T\phi)(t) = \int_0^1 G(t,s)H(s,\phi(s),\phi^{'}(s))ds + \psi(t).$$

Observe that fixed points of T are solutions of the BVP (5.6.1) and (5.6.2), Define the set E_1 by

$$E_1 = \{\phi \in E_0 : \sup_J \|\phi(t)\| \le PL + Q, \sup_J \|\phi^{'}(t)\| \le L + S].$$

Clearly E_1 is bounded, closed and convex subset of E_0, $T(E_1) \subseteq E_1$ and T is continuous on E_1. If we can show that $\alpha(T(A)) \le 2Pk\alpha(A)$ for every $A \subseteq E_1$, then Theorem 5.6.1 implies that T has a fixed point and that fixed point is a solution of the BVP (5.6.1) and (5.6.2).

Consider $A \subseteq E_1$ and observe that for $\phi \in A$

$$(T\phi)^{''}(t) = H(t,\phi(t),\phi^{'}(t))$$

and hence $\sup_J \|(T\phi)''(t)\| \leq L$, which means $(TA)'$ is equicontinuous. Applying Lemma 1.6.3 for $\epsilon > 0$ there exists $\bar{t} \in J$ or $\bar{\bar{t}} \in J$ with

$$\alpha(T(A)) \leq \alpha(TA(\bar{t})) + \epsilon, \tag{5.6.9}$$

or

$$\alpha(T(A)) \leq \alpha(TA)'(\bar{\bar{t}})) + \epsilon. \tag{5.6.10}$$

Let us first consider the case (5.6.9). Using the properties of α, given in Theorem 1.6.1, we see that

$$\alpha(TA) \leq \alpha(TA(\bar{t})) + \epsilon$$

$$= \alpha(\{\int_0^1 G(\bar{t},s)H(s,\phi(s),\phi'(s))\mathrm{d}s + \psi(\bar{t})|\phi \in A\}) + \epsilon$$

$$= \alpha(\{\int_0^1 G(\bar{t},s)H(s,\phi(s),\phi'(s))\mathrm{d}s|\phi \in A\}) + \epsilon$$

$$\leq \alpha(\overline{co}\{G(\bar{t},s)H(s,\phi(s),\phi'(s))|\phi \in A, s \in J\}) + \epsilon$$

$$+\alpha(\{G(\bar{t},s)H(s,\phi(s),\phi'(s))|\phi \in A, s \in J\}) + \epsilon.$$

Now by (i), we have

$$\alpha(TA) \leq (\max_J |G(\bar{t},s)|)\alpha(\{H(s,\phi(s),\phi'(s))|\phi \in A, s \in J\}) + \epsilon$$

$$\leq P\alpha(\{H(s,\phi(s),\phi'(s))|\phi \in A, s \in J\}) + \epsilon$$

$$\leq P\alpha(H(J \times A(J) \times A'(J))) + \epsilon \leq Pk\max(\alpha(A(J)), \alpha(A'(J))) + \epsilon.$$

This implies, by Lemma 1.6.4,

$$\alpha(TA) \leq Pk\max(\alpha(A), 2\alpha(A)) + \epsilon \leq 2Pk\alpha(A) + \epsilon.$$

Since ϵ is arbitrary, it follows that

$$\alpha(TA) \leq 2kP\alpha(A).$$

Consider the alternative (5.6.10). Proceeding as before, we obtain

$$\alpha(TA) \leq \alpha\{(TA)'(\bar{\bar{t}})\} + \epsilon$$

$$= \alpha(\{\int_0^1 G_t(\bar{\bar{t}},s)H(s,\phi(s),\phi'(s))\mathrm{d}s + \psi'(\bar{\bar{t}})|\phi \in A\}) + \epsilon$$

$$= \alpha(\{\int_0^1 G_t(\bar{\bar{t}},s)H(s,\phi(s),\phi'(s))\mathrm{d}s|\phi \in A\}) + \epsilon$$

$$\leq (\overline{co}\{G_t(\bar{\bar{t}},s)H(s,\phi(s),\phi'(s))\mathrm{d}s|\phi \in A, \quad s \in J\}) + \epsilon$$

$$= \alpha(\{G_t(\bar{\bar{t}}, s)H(s, \phi(s), \phi^{'}(s))ds | \phi \in A, \quad s \in J\}) + \epsilon$$

Remark 1.6.1, (i) and the fact $\sup_{J \times J} |G_t(t, s)| \leq 1$ yields

$$\alpha(TA) \leq (\max_j |G_t(\bar{\bar{t}}, s)|)\alpha(\{H(s, \phi(s), \phi^{'}(s)), s \in J, \phi \in A\}) + \epsilon$$

$$\leq \alpha(\{H(s, \phi(s), \phi^{'}(s)) | s \in J, \phi \in A\}) + \epsilon$$

$$\leq \alpha(H\{J \times A(J) \times A^{'}(J)\}) + \epsilon \leq k\max[\alpha(A(J)), \alpha(A^{'}(J))] + \epsilon.$$

In view of Lemma 1.6.4, this implies that

$$\alpha(TA) \leq k\max[\alpha(A), 2\alpha(A)] + \epsilon.$$

Thus, as before, we get

$$\alpha(TA) \leq 2k\alpha(A) \leq 2kP\alpha(A).$$

Therefore, in either case, we obtain

$$\alpha(TA) \leq 2kP\alpha(A)$$

and the proof is complete.

Remark 5.6.1. . *By Remark 1.6.1, we see that*

$$\alpha(H(J \times A \times B)) = \max(\alpha(J), \alpha(A), \alpha(B)) = \max(\alpha(A), \alpha(B))$$

since $\alpha(J) = 0$. *Consequently, it is instructive to note that the compactness-like condition assumed in Theorem 5.6.2 is actually equivalent to*

$$\alpha(H(J \times A \times B)) \leq k\alpha(J \times A \times B).$$

Furthermore, this assumption also implies that H *maps bounded sets into bounded sets.*

If assumption (ii) of Theorem 5.6.2 is dispensed with, then we can prove a result which gives existence in the small. This is precisely the next result in which the fact that H *maps bounded sets into bounded sets is fully utilized.*

Theorem 5.6.3. *Assume that hypothesis (i) of Theorem 5.6.2 holds. For any given* $M > 0$, *let* $N > 0$ *be such that* $\|H(t, x, y)\| \leq N$ *for all* $(t, x, y) \in J \times \{x : \|x\| \leq M\} \times \{y : \|y\| \leq M\}$. *If* $t_1, t_2 \in J$ *are such that* $0 < t_2 - t_1 < \dfrac{M}{2PN}$. *then the BVP*

$$x^{''} = N(t, x, x^{'}), \tag{5.6.11}$$

$$ax(t_1) - bx^{'}(t_1) = x_0, \quad \textit{and} \quad cx(t_2) + dx^{'}(t_2) = x_1, \tag{5.6.12}$$

has a solution $x \in C^2([t_1, t_2], E)$ *provided* $\max[\|x_0\|, \|x_1\|](\frac{a+b+c+d}{ad+bc}) \leq \frac{M}{2}$ *and* $k < \frac{1}{2P}$.

Proof. Let $\hat{G}(t,s)$ be the Green's function associated with the scalar BVP

$$y^{''} = h(t); \quad ay(t_1) - by^{'}(t_1) = 0 \quad \text{and} \quad cy(t_2) + dy^{'}(t_2) = 0$$

and let $\hat{\psi}(t)$ be the unique function satisfying

$$\hat{\psi}^{''} = 0, \quad a\hat{\psi}(t_1) - b\hat{\psi}'(t_1) = x_0 \quad \text{and} \quad c\hat{\psi}(t_2) + d\hat{\psi}'(t_2) = x_1,$$

where $t_1, t_2 \in J$. It is easily verified that

$$\sup[|\hat{G}(t,s)| : t, s \in [t_1, t_2]] \leq P,$$

$$\sup[|\hat{G}_t(t,s)| : t, s \in [t_1, t_2]] \leq 1,$$

$$\sup[\|\hat{\psi}(t)\| : t \in [t_1, t_2]] \leq Q,$$

and $\sup[\|\hat{\psi}^{'}(t)\| : t \in [t_1, t_2]] \leq S$ where P, Q, S are the same constants given in (5.6.7) and (5.6.8).

Consider $E_0 = C^1[[t_1, t_2], E]$ with $\|\psi\| = \max\{\sup_{[t_1,t_2]} \|\psi(t)\|, \quad \sup_{[t_1,t_2]} \|\psi^{'}(t)\|\}$ and define $E_1 = \{\phi \in E_0 : \sup_{[t_1,t_2]} \|\phi(t)\| \leq M, \sup_{[t_1,t_2]} \|\phi^{'}(t)\| \leq M\}$. The operator $T : E_0 \to E_0$ is defined by

$$(T\phi)(t) = \int_{t_1}^{t_2} \hat{G}(t,s) H(s, \phi(s), \phi^{'}(s) ds + \hat{\psi}(t).$$

For $\phi \in E_1$,

$$\|T\phi(t)\| \leq \int_{t_1}^{t_2} |\hat{G}(t,s)| \|H(s, \phi(s), \phi^{'}(s))\| ds + \|\hat{\psi}(t)\|$$

$$\leq PN(t_2 - t_1) + Q \leq \frac{M}{2} + (\frac{a+b+c+d}{ad+bc}) \max[\|x_0\|, \|x_1\|] \leq \frac{M}{2} + \frac{M}{2} = M,$$

and

$$\|(T\phi)^{'}(t)\| \leq \int_{t_1}^{t_2} |\hat{G}_t(t,s)| \|H(s, \phi(s), \phi^{'}(s))\| ds + \|\psi^{'}(t)\|$$

$$\leq (t_2 - t_1)N + (\frac{a+c}{ad+bc}) \max\{\|x_0\|, \|x_1\|\} \leq \frac{M}{3} + \frac{M}{2} = M.$$

Thus $T(E_1) \subseteq E_1$. Also, E_1 is closed, bounded, convex subset of E_0 and T is continuous on E_1. As before , we can obtain $\alpha(TA) \leq 2kP(t_2 - t_1)\alpha(A)$, where the factor $(t_2 - t_1)$ arises when the integral is replaced by the closure of the convex hull, and so

$$\alpha(TA) \leq 2kP\alpha(A).$$

Remark 5.6.2. *Since we actually get $\alpha(TA) \leq 2kP(t_2 - t_1)\alpha(A)$, it is enough to assume in Theorem 5.6.3 that* $k < \dfrac{1}{2P(t_2 - t_1)}$.

If we wish to dispense with assumption (ii) of Theorem 5.6.2 and to prove an existence theorem in the large we have to utilize Leray-Schauder's alternative or, equivalently, the modified function approach. We now develop the modified function approach which allows us to remove the restrictive boundedness condition (ii) of Theorem 5.6.2.

Let us list the following hypotheses for convenience:

(H_1) $f \in C(J \times R \times R, R)$, $W \in C^2(J, R)$ with $W(t) \geq 0$, $t \in J = [0, 1]$ and for $t \in (0, 1)$,

$$W''(t) \leq f(t, W(t), W'(t)),$$

such that $\alpha_0 W(0) - \beta_0 W'(0) \geq \gamma_0$ and $\alpha_1 W(1) + \beta_1 W'(1) \geq \gamma_1$ where $\alpha_0, \alpha_1 \geq 0$, $\beta_0, \beta_1 > 0$;

(H_2) $z \in C^2(J, R)$ with $z(t) > 0$, $t \in J$ and for each $\lambda > 0$, $\lambda z'' < f(t, W(t) + \lambda z(t), W'(t) + \lambda z'(t)) - f(t, W(t), W'(t))$, such that $\alpha_0 z(0) - \beta_0 z'(0) > 0$ and $\alpha_1 z(1) + \beta_1 z'(1) > 0$;

(H_3) the left maximal solution $r(t, 1, \eta_1)$ and the right minimal solution $\rho(t, 0, \eta_0)$ of $v' = \hat{f}(t, v)$ exist on J, where $\hat{f}(t, v) = \min_{0 \leq u \leq B_0} f(t, u, v)$, $B_0 = \max_J W(t)$,

$$\eta_0 = \frac{-\gamma_0}{\beta_0} \text{ and } \eta_1 = \frac{\gamma_1}{\beta_1};$$

(H_4) $V \in C^2(J \times E, R^+)$ such that $V(t, x) \to \infty$ as $\|x\| \to \infty$ uniformly in $t \in J$;

(H_5) for $0 \leq \delta \leq 1$,

$$V''_{\delta H}(t, x) \geq f(t, V(t, x), V'(t, x)) + \sigma\delta\|H(t, x, x')\|,$$

where $\sigma > 0$, $V''_{\delta H}(t, x) = U(t, x, x') + \delta V_x(t, x)H(t, x, x')$, and

$$U(t, x, x') = V_{tt}(t, x) + 2V_{tx}(t, x)x' + V_{xx}(t, x)(x', x').$$

Here we have used the known facts that if $V \in C^2(J \times E, R_+)$, then $V'(t, x) = V_t(t, x) + V_x(t, x)x'$ and $V''(t, x) = U(t, x, x') + V_x(t, x)x''$, and $V_{xx}(t, x)$ occuring in U is bilinear operator mapping $E \times E$ into $L(E \times E, R)$;

(H_6) the boundary conditions (5.6.2) imply that

$$\alpha_0 V(0, x(0)) - \beta_0 V'(0, x(0)) \leq \gamma_0 \quad \text{and} \quad \alpha_1 V(1, x(1)) + \beta_1 V'(1, x(1)) \leq \gamma_1.$$

We are now in a position to prove the following theorem concerning the existence in the large.

Theorem 5.6.4. *Let hypothesis (i) of Theorem 5.6.2 hold. Assume further that* (H_1) - (H_6) *hold. Then there exists a solution* $x \in C^2(J, E)$ *of BVP (5.6.1) and (5.6.2) provided* $k < \dfrac{1}{2P}$.

Proof. Let $\delta(t, x, y) \in C[J \times E \times E, J]$ satisfy

$$\delta(t,x,y) = \begin{cases} 1 & \text{if} \quad \|x\| \le B \quad \text{and} \quad \|y\| \le M \\ 0 & \text{if} \quad \|x\| \ge B+1 \quad \text{or} \quad \|y\| \ge M+1, \end{cases}$$

where the constants B and M are to be specified later. Define the modified function $\hat{H}(t, x, y)$ by

$$\hat{H}(t,x,y) = \delta(t,x,y)H(t,x,y).$$

Surely $\hat{H} \in C(J \times E \times E, E)$. Now, by (i) and Remark 1.6.1, we see that

(a) for A, B bounded subsets of E,

$$\alpha(\hat{H}(J\times A\times B)) \le \alpha[\delta(J\times A\times B)H(J\times A\times B)] \le (\max_{J\times E\times E} \delta)\alpha(H(J\times A\times B))$$

$$\le k\max(\alpha(A), \alpha(B))$$

(b) $\|\hat{H}(t,x,y)\| = \|\delta(t,x,y)H(t,x,y)\| \le \max|H(t,x,y)|t \in J, y, \|x\| \le M+1 \le \hat{M}$, (constant), since H maps bounded sets to bounded sets by (i).

Thus $\hat{H}$ satisfies the hypotheses of Theorem 5.6.2. Consequently there exists $x \in C^2(J, E)$ such that

$$x'' = \hat{H}(t, x(t), x'(t)),$$

and $ax(0) - bx(1) = x_0$, $cx'(0) + dx'(1) = x_1$. Setting $m(t) = V(t, x(t))$ and using (H_5) we have

$$m''(t) \ge f(t, m(t), m'(t)) + \sigma\|x''(t)\| \ge f(t, m(t), m'(t)). \tag{5.6.13}$$

By (H_6), it follows that $\alpha_0 m(0) - \beta_0 m'(0) \le \gamma_0$, $\alpha_1 m(1) + \beta_1 m'(1) \le \gamma_1$. Hence by Theorem 1.7.4, we get

$$m(t) \le W(t) \quad \text{for} \quad t \in J. \tag{5.6.14}$$

Using (H_4), we know that there exists a B which depends on B_0 such that $V(t, x(t)) = m(t) \le W(t) \le B_0$ implies $\|x(t)\| \le B$. This is the B, used in the definition of δ, which is independent of $x(t)$.

Consider $\hat{f}$ from (H_3) and observe that

$$m''(t) \ge f(t, m(t), m'(t)) \ge \hat{f}(t, m'(t))$$

Hence, defining $v(t) = m'(t)$, we have

$$v' \ge \hat{f}(t, v(t)).$$

Also, notice that in view of (H_6), $v(0) \geq \frac{-\gamma_0}{\beta_0} = \eta_0$ and $v(1) \leq \frac{\gamma_1}{\beta_0} = \eta_1$. From the theory of differential inequalities we therefore obtain

$$v(t) \leq r(t, 1.\eta_1).$$

$$v(t) \geq \rho(t, 0, \eta_0), \quad for \quad t \in J$$

where $r(t, 1, \eta_1)$ and $\rho(t, 0, \eta_0)$ are the left maximal and right minimal solutions of $\xi' = \hat{f}(t, \xi)$ whose existence is assured by (H_3).

Let $B_1 = \max\{\max_J |r(t, 1, \eta_1)|, \min_J |\rho(t, 0, \eta_0)|\}$, then it follows that $|m'(t)| = |v(t)| \leq B_1$. Now define $N_0 = \min\{f(t, u, y)|t \in J, \quad 0 \leq u \leq B_0, \|y\| \leq B_1\}$ and note from equations (5.6.13), (5.6.14) and the definition of B_0 and B_1, that $m''(t) \geq N_0 + \sigma\|x''(t)\|$.

For $0 \leq s \leq t \leq 1$

$$2B_1 \geq m'(t) - m'(s) \geq \int_s^t m''(\xi)d\xi \geq N_0(t-s) + \sigma\|x'(t) - x'(s)\|,$$

and so $2B_1 + |N_0| \geq \sigma\|x'(t) - x'(s)\|$. Similarly for $0 \leq t \leq s \leq 1$, we have

$$2B + |N_0| \geq \sigma\|x'(t) - x'(s)\|.$$

Let $t \in [0.1]$ and integrate from 0 to 1 to obtain

$$\frac{1}{\sigma}(2B_1 + |N_0|) \geq \int_0^1 \|x'(t) - x'(s)\|ds \geq \|\int_0^1 (x'(t) - x'(s))ds\|$$

$$= \|x'(t) - x(1) + x(0)\|.$$

But $\|x(t)\| \leq B$ on J and hence

$$M \equiv \frac{1}{\sigma}(2B_1 + |N_0|) + 2B \geq \|x'(t)\| \quad on \quad J.$$

This constant M is independent of $x(t)$ and is to be used in the definition of δ. This, in view of the definition of $\hat{H}$, shows $\hat{H} \equiv H$ and so $x(t)$ is actually a solution of (5.6.1) and (5.6.2). The proof is therefore complete.

5.7 Fractional Differential Equations in a Banach Space

In this section, we discuss the theory of fractional differential equations in a Banach Space utilizing the initial time $t_0 \geq 0$. We prove general existence and uniqueness, continuous dependence, fractional differential inequalities in cones and flow invariance.

Let E be a real Banach Space with the norm $\|\cdot\|$. Let $0 < q < 1$ and $p = 1-q$. We let $C_p([t_0, t_0+a], E) = [u : C((t_0, t_0+a), E)$ and $(t-t_0)^{1-q}u(t) \in$

$C([t_0, t_0 + a], E)$. Let us consider the initial value problem (IVP) for fractional differential equations in E given by

$$D^q x = f(t, x), \quad x(t)(t - t_0)^{1-q}|_{t=t_0} = x^0, \tag{5.7.1}$$

where $f \in C(R_0, E)$ with $R_0 = [(t, x) : t_0 \leq t \leq t_0 + a$ and $\|x - x_0\| \leq b], D^q x$ is the fractional derivative of x of the order $0 < q < 1$ and $x^0(t) = \frac{x^0(t-t_0)^{q-1}}{\Gamma(q)}$. Since f is assumed to be continuous, the IVP (5.7.1) is equivalent to the following fractional Volterra integral

$$x(t) = x^0(t) + \frac{1}{\Gamma(q)} \int_{t_0}^{t} (t - s)^{q-1} f(s, x(s)) ds, \ t_0 \leq t \leq t_0 + a; \tag{5.7.2}$$

that is, every solution of (5.7.2) is a solution of (5.7.1) and vice versa. Here and in what follows Γ is the Gamma function.

We are now in a position to prove the following general existence and uniqueness result.

Theorem 5.7.1. *Assume that*

(a) $f \in C(R_0, E)$ *and* $\|f(t, x)\| \leq M_0$ *on* R_0;

(b) $g \in C([t_0, t_0 + a] \times [0, 2b], R_+), g(t, u) \leq M_1$ *on* $[t_0, t_0 + a] \times [0, 2b], g(t, 0) \equiv 0, g(t, u)$ *is nondecreasing in* u *for each* t *and* $u(t) \equiv 0$ *is the only solution of*

$$D^q u = g(t, u), \quad u(t)(t - t_0)^{1-q}|_{t=t_0} = 0 \quad on \quad [t_0, t_0 + a]; \tag{5.7.3}$$

(c) $\|f(t, x) - f(t, y)\| \leq g(t, \|x - y\|)$ *on* R_0.

Then, the successive approximations defined by

$$x_{n+1}(t) = x^0(t) + \frac{1}{\Gamma(q)} \int_{t_0}^{t} (t - s)^{q-1} f(s, x_n(s)) ds, \quad n = 0, 1, 2, \ldots, \tag{5.7.4}$$

on $[t_0, t_0 + a]$, where $\alpha = min(a, [\frac{b\Gamma(q+1)}{M}]^{\frac{1}{q}}), M = max(M_0, M_1)$, are continuous and converge uniformly to the unique solution $x(t)$ of the IVP (5.7.1) on $[t_0, t_0 + \alpha]$.

Proof. For $t_0 \leq t_1 \leq t_2 \leq t_0 + \alpha$, we find

$$\begin{aligned} &\|x_1(t_1) - x^0(t_1) - x_1(t_2) + x^0(t_2)\| \\ \leq\ & \frac{M}{\Gamma(q)} \left| \int_{t_0}^{t_1} [t_1 - s)^{q-1} - (t_2 - s)^{q-1}] ds \right| \end{aligned}$$

$$\begin{aligned} &\leq \frac{M}{\Gamma(q+1)}[(t_1-t_0)^q-(t_2-t_0)^q+2(t_2-t_1)^q] \\ &\leq \frac{2M}{\Gamma(q+1)}[(t_2-t_1)^q \leq \epsilon, \end{aligned}$$

provided $|t_2-t_1|<\delta$, where $\delta=[\frac{\epsilon\Gamma(q+1)}{2M}]^{\frac{1}{q}}$, proving that $x_1(t)$ is continuous on $[t_0,t_0+\alpha]$. Similarly,

$$\|x_1(t_1)-x^0(t_1)\| \leq \frac{1}{\Gamma(q)}\int_{t_0}^{t}(t-s)^{q-1}\|f(s,x_0)\|ds \leq \frac{M(t-t_0)^q}{\Gamma(q+1)} \leq \frac{M\alpha^q}{\Gamma(q+1)} \leq b$$

Hence it is easily seen by induction that the successive approximations are continuous and satisfy $\|x_n(t)-x_0\| \leq b, n=1,2,3,\ldots$

We shall next define the successive approximations for the IVP (5.7.3) as follows:

$$u_0(t)=\frac{M(t-t_0)^q}{\Gamma(q+1)},$$

$$u_{n+1}=\frac{1}{\Gamma(q)}\int_{t_0}^{t}(t-s)^{q-1}g(s,u_n(s))ds, t_0 \leq t \leq t_0+\alpha. \tag{5.7.5}$$

Since $g(t,u)$ is assumed to be nondecreasing in u for each t, an easy induction shows that the successive approximations (5.7.5) are well defined and satisfy

$$0 \leq u_{n+1}(t) \leq u_n(t), \quad t_0 \leq t \leq t_0+\alpha.$$

Moreover, $|D^q u_n(t)| = g(t,u_{n-1}(t) \leq M_1$ and therefore, we can conclude by Ascoli-Arzela theorem and monotonicity of the sequence $\{u_n(t)\}$ that $\lim_{n\to\infty} u_n(t) = u(t)$ uniformly on $[t_0,t_0+\alpha]$. It is also clear that $u(t)$ satisfies the IVP (5.7.3) and hence by (b) $u(t)\equiv 0$ on $[t_0,t_0+\alpha]$.

Now from the earlier estimate

$$\|x_1(t)-x^0(t)\| \leq \frac{M(t-t_0)^q}{\Gamma(q+1)} = u_0(t).$$

Assume that $\|x_k(t)-x_{k-1}(t)\| \leq u_{k-1}(t)$ for some given k. Since

$$\begin{aligned} \|x_k(t)-x_{k-1}(t)\| &= \frac{1}{\Gamma(q)}\|\int_{t_0}^{t}(t-s)^{q-1}f(s,x_k(s))ds - \int_{t_0}^{t}(t-s)^{q-1}f(s,x_{k-1}(s))ds \\ &\leq \frac{1}{\Gamma(q)}\int_{t_0}^{t}(t-s)^{q-1}\|f(s,x_k(s))-f(s,x_{k-1}(s))\|ds, \end{aligned}$$

using condition (c) and the monotone character of $g(t,u)$, we get

$$\|x_k(t)-x_{k-1}(t)\| \leq \frac{1}{\Gamma(q)}\int_{t_0}^{t}(t-s)^{q-1}g(s,\|x_k(s))-g(s,x_{k-1}(s)\|)ds = u_k(t).$$

Thus by induction, the inequality

$$\|x_k(t) - x_{k-1}(t)\| \le u_k(t), \ t_0 \le t \le t_0 + \alpha.$$

holds for all n. Also

$$\begin{aligned}&\|D^q x_{n+1}(t) - D^q x_n(t)\| \le \|f(t, x_n(t)) - f(t, x_{n-1}(t)\| \\ &\le g(t, \|x_n(t) - x_{n-1}(t)\|) \le g(t, u_{n-1}(t)).\end{aligned}$$

Let $n \le m$. Then we can easily obtain

$$\begin{aligned}&D^{+q}\|x_n(t) - x_m(t)\| \le \|D^q x_n(t) - D^q x_m(t)\| \\ &\le g(t, u_{n-1}(t)) + g(t, u_{m-1}(t)) + g(t, \|x_n(t) - x_m(t))\|).\end{aligned}$$

Since $u_{n+1}(t) \le u_n(t)$ for all n, it follows that

$$D^{+q}\|x_n(t) - x_m(t)\| \le g(t, \|x_n(t) - x_m(t)\|) + 2g(t, u_{n-1}(t)),$$

where D^{+q} denotes the Dini derivative corresponding to D^+. An application of comparison result Theorem 1.7.1 gives

$$\|x_n(t) - x_m(t)\| \le r_n(t), \ t_0 \le t \le t_0 + \alpha,$$

where $r_n(t)$ is the maximal solution of the IVP

$$D^q v = g(t, v) + 2g(t, u_{n-1}(t)), \ v(t)(t - t_0)^{1-q}|_{t=t_0} = 0,$$

for each n. Since as $n \to \infty$, $2g(t, u_{n-1}(t)) \to 0$ uniformly on $[t_0, t_0 + \alpha]$, it follows by lemma 1.7.1 that $r_n(t) \to 0$ uniformly on $[t_0, t_0 + \alpha]$. This implies that $\{x_n(t)\}$ converges to $x(t)$ and it is now easy to show that $x(t)$ is a solution of IVP (5.7.1).

To show that this solution $x(t)$ is unique, let $y(t)$ be another solution of the IVP (5.7.1) on $[t_0, t_0 + \alpha]$. Define $m(t) = \|x(t) - y(t)\|$ and note that $m(t_0) = 0$. Then $D^{+q}m(t) \le \|D^q x(t) - D^q y(t)\| \le \|f(t, x(t)) - f(t, y(t))\| \le g(t, m(t))$, using condition (c). Again applying the comparison result Theorem 1.7.1, we have

$$m(t) \le r(t), \quad t_0 \le t \le t_0 + \alpha,$$

where $r(t)$ is the maximal solution of IVP (5.7.1). By assumption (b), $r(t) \equiv 0$ and this proves that $x(t) = y(t)$ on $[t_0, t_0 + \alpha]$. Hence the proof is complete.

Corollary 5.7.1. *The function $g(t, u) = Lu, L > 0$ is admissible in Theorem 5.7.1.*

Let us note first that when the initial time is changed the corresponding fractional differential equation becomes different because the notion of fractional derivative is nonlocal. We are therefore content with proving the continuous dependence of solutions $x(t, t_0, x_0)$ of IVP with respect to x_0 only.

Theorem 5.7.2. *Let $f \in C(R_+ \times E, E)$ and for $(t,x) \in R_+ \times E$,*

$$\|f(t,x) - f(t,y)\| \leq g(t, \|x-y\|) \tag{5.7.6}$$

where $g \in C(R_+^2, R_+)$. Assume that $u(t) \equiv 0$ is the unique solution of the fractional differential equation

$$D^q u = g(t,u) \tag{5.7.7}$$

with $u(t)(t-t_0)^{1-q}|_{t=t_0} = 0$. Then, if the solutions $u(t,t_0,u_0)$ where $u^0 = u(t)(t-t_0)^{1-q}|_{t=t_0} = 0$ of (5.7.7) are continuous with respect to the initial condition u^0, the solutions $x(t,t_0,x^0)$ of (5.7.1) are unique and continuous relative to x^0.

Proof. Since uniqueness follows from Theorem 5.7.1, we have to consider continuity part only. To this end, let $x(t,t_0,x^0)$, $y(t,t_0,y^0)$ be the two solutions of (5.7.1) through (t_0,x^0), (t_0,y^0) respectively. Defining $m(t) = \|x(t,t_0,x^0) - y(t,t_0,y^0)\|$, condition (5.7.6) implies the inequality

$$D^{+q} m(t) \leq g(t, m(t)),$$

and by Theorem 1.7.1, we get $m(t) \leq r(t,t_0,\|x^0 - y^0\|), t \geq t_0$, where $r(t,t_0,u^0)$ is the maximal solution of (5.7.7) such that $u^0 = \|x^0 - y^0\|$. Since the solutions $u(t,t_0,u^0)$ are assumed to be continuous relative to u^0, it follows that $\lim_{x^0 \to y^0} r(t,t_0,\|x^0 - y^0\|) = r(t,t_0,0) \equiv 0$ by hypothesis. It then follows that $\lim_{x^0 \to y^0} x(t,t_0,x^0) = y(t,t_0,y^0)$, proving the continuity with respect to x^0. The proof is complete.

Next we consider the flow invariance and inequalities in cones. The definition of the fractional derivative of arbitrary order q, $0 < q < 1$ of a function $x \in C([t_0,\infty), E)$ is given by

$$D_0^q x(t) = \lim_{h \to 0} h^{-q} \sum_{\eta=0}^{n} (-1)^\eta \binom{q}{\eta} x(t - \eta h) = \lim_{h \to 0} x_h^q(t), \tag{5.7.8}$$

where

$$x_h^{(q)}(t) = h^{-q} \sum_{\eta=0}^{n} (-1)^\eta \binom{q}{\eta} x(t - \eta h). \tag{5.7.9}$$

This implies, on expanding

$$\begin{aligned} x_h^{(q)}(t) &= \tfrac{1}{h^q}[x(t) - qx(t-h) + \tfrac{q(q-1)}{2!}x(t-2h) - \tfrac{q(q-1)(q-2)}{3!}x(t-3h) + \ldots] \\ &= \tfrac{1}{h^q}[x(t) - S(t,h,q)]. \end{aligned} \tag{5.7.10}$$

Let us consider the IVP for fractional differential equation

$$D^q x = f(t,x), x(t)(t-t_0)^{1-q}|_{t=t_0} = x^0 \in F, \tag{5.7.11}$$

where $f \in C([t_0,\infty) \times E, E)$ and $F \subset E$ is a closed set. Let

$$\lim_{n\to\infty} \frac{1}{h^q} d[x - h^q f(t,x), F] = 0, \tag{5.7.12}$$

where $d(x,F) = \inf_{y\in F} \|x-y\|$. The set F is said to be flow invariant relative to f, if every solution $x(t)$ of IVP (5.7.1) on $[t_0,\infty)$ is such that $x(t) \in F$ for $t_0 \le t < \infty$. A set $A \subset E$ is called a distance set, if to each $x \in E$ there corresponds a point $y \in A$ such that $d(x,A) = \|x-y\|$. A function $g \in C([t_0,\infty) \times R_+, R_+)$ is said to be a uniqueness function, if the following holds: if $m \in C([t_0,\infty), R_+)$ is such that $m(t)(t-t_0)^{1-q}|_{t=t_0} \le 0$ and $D^q m(t) \le g(t, m(t))$, whenever $m(t) > 0$, then $m(t) \le 0$ for $t_0 \le t < \infty$.

We are now in a position to prove the following result on flow invariance of F.

Theorem 5.7.3. *Let $F \subset E$ be a closed and distance set. Assume further that*

(i) $\lim_{h\to 0} \frac{1}{h^q} d[x - h^q f(t,x), F] = 0, t \in [t_0,\infty), x \in \partial F$

(ii) $\|f(t,x) - f(t,y)\| \le g(t, \|x-y\|), x \in E - F, y \in \partial F$, *where $g(t,u)$ is a uniqueness function.*

Then F is flow invariant with respect to f.

Proof. Let $x(t)$ be a solution of (5.7.11). Suppose that $x(t) \in F$ for $t_0 \le t \le t_0 + \alpha$, where $t_0 + \alpha < \infty$ is maximal, that is, $x(t)$ leaves the set F at $t = t_0 + \alpha$ for the first time. Let $t_1 \in (t_0 + \alpha, \infty)$ and $x(t_1) \notin F$ and let $y_0 \in \partial F$ be such that $d(x(t_1), F) = \|x(t_1) - y_0\|$. Set for $t \in [t_0,\infty), m(t) = d[x(t), F]$ and $v(t) = \|x(t) - y_0\|$. For sufficiently small $h > 0$, letting $x = x(t_1)$, we have

$$\begin{aligned} d[S(t_1,h,q),F] \ge\ & d[x - h^q f(t_1,x), F] - \epsilon(h^q) \\ =\ & d[x - y_0 - h^q(f(t_1,x) - f(t_1,y_0)), F] \\ & -d[x - y_0 - h^q(f(t_1,x) - f(t_1,y_0)), F] \\ & +d[x - h^q f(t_1,x), F] - \epsilon(h^q) \\ \ge\ & d[x - y_0 - h^q(f(t_1,x) - f(t_1,y_0)), F] \\ & -d[y_0 - h^q(f(t_1,y_0), F] - \epsilon(h^q) \\ \ge\ & d[x - y_0, F) - h^q\|f(t_1,x) - f(t_1,y_0)\| - \epsilon(h^q). \end{aligned} \tag{5.7.13}$$

Since $m(t_1) = v(t_1) > 0$ and $m(t_1) = d[x,F] = \|x(t_1 - y_0)\|$, we find

$$\|x(t_1) - y_0\| - d[S(t_1,h,q),F] \le h^q\|f(t_1,x) - f(t_1,y_0)\| - \epsilon(h^q),$$

which yields the inequality

$$D^q m(t_1) \le g(t_1, m(t_1)).$$

This gives, in view of the facts that g is the uniqueness function and $m(t)(t-t_0)^{1-q}|_{t=t_0} = 0$, the relation $m(t) \leq 0, t_0 \leq t < \infty$. But, $d[x(t_1), F] = m(t_1) > 0$, which is a contradiction. Hence the set F is flow invariant relative to $f(t, x)$ and the proof is complete.

We shall next develop the theory of fractional differential inequalities. To do this, we shall use the concept of a cone which induces a partial order in E.

Theorem 5.7.4. *Let K ne a cone with nonempty interior. Assume that*

(a) $u, v \in C(R_+, E)$ such that $D^q v, D^q u$ exist, $f \in C(R_+ \times E, E)$ and $f(t, x)$ is quasimonotone nondecreasing in x for each $t \in R_+$;

(b) $D^q u(t) - f(t, u(t)) < D^q v(t) - f(t, v(t)),\ t \in [t_0, \infty)$.

Then $v^0 < w^0$ implies that $u(t) < v(t), t_0 \leq t < \infty$.

Proof. Suppose that the assertion of the Theorem is false. Then there exist a $t_1 > t_0$ such that

$$v(t_1) - u(t_1) \in \partial K$$

and

$$v(t) - u(t)] \in K^0,\ t \in [t_0, t_1).$$

By Lemma 2.2.1, there exists a $c \in K_0^*$ with $c(v(t_1)) - c(u(t_1)) = 0$. Setting $m(t) = c(v(t) - u(t))$, we see that $m(t) > 0$ for $t_0 \leq t \leq t_1$ and $m(t_1) = 0$. Consequently, we get $D^q m(t_1) \leq 0$. At $t = t_1$, we have $u(t_1) \leq v(t_1)$ and $c(u(t_1)) = c(v(t_1))$. Hence using quasimonotone property of f and (b), it follows that

$$D^q m(t_1) = c(D^q v(t_1) - D^q u(t_1)) > c(f(t_1, v(t_1)) - f(t_1, u(t_1)).$$

This contradiction proves the Theorem.

Remark 5.7.1 Observe that Theorem 5.7.3 is true when $F = K$. Although K^0 is not assumed to have nonempty interior, Theorem 5.7.3 requires that K must be a distance set. This, however, is a weaker assumption because the cones in L^p - space are distance sets whose interior is empty. We note also that every closed convex set in a reflexible Banach space is a distance set.

5.8 Integro-differential Equations

In this section, we investigate the initial value problem of first-order nonlinear integro-differential equation of Volterra type, namely

$$x' = H(t, x, Tx), x(t_0) = x_0 \tag{5.8.1}$$

in a real Banach space E, where

$$(Tx)(t) = \int_{t_0}^{t} K(t,s)x(s)ds, \tag{5.8.2}$$

$$\begin{aligned}
&K \in C(J \times J, R), \ |K(t,s)| \leq K_1 \ \text{ on } J \times J, \\
&H \in C(J \times \Omega \times \Omega, E), \\
&J = [t_0, t_0 + a], \ \ t_0 \geq 0, \\
&\Omega = B(0, N) = \{\|x\| < N : x \in E\}, \\
&x_0 \in \Omega
\end{aligned}$$

The problem (5.8.1) is equivalent to the following integral equation:

$$x(t) = x_0 + \int_{t_0}^{t} H(s, x(s), (Tx)(s))ds. \tag{5.8.3}$$

We obtain an existence theorem using Darbo's fixed point theorem and an existence and uniqueness theorem by means of the classical method of successive approximations.

Let us begin by prove the following existence result.

Theorem 5.8.1. *Assume that*

(A_1) $K(t,s) \in C(J \times J, R)$ *and* $|K(t,s)| \leq K_1(K_1 a \geq 1)$ *for* $(t,s) \in J \times J$;

(A_2) $H(t,x,y) \in C(J \times \Omega \times \Omega, E), \Omega_1 = B(0, K_1 a N) \subset E$;

(A_3) $\alpha(H(I \times B_1 \times B_2)) \leq \lambda \ \max[\alpha(B_1), \alpha(B_2)]$ *for every bounded subset* $B_1 \subset \Omega$, $B_2 \subset \Omega_1$, *and every interval* $I \subset J$, *where* λ *is some positive number and* $\alpha(\cdot)$ *is Kuratowski's measure of noncompactness.*

Then there exists a $\gamma > 0$ *such that* (5.8.1) *has a solution* $x(t)$ *for* $t \in J_0 = [t_0, t_0 + \gamma]$.

Proof. Let $\eta = \sup\{\epsilon : B(x_0, \epsilon) \subset \Omega\}$. It follows by the assumption (A_3) that $H(t,x,y)$ maps bounded sets into bounded sets, and hence there exists a constant $M > 0$, such that

$$\|H(t,x,y)\| \leq M \ \text{ for } t \in J, x, y \in B(0, \|x_0\| + \frac{1}{2}\eta) = \Omega_0.$$

Let $\gamma = \min(a, \eta/2M, 1/K_1, 1/2\lambda)$. Define the operator $H = C(J_0, E) \to C(J_0, E)$ by

$$(H\phi)(t) = x_0 + \int_{t_0}^{t} H(s, \phi(s), (T\phi)(s))ds.$$

Define the set A by

$$A = \{\phi \in C(J_0, \bar{\Omega}_0) : \max_{J_0} |\phi(t) - x_0| \leq \tfrac{1}{2}\eta, \ |\phi(t_1) - \phi(t_2)| \\ \leq M|t_1 - t_2|, \text{ for } t_1, t_2 \in J_0\}.$$

It is clear that A is closed, bounded and convex. If $\phi \in A$ then, by the assumptions $(A_1), (A_2)$ and the choice of γ, we have

$$|(T\phi)(t)| = |\int_{t_0}^{t} K(t,s)\phi(s)ds| \leq K_1\gamma \max_{s\in[t_0,t]} |\phi(s)| \leq \|x_0\| + \frac{1}{2}\eta, \ t \in J_0.$$

Hence

$$\|(H\phi)(t) - x_0\| \leqslant \int_{t_0}^{t} \|H(s, \phi(s), (T\phi)(s))\|ds \leq M(t - t_0) \leq \frac{1}{2}\eta, \ t \in J_0$$

and

$$\|(H\phi)'(t)\| = \|H(t, \phi(t), (T\phi)(t))\| \leq M, \quad t \in J_0.$$

Therefore

$$\|(H\phi)(t_1) - (H\phi)(t_2)\| \leq M|t_1 - t_2|, \quad t_1, t_2 \in J_0.$$

Thus $HA \subset A$, and clearly $H \in C(A, A)$. Now note that A is a bounded equicontinuous set. Let $B \subset A$; then, using the properties of α, assumption (A_3) and $\gamma\lambda \leq \frac{1}{2}$, we have

$$\begin{aligned} \alpha((HB)(t)) &= \alpha\left(\int_{t_0}^{t} H(s, \phi(s), (T\phi)(s))ds + x_0 \mid \phi \in B\right) \\ &\leq \gamma\alpha(\overline{co}\{H(s, \phi(s), (T\phi)(s)) \mid \phi \in B, s \in J_0\}) \\ &\leq \gamma\alpha(H(J_0 \times B(J_0) \times (TB)(J_0))) \\ &\leq \gamma\lambda \max[\alpha[B(J_0)), \gamma K_1 \alpha(B(J_0))] \\ &\leq \tfrac{1}{2}\alpha(B), \end{aligned}$$

where

$$B(J_0) = \bigcup_{s\in J_0} \{\phi(s), \phi \in B\}$$

and

$$\begin{aligned} \alpha((TB)(J_0)) &= \alpha\left(\bigcup_{t\in J_0} \int_{t_0}^{t} K(t,s)\phi(s)ds, \phi \in B\right) \\ &\leq K_1\gamma\alpha(B(J_0)). \end{aligned}$$

Since HB is also a bounded equicontinuous set, we have

$$\alpha(H(B)) = \sup_{J_0} \alpha((HB)(t) \leq \frac{1}{2}\alpha(B).$$

Consequently, by Darbo's fixed point theorem, there exists a fixed point x of H. Clearly such a fixed point is a solution of (5.8.1). The proof is complete. □

Remark 5.8.1 If $E = R^n$ then the assumption (A_3) in Theorem 5.8.1 is superfluous.

We need the following lemmas before we proceed further.

Lemma 5.8.1. *Assume that*

(L_1) $g \in C(R_0, R)$, *where* $R_0 = [t_0, t_0+a) \times (-q, q) \times (-K_1 aq, K_1 aq)$, $|g(t, u, v)| \leq M_0$, *for* $(t, u, v) \in R_0$, *and* $g(t, u, v)$ *is nondecreasing in* v *for each* (t, u);

(L_2) $[t_0, t_0 + a)$ *is the largest interval in which the maximal solution* $\gamma(t, t_0, u_0)$ *of*

$$u' = g(t, u, Su), \quad u(t_0) = u_0 \tag{5.8.4}$$

exists, where $u_0 \in (-q, q)$ *and*

$$(Su)(t) = \int_{t_0}^{t} N(t, s)u(s)ds,$$

$N(t, s) \in C(J \times J, R^+)$ *and* $N(t, s) \leq K_1$, *for* $(t, s) \in J \times J$.

Suppose further that $J_1 = [t_0, t_1] \subset [t_0, t_0 + a]$.

Then there is an $\epsilon_0 > 0$ *such that for* $0 < \epsilon < \epsilon_0$, *the maximal solution* $\gamma(t, \epsilon)$ *of*

$$u' = g(t, u, Su) + \epsilon, \quad u(t_0) = u_0 + \epsilon, \tag{5.8.5}$$

exists on J_1 *and*

$$\lim_{\epsilon \to 0} \gamma(t, \epsilon) = \gamma(t, t_0, u_0)$$

uniformly on J_1.

Proof. let $\triangle$ be an open interval $\triangle \subset (-q, q)$, and $(t, \gamma(t, t_0, u_0)) \in J \times \triangle$ for $t \in J_1$. We can choose a $b > 0$ such that for $t \in J_1$, the rectangle $R_t^\epsilon = [(s, u) : s \in [t, t + b]$ and $|u - (\gamma(t) + \epsilon)| \leq b$ is included in $J \times \triangle$ for $\epsilon \leq \frac{1}{2}b$. Note that

$$(Su)(t) \leq \int_{t_0}^{t} N(t, s)|u(s)|ds \leq K_1|t - t_0| \max_{s \in [t_0, t]} |u(s)| < K_1 aq$$

and $|g(t, u, Su)| \leq M_0$ for $(t, u(t), (Su)(t)) \in R_0$. Then it is evident that

$$|g(t, u, Su) + \epsilon| \leq M_0 + \frac{1}{2}b.$$

Let $\eta = \min(b, 2b/(2M_0 + b), 1/K_1)$. Note that η does not depend upon ϵ. Then using the special case of Theorem 5.8.1, we can show that (5.8.5) has the maximal solution $\gamma(t, \epsilon)$ and

$$\lim_{\epsilon \to 0} \gamma(t, \epsilon) = \gamma(t, t_0, u_0)$$

uniformly on $[t_0, t_0 + \eta]$. The rest of the proof is standard.

Lemma 5.8.2. *Assume that* (L_1) *and* (L_2) *hold, and suppose further that:*

(L_3) $m \in C(t_0, t_0 + a), R)$ *is such that* $(t, m(t)) \in J \times (-q, q)$ *for* $t \in [t_0, t_0 + a), m(t_0) \leq u_0$ *and for a fixed Dini derivative*

$$D_- \, m(t) \leq g(t, m(t), (Sm)(t)), \quad t \in [t_0, t_0 + a).$$

Then

$$m(t) \leq \gamma(t, t_0, \epsilon) \quad t \in [t_0, t_0 + a). \tag{5.8.6}$$

Proof. First of all, we prove that $m(t) \leq \gamma(t, \epsilon)$, on $[t_0, t_0 + a)$, where $\gamma(t, \epsilon)$ is the maximal solution of (5.8.5). In fact, if this assertion is false then the set $Z = [t \in [t_0, t_0 + a) : m(t) > \gamma(t, \epsilon)]$ is nonempty. Define $t_1 = \inf Z$. Since $m(t_0) \leq u_0 < \gamma(t_0, \epsilon) + \epsilon,\ t_1 > t_0$. Furthermore $m(t_1) = \gamma(t_1, \epsilon)$ and $m(t) \leq \gamma(t, \epsilon), t \in [t_0, t_1)$. Then

$$D_- \, m(t_1) \geq D_- \, \gamma(t_1, \epsilon).$$

On the other hand, since $N(t, s) \geq 0, (Sm)(t) \leq (S\gamma)(t), t \in [t_0, t_1)$, and using the monotonicity of g, we obtain

$$\begin{aligned} D_- \, m(t_1) \quad & \leq g(t_1, m(t_1), (Sm)(t_1)) \\ & \leq g(t_1, \gamma(t_1, \epsilon), (S\gamma)(t_1)) \\ & < D_- \, \gamma(t_1, \epsilon). \end{aligned}$$

This contradiction proves that Z is empty. Thus

$$m(t) \leq \gamma(t, \epsilon) \quad \text{on } [t_0, t_0 + a).$$

Finally, by Lemma 5.8.1, we have, as $\epsilon \to 0$,

$$m(t) \leq \gamma(t, t_0, \epsilon) \quad t \in [t_0, t_0 + a).$$

The proof of the lemma is complete. □

We are now in a position to prove our main result.

Theorem 5.8.2. *Assume that*

$(A_1)^*$ $K(t, s) \in C(J \times J, R)$ *and* $|K(t, s)| \leq K$ *for* $(t, s) \in J \times J$;

$(A_2)^*$ $H(t, x, y) \in C(J \times \Omega \times \Omega, E)$, *and* $\|H(t, x, y)\| \leq M_1$ *for* $(t, x, y) \in J \times \Omega \times \Omega$;

$(A_4)^*$ $g \in (J \times [0, 2N) \times [0, 2N_1], R^+), N_1 = KaN, g(t, u, v) \leq M_0$ *on* $J \times [0, 2N] \times [0, 2N_1], g(t, 0, 0) \equiv 0, g(t, u, v)$ *is nondecreasing in* v *for fixed* t *and* u, *and* *nondecreasing in u for fixed t and v, and $u \equiv 0$ is the unique solution of the scalar integro-differential equation*

$$u'(t) = g(t, u, Su), \quad u(t_0) = 0 \tag{5.8.7}$$

on $[t_0, t_0 + a]$, where

$$(Su)(t) = \int_{t_0}^{t} |K(t,s)|u(s)ds;$$

$(A_5)^*$ $\|H(t, x_1, y_1) - H(t, x_2, y_2)\| \leq g(t, \|x_1 - x_2\|, \|y_1 - y_2\|)$ *on* $J \times \Omega \times \Omega$.

Then there exists a $\gamma > 0$ such that (5.8.1) has an unique solution for $t \in [t_0, t_0 + \gamma]$.

Proof. We shall prove the theorem by the method of successive approximations. Let $\eta = \sup\{\epsilon : B(x_0, \epsilon) \subset \Omega\}$, $M = \max(M_0, M_1), \gamma = \min(a, \eta/2M, 1/K_1)$ and $J_0 = [t_0, t_0 + \gamma]$. The successive approximations are defined by

$$\begin{cases} x_0(t) = x_0 \\ x_n(t) = x_0 + \int_{t_0}^{t} H(s, x_{n-1}(s), (Tx_{n-1})(s))ds, \ (n = 1, 2, \ldots). \end{cases} \tag{5.8.8}$$

If $x \in C(J_0, \Omega)$, then, by the assumption $(A_1)^*$

$$|(Tx((t)| \leq \int_{t_0}^{t} |K(t,s)|\|x(s)\|ds \leq K_1\gamma \max_{s\in[t_0,t]} \|x(s)\| < N;$$

that is, T maps $C(J_0, \Omega) \to \Omega$. By the assumption $(A_2)^*$ and the choice of γ, we have

$$\|x_n(t) - x_0\| \leq M|t - t_0| \leq \frac{1}{2}\eta \text{ on } J_0,$$

and hence

$$\|x_n(t)\| \leq \|x_0\| + \frac{1}{2}\eta \leq N.$$

Thus the successive approximations (5.8.8) are well defined and continuous on $[t_0, t_0 + \gamma]$.

We also define the successive approximations for the problem (5.8.7) as follows

$$\begin{cases} u_0(t) = M(t - t_0) \leq \frac{1}{2}\eta < N, \\ u_{n+1}(t) = \int_{t_0}^{t} g(s, u_n(s), (Su_n)(s))ds, \quad t \in J_0, \quad n = 1, 2, \ldots. \end{cases} \tag{5.8.9}$$

Note that

$$u_1(t) = \int_{t_0}^{t} g(s, u_0(s), (Su_0)(s)ds \leq M(t - t_0) = u_0(t)$$

by the assumption (A_4) and $0 \leq |K(t,s)| \leq K_1$ for $(t,s) \in J \times J$. By induction, it is easy to prove that the successive approximations are well defined and satisfy

$$0 \leq u_{n+1}(t) \leq u_n(t), \quad t \in J_0.$$

Since $|u_n(t)| \leq M$, we conclude by the Ascoli-Arzela theorem and the monotonicity of the sequence $\{u_n(t)\}$ that $\lim_{n\to\infty} u_n(t) = u(t)$ uniformly on J_0. It is also clear that $u(t)$ satisfies (5.8.7). Hence by $(A_4), u(t) \equiv 0$ on J_0.

Now since

$$\|x_1(t) - x_0\| \leq \int_{t_0}^{t} H(s, x_0(s), (Tx_0)(s))|ds \leq M(t - t_0) = u_0(t),$$

we assume that $\|x_k(t) - x_{k-1}(t)\| \leq u_{k-1}(t)$ for a given k. Then using the assumption (A_4) and (A_5), we have

$$\begin{aligned}
\|x_{k+1}(t) - x_k(t)\| &\leq \int_{t_0}^{t} \|H(s, x_k(s), (Tx_k)(s)) - H(s, x_{k-1}(s), (Tx_{k-1})(s))\|ds \\
&\leq \int_{t_0}^{t} g(s, \|x_k(s) - x_{k-1}(s)\|, |(Sx_k)(s) - (Sx_{k-1})(s)|)ds \\
&\leq \int_{t_0}^{t} g(s, u_{k-1}(s), |Su_{k-1})(s)|)ds = u_k(t)
\end{aligned}$$

Thus, by induction, the inequality

$$\|x_{n+1}(t) - x_n(t)\| \leq u_n(t), \ t \in J_0,$$

is true for all n. Also

$$\begin{aligned}
\|x'_{n+1}(t) - x'_n(t)\| &= \|H(t, x_n(t), (Tx_n)(t)) - H(t, x_{n-1}(t), (Tx_{n-1})(t))\| \\
&\leq g(t, \|x_n(t) - x_{n-1}(t)\|, |(Sx_n)(t) - (Sx_{n-1})(t))|) \\
&\leq g(t, u_{n-1}(t), (Su_{n-1})(t)).
\end{aligned}$$

Let $n \leq m$. Then we have

$$\begin{aligned}
\|x'_n(t) - x'_m(t)\| &= \|x'_{n+1}(t) - x'_n(t)\| + \|x'_{m+1}(t) - x'_m(t)\| \\
&+ \|x'_{n+1}(t) - x'_{m+1}(t)\| \\
&\leq g(t, u_{n-1}(t), (Su_{n-1})(t)) + g(t, u_{m-1}(t), (Su_{m-1})(t)) \\
&+ g(t, \|x_n(t) - x_m(t)\|, |(Sx_n)(t) - (Sx_m)(t)|).
\end{aligned}$$

Since $u_{n+1}(t) \leq u_n(t)$ for all n, it follows that

$$\begin{aligned} D^+(\|x_n(t) - x_m(t)\|) &\leq g(t, \|x_n(t) - x_m(t)\|, |(S(x_n - x_m)(t)|) \\ &+2g(t, u_{n-1}(t), (Su_{n-1})(t)). \end{aligned}$$

Using Lemma 5.8.2, we have

$$\|x_n(t) - x_m(t)\| \leq \gamma_n(t), \quad t \in J_0$$

where $\gamma_n(t)$ is the maximal solution of

$$v' = g(t, v, Sv) + 2g(t, u_{n-1}(t), (Su_{n-1})(t)), \quad v(0) = 0$$

for each n. Since, as $n \to \infty, 2g(t, u_{n-1})(t)), (Su_{n-1})(t)) \to 0$ uniformly on J_0, it follows by Lemma 5.8.2 that $\gamma_n(t) \to 0$ uniformly on J_0. This implies that $x_n(t)$ converges uniformly to $x(t)$, and it is now easy to show that $x(t)$ is a solution of (5.8.1) by standard arguments.

To show that this solution is unique, let $y(t)$ be another solution of (5.8.1) existing on J_0. Define $m(t) = \|x(t) - y(t)\|$ and note that $m(t_0) = 0$ using the assumption (A_5); then

$$\begin{aligned} D^+m(t) &\leq \|x(t) - y(t)\| \\ &= \|H(t, x(t), (Tx)(t)) - H(t, y(t), (Ty)(t))\| \\ &\leq g(t, m(t), (Sm)(t)). \end{aligned}$$

Again applying Lemma 5.8.2, we have

$$m(t) \leq \gamma(t), \quad t \in J_0$$

where $\gamma(t)$ is the maximal solution of (5.8.7). But by the assumption $(A_4), \gamma(t) \equiv 0$ and this proves that $x(t) \equiv y(t)$. Hence the limit of the successive approximations is the unique solution of (5.8.1). The proof is complete. □

Corollary 5.8.1. *Assume that* $(A_1)^*$ *and* $(A_2)^*$ *hold and suppose further that*

(A_6) $\|H(t, x_1, y_1) - H(t, x_2, y_2)\| \leq L(\|x_1 - x_2\| + \|y_1 - y_2\|)$ *for any* (t, x_1, y_1), $(t, x_2, y_2) \in J \times \Omega \times \Omega$, *where* L *is a Lipschitz constant.*

Then there exists a $\gamma > 0$ *such that the problem* (5.8.1) *has a unique solution on* J_0.

Using the monotone iterative method, we discuss the existence of maximal and minimal solutions of the initial value problem of the nonlinear integro-differential equation

$$x'(t) = H(t, x, Sx), \quad x(0) = x_0, \tag{5.8.10}$$

in a real Banach space E, where $(Sx)(t) = \int_0^t S(t, s)x(s)ds$, for $x \in C(I, E), x_0 \in E, S(\cdot, \cdot) \in C(I \times I, R_+), S(t, s) \leq S_0$ for $(t, s) \in I \times I, H \in C(I \times E \times E, E), I = [0, \tau]$, and $\tau > 0$ is finite.

Let K be cone of E, then the cone K induces a partial ordering on E defined by $u \leq v$ if $v - u \in K$.

Let E^* denote the set of continuous linear functionals on E. Given a cone K, we let

$$K^* = \{\phi \in E : \phi(u) \geq 0 \quad \text{for all } u \in K\}.$$

Recall that a cone K is said to be normal if there exists a real number $L > 0$ such that $0 \leq u \leq v$ implies $\|u\| \leq L\|v\|$, where L is independent of u and v. We shall assume in this section that K is a normal cone.

Definition 5.8.1. *The function $w \in C^1(I, E)$ is called an upper solution of (5.8.10) if*

$$w'(t) \geq H\left(t, w(t), \int_0^t S(t,s)w(s)ds\right), \quad w(0) \geq x_0. \tag{5.8.11}$$

Similarly, a function $v \in C^1(I, E)$ is called a lower solution of (5.8.10) if

$$v'(t) \geq H\left(t, v(t), \int_0^t S(t,s)v(s)ds\right), \quad v(0) \geq x_0. \tag{5.8.12}$$

Definition 5.8.2. *The functions ρ and $r \in C^1(I, E)$ are called minimal and maximal solutions of (5.8.10) respectively if every solution $x \in C^1(I, E)$ of (5.8.10) satisfies the relation $\rho(t) \leq x(t) \leq r(t)$ for $t \in I$.*

We need the following lemmas.

Lemma 5.8.3. *Let E_1 be s separable Banach space and $\beta(\cdot)$ the Hausdorff measure of noncompactness on E_1. Let $\{x_n\}$ be a sequence of continuous functions from $J = [a, b]$ to E_1 such that there is some function $u \in L^1(a, b)$ with $\|x_n(t)\| \leq u(t)$ on J. Let $\psi(t) = \beta(\{x_n(t)\}_{n=1}^{\infty})$. Then $\psi(t)$ is integrable on J and*

$$\left(\left\{\int_a^b x_n(t)dt\right\}_{n=1}^{\infty}\right) \leq \int_a^b \psi(t)dt.$$

Lemma 5.8.4. *Let $\{x_n\}$ be a sequence of continuously differentiable functions from $J = [a, b]$ to E such that there is some $u \in L^1(a, b)$ with $\|x_n(t)\| \leq u(t)$ and $\|x_n'(t)\| \leq u(t)$ on J. Let $\psi(t) = \beta(\{x_n(t)\}_{n=1}^{\infty})$. Then $\psi(t)$ is absolutely continuous and*

$$\psi'(t) \leq 2\beta(\{x'_{n(t)}\}_{n=1}^{\infty}) \quad a.e. \text{ on } J.$$

We make the following assumptions for convenience.

(A_1) The functions $v_0, w_0 \in C^1(I, E)$ with $v_0(t) \le w_0(t)$ on I are lower and upper solutions of (5.8.10) respectively. We define the conical segments

$$[v_0, w_0] = \{u \in E : v_0(t) \le u \le w_0(t),\ t \in I\}.$$

Similarly,

$$[Sv_0, Sw_0] = \{u \in E : (Sv_0)(t) \le u \le (Sw_0)(t),\ t \in I\}.$$

(A_2) $H(t,x,u) - H(t,\bar{x},\bar{u}) \ge -M(x-\bar{x}) - N(u-\bar{u})$ whenever $x, \bar{x} \in [v_0, w_0], u, \bar{u} \in [Sv_0, Sw_0]$ and $x \ge \bar{x}, u \ge \bar{u}$, where $M > 0, N \ge 0$ are constants with $NS_0\tau(e^{M\tau} - 1) \le M$.

(A_3) For any bounded set $B \subset [v_0, w_0], B_1 \subset [Sv_0, Sw_0], t \in I, \beta(H(t, B, B)) \le \lambda \max(\beta(B), \beta(B_1))$ for some $\lambda > 0$, where $\beta(\cdot)$ denotes the Hausdorff measure of noncompactness. Then we have the following main result.

Theorem 5.8.3. *Let the cone K be normal and the assumptions $(A_1) - (A_3)$ hold. Then, for any $x_0 \in [v_0(0), w_0(0)]$, there exist monotone sequences $\{v_n\}$ and $\{w_n\}$ that converge uniformly and monotonically to the minimal and maximal solutions $\rho(t)$ and $r(t)$ respectively of the problem* (5.8.10) *in $[v_0(0), w_0(0)]$, and*

$$v_0 \le v_1 \le v_2 \le ... \le v_n \le \rho \le u \le r \le w_n \le ... \le w_2 \le w_1 \le w_0 \quad on \quad I.$$

The proof of Theorem 5.8.3 will be completed by the following five lemmas.

Lemma 5.8.5. *Let $y(t) \in C^1(I, R)$ be such that*

$$y'(t) \le -My(t) - N\int_0^t S(t,s)y(s)ds, \quad y(0) \le 0, \tag{5.8.13}$$

where $M > 0, N \ge 0, S(\cdot,\cdot) \in C(I \times I, R^+), S(t,s) \le S_0$ for $(t,s) \in I \times I$. Suppose further that $NS_0\tau(e^{M\tau} - 1) \le M$. Then $y(t) \le 0, t \in I$.

Proof. Set $m(t) = y(t)e^{Mt}$, so that the inequality (5.8.13) reduces to

$$m'(t) \le -N\int_0^t S^*(t,s)m(s)ds, \quad m(0) \le 0, \tag{5.8.14}$$

where $S^*(t,s) = S(t,s)e^{M(t-s)}$. It is enough to prove that $m(t) \le 0$, for $t \in I$. If this is false then there exists $t_1 \in (0, \tau], m(t_1) > 0$. If $m(t) \ge 0$ for $t \in [0, t_1]$ then from (5.8.14) we get $m'(t) \le 0, t \in [0, t_1]$; that is, if $m(t)$ is nonincreasing in $[0, t_1]$ then $m(t_1) \le m(0) \le 0$. This is a contradiction. So there exists

$t_2 \in [0,t_1], m(t_2) = \min_{t\in[0,t_1]} m(t) = -\mu < 0$. By the mean value theorem on $[t_2,t_1]$; there exists $t_3 \in (t_2,t_1)$ such that

$$m'(t_3) = \frac{m(t_1)+\mu}{t_1-t_2} > \frac{\mu}{\tau}$$

On the other hand, it follows from (5.8.14) that

$$\begin{aligned} m'(t) &\le -N\int_0^{t_3} S^*(t_3,s)m(s)ds \le \frac{NS_0\mu}{M}(e^{Mt_3}-1) \\ &\le \tfrac{NS_0\mu}{M}(e^{M\tau}-1) \end{aligned}$$

Then from

$$\frac{NS_0\mu}{M}(e^{M\tau}-1) > \frac{\mu}{\tau},$$

we get

$$NS_0\tau(e^{M\tau}-1) > M.$$

This is contradiction to the assumption $NS_0\tau(e^{M\tau}-1) \le M$. The proof of Lemma 5.8.5 is complete. □

Lemma 5.8.6. *Let $y(t) \in C^1(I,R)$, $y(0) \le 0$, satisfy*

$$y'(t) \le A_1y(t) + A_2\int_0^t y(s)ds \quad a.e. \quad on\ I = [0,\tau], \tag{5.8.15}$$

where $A_1, A_2 > 0$. Then $y(t) \le 0$, for any $t \in I$.

Proof. If the conclusion is false we choose $\eta > 0$ such that $(A_1 + A_2\eta)\eta < 1$ and $n\eta = \tau$ for some integer n. First of all, we prove that $y(t) \le 0$ for $t \in [0,\eta]$. If this is not true then there exists $t_1 \in [0,\eta]$ such that $y(t_1) = \mu = \max_{t\in[0,\eta]} y(t) > 0$. Define

$$\sum = \left\{ t \in (0,t_1] : y'(t) \text{ exists and } y'(t) \le A_1y(t) + A_2\int_0^t y(s)ds \right\}.$$

So mes$(\sum) = t_1$. Then it is certain that there exists a $t_2 \in \sum$ such that $y'(t_2) \ge \mu/t_1$. Otherwise, if $y'(t) \ge \mu/t_1$ for all $t \in \sum$ then we get

$$\mu \le \int_0^{t_1} y'(t)dt = \int_{\Sigma} y'(t)dt < \mu.$$

This is a contradiction. On the other hand, from (5.8.15), we have

$$\begin{aligned} \tfrac{\mu}{t_1} &\le y'(t_2) \le A_1y(t_2) + A_2\textstyle\int_0^{t_2} y(t)dt \\ &\le A_1\mu + A_2\mu\eta, \end{aligned}$$

and hence

$$1 \leq (A_1 + A_2\eta)\eta.$$

This is a contradiction to the choice of η. Thus we have proved that $y(t) \leq 0$, for $t \in [0, \eta]$. We can further prove that $y(t) \leq 0$ for $t \in [\eta, 2\eta]$, by repeating the above process. This argument can be repeated until $n\eta = \tau$. So we get $y(t) \leq 0$, for $t \in [0, \tau]$. The proof of Lemma 5.8.6 is complete. □

In order to prove Theorem 5.8.3, let us consider the following linear initial value problem:

$$x'(t) = -Mx - N\int_0^t S^*(t,s)x(s)ds + \sigma(t), \quad x(0) = x_0, \tag{5.8.16}$$

where

$$\sigma(t) = H(t, \eta(t), (S\eta)(t)) + M\eta(t) + N(S\eta)(t)$$

for $\eta \in C(I, E)$ such that $\eta(t) \in [v_0, w_0]$ for all $t \in I$ and $x_0 \in E$ such that $v_0(0) \leq x_0 \leq w_0(0)$.

Lemma 5.8.7. *For any $\eta \in C(I, E)$ such that $\eta(t) \in [v_0, w_0]$ for $t \in I$, the problem* (5.8.16) *possesses a unique solution on I.*

Proof. It is easy to check that the problem of (5.8.16) is equivalent to the following equation:

$$x(t) = x_0 e^{-Mt} + \int_0^t \sigma(s)e^{M(s-t)}ds - Ne^{-Mt}\int_0^t \left[e^{Ms}\int_0^s S(s,\xi)x(\xi)d\xi \right] ds. \tag{5.8.17}$$

By changing the order of integration, (5.8.17) reduces to the following linear integral equation of Volterra type:

$$x(t) = x_0 e^{-Mt} + \int_0^t \sigma(s)e^{M(s-t)}ds + \int_0^t S^*(t,\xi)x(\xi)d\xi. \tag{5.8.18}$$

where

$$S^*(t,\xi) = -Ne^{-Mt}\int_\xi^t S(s,\xi)e^{Ms}ds \in C(I \times I, R).$$

Now, it is easy to see that (5.8.18) has a unique solution on I. The proof of Lemma 5.8.7 is therefore complete. □

For each $\eta \in C(I, E)$ such that $v_0(t) \leq \eta(t) \leq w_0(t)$ on I, we define the mapping A by $A\eta = x$, where x is the unique solution of (5.8.16) corresponding to η. This mapping will be used to define the sequences that converge to the minimal and maximal solution of (5.8.10)

Lemma 5.8.8. *Assume that the hypotheses (A_1) and (A_2) hold. Then*

(i) $v_0 \leq Av_0$ *and* $w_0 \geq Aw_0$;

(ii) A *is a monotone operator on the segment* $[v_0, w_0]$; *that is, if* $\eta_1(t), \eta_2(t) \in [v_0, w_0]$ *for* $t \in I$ *with* $\eta_1 \leq \eta_2$ *then* $A\eta_1 \leq A\eta_2$ *for* $t \in I$.

Proof. To prove (i), set $Av_0 = v_1$, where v_1 is the unique solution of (5.8.16) corresponding to v_0. For any $\phi \in K^*$, set $y(t) = \phi(v_0(t) - v_1(t))$, so that $y(0) \leq 0$, and $y'(t) = \phi(v_0'(t) - v'(t))$. Then we have

$$\begin{aligned} y'(t) &\leq \phi\left(H(t, v_0, Sv_0) - H(t, v_0, Sv_0) + M(v_1 - v_0) + N\int_0^t S(t,s)(v_1 - v_0)ds\right) \\ &= -My - N\int_0^t S(t,s)y(s)ds. \end{aligned}$$

By Lemma 5.8.5, it follows that $y(t) \leq 0$ on I. Since $\phi \in K^*$ is arbitrary, this implies that $v_0(t) \leq v_1(t), t \in I$. Thus we have proved that $v_0 \leq Av_0$ for $t \in I$. Similar argument shows that $w_0 \geq Aw_0$ for $t \in I$, proving (i).

To prove (ii), let $\eta_1, \eta_2 \in C(I, E)$ be such that $\eta_1(t), \eta_2(t) \in [v_0, w_0]$ for $t \in I$ and $\eta_1 < \eta_2$. Suppose that $\xi_1 = A\eta_1$ and $\xi_2 = A\eta_2$. Let $Z(t) = \phi(\xi_1(t) - \xi_2(t))$, so that $Z(0) \leq 0$, where $\phi \in K^*$. Then

$$\begin{aligned} Z'(t) &= \phi(\xi_1'(t) - \xi_2'(t)) \\ &= \phi(H(t, \eta_1(t), (S\eta_1)(t)) - M(\xi_1(t) - \eta_1(t)) - N((S\xi_1(t) - (S\eta_1(t)) \\ &\quad -H(t, \eta_2(t), (S\eta_2)(t)) - M(\xi_2(t) - \eta_2(t)) - N((S\xi_2(t) - (S\eta_2(t)) \\ &\leq -MZ(t) - N\int_0^t S(t,s)Z(s)ds, \end{aligned}$$

using (A_2). By Lemma 5.8.5, we get $Z(t) \leq 0$ on I. Since $\phi \in K^*$ is arbitrary, this implies that $\xi_1(t) \leq \xi_2(t)$ on I. That is $A\eta_1 \leq A\eta_2$. The proof of Lemma 5.8.6 is therefore complete. □

We now define the sequences $\{v_n\}$ and $\{w_n\}$ as follows:

$$v_n = Av_{n-1}, \quad w_n = Aw_{n-1}.$$

It is easy to see from Lemma 5.8.6 the sequences $\{v_n\}$ and $\{w_n\}$ are monotone nondecreasing and nonincreasing respectively, and $v_n \leq w_n$ and $v_n(t), w_n(t) \in [v_0, w_0]$ for all $t \in I$. Now we show that there exist subsequences of $\{v_n\}$ and $\{w_n\}$ that converge uniformly on I.

Lemma 5.8.9. *Suppose that the assumptions (A_1) - (A_3) hold. Then the sequences $\{v_n\}$ and $\{w_n\}$ are uniformly bounded, equicontinuous and relatively compact on I.*

Proof. Since the cone K is assumed to be normal, it follows from $v_n(t), w_n(t) \in [v_0, w_0]$ for all $t \in I$, that $\{v_n\}$ and $\{w_n\}$ are uniformly bounded.

By means of the assumption (A_2) and $v_n(t), w_n(t) \in [v_0, w_0], (Sv_n)(t), (Sw_n)(t) \in [Sv_0, Sw_0]$, for all $t \in I$ and for each n, we have $H(t, w_0, Sw_0) - H(t, v_n, Sv_n) \geq -M(w_0 - v_n) - N(Sw_0 - Sv_n)$. Hence

$$H(t, v_n, Sv_n) \leq H(t, w_0, Sw_0) + M(w_0 - v_0) + N(Sw_0 - Sv_0).$$

Similarly, for each n, we have

$$H(t, v_n, Sv_n) \geq H(t, v_0, Sv_0) - M(w_0 - v_0) - N(Sw_0 - Sv_0).$$

Then it follows that $H(t, v_n, Sv_n)$ is uniformly bounded. Since both $\{v_n\}$ and $\{w_n\}$ are uniformly bounded and

$$v_n' = H(t, v_{n-1}, Sv_{n-1}) - M(v_n - v_{n-1}) - N(Sv_n - Sv_{n-1}),$$

this implies the equicontinuity of the sequence $\{v_n\}$. A similar argument can be used to prove that the sequence $\{w_n\}$ is equicontinuous.

Now we set $B(t) = \{v_n(t)\}_{n=0}^{\infty}, B'(t) = \{v_n'(t)\}_{n=0}^{\infty}$ and $\psi(t) = \beta(B(t))$, and since $\{v_n\}$,$\{v_n'\}$ are uniformly bounded, it is easy to see that the sequence $\{v_n\}$ satisfies the assumptions of Lemma 5.8.3 and 5.8.4. Note that

$$\begin{aligned}
&\beta\Big(\Big\{\int_0^t S(t,s)v_n(s)\mathrm{d}s\Big\}_{n=0}^{\infty}\Big) \\
&\leq \beta_{E_1}\Big(\Big\{\int_0^t S(t,s)v_n(s)\mathrm{d}s\Big\}_{n=0}^{\infty}\Big) \\
&\leq S_0 \int_0^t \beta_{E_1}(\{v_n(s)\}_{n=0}^{\infty})\mathrm{d}s \leq 2S_0 \int_0^t \psi(s)\mathrm{d}s
\end{aligned} \tag{5.8.19}$$

using the Lemma 5.8.5, where E_1 is the closed subspace of E spanned by $B(t) = \{v_n(t)\}_{n=0}^{\infty}, t \in I)$. Hence E_1 is separable (β_{E_1} denotes the Hausdorff measure of non-compactness on E_1). Using (5.8.19), the assumption (A_3) and Lemma 5.8.4, $\psi(t)$ is absolutely continuous and satisfies

$$\begin{aligned}
\psi'(t) &\leq 2\beta\big(B'(t)\big) \\
&= 2\beta\Big(\Big\{H\big(t, v_n(t), \int_0^t S(t,s)v_n(s)\mathrm{d}s\big) - M\big(v_n(t) - v_{n-1}(t)\big) \\
&- N\Big[\int_0^t S(t,s)v_n(s)\mathrm{d}s - \int_0^t S(t,s)v_{n-1}(s)\mathrm{d}s\Big]\Big\}_{n=0}^{\infty}\Big) \\
&\leq 2\lambda \max\Big[\psi(t), 2S_0 \int_0^t \psi(s)\mathrm{d}s\Big] + 4M\psi(t) + 8NS_0 \int_0^t \psi(s)\mathrm{d}s \\
&\leq A_1\psi(t) + A_2 \int_0^t \psi(s)\mathrm{d}s
\end{aligned} \tag{5.8.20}$$

where $A_1 = 2\lambda + 4M > 0, A_2 = 4S_0\lambda + 8NS_0 > 0$ and $\psi(0) = 0$. From (5.8.20), applying Lemma 5.8.6 and $\psi(0) \geq 0$, we have $\psi(0) = 0, t \in I$. Thus it implies the relative compactness of the sequence $\{v_n(t)\}$ for each $t \in I$. Similarly, $\{w_n(t)\}$ is relatively compact for each $t \in I$. The proof of Lemma 5.8.9 is complete. □

We now apply Ascoli's theorem to the sequences $\{v_n\}$ and $\{w_n\}$ to obtain subsequences $\{v_{n_k}\}$ and $\{w_{n_k}\}$ that converge uniformly on I. Since the sequences $\{v_n\}$ and $\{w_n\}$ are monotone and K is normal, this then shows that the full sequences converge uniformly and monotonically to continuous functions, namely $\lim_{n\to\infty} v_n(t)\} = \rho(t)$ and $\lim_{n\to\infty} w_n(t)\} = r(t)$ on I. It then follows easily from (5.8.16) that $\rho(t)$ and $r(t)$ are solutions of the problem (5.8.10) on I.

Finally, we show that $\rho(t), r(t)$ are minimal and maximal solutions of (5.8.10). To this end, let $x(t)$ be any solution of (5.8.10)on I such that $x(t) \in [v_0, w_0]$ for all $t \in I$. Assume that $v_n \leq x \leq w_n$ on I. Set $p(t) = \phi(v_{n+1}(t) - x(t))$. Thus $p(0) = 0$, where $\phi \in K^*$. Then by (A_2) and the assumption that $v_n \leq x$, we have

$$\begin{aligned} p'(t) &= \phi(H(t, v_n(t), Sv_n(t)) - M(v_{n+1} - v_n) - N(Sv_{n+1} - Sv_n) \\ &\quad -H(t, x_n(t), Sx(t))) \\ &\leq -Mp(t) - N(Sp)(t). \end{aligned}$$

This implies that $v_{n+1} \leq x$ on I, using Lemma 5.8.5. Similarly, we can show that $x \leq w_{n+1}$ on I. Since $x(t) \in [v_0, w_0]$ for all $t \in I$, we have, by induction, $v_n \leq x \leq u_n$ on I for all n. Thus we obtain, taking the limits as $n \to \infty, \rho(t) \leq x(t) \leq r(t)$ on I, proving that $\rho(t)$ and $r(t)$ are minimal and maximal solutions of (5.8.10) on I. The proof is therefore complete. □

5.9 Notes and Comments

The contents of Section 5.2 are taken from Drici [1], while the material of Section 5.3 is adopted from Eisenfeld and Lakshmikantham [1,2,3]. Sections 5.4 to 5.6 consists of the work taken from Lakshmikntham and Leela [1]. The material of Section 5.7 is adopted from Lakshmikantham, Leela and Vasundhara Devi [1], while the contents of Section 5.8 are taken from Lakshmikantham and Rama Mohan Rao [1]. See also, Lakshmikantham and Leela [1], Bernfeld and Lakshmikantham [1].

References

Bailey, F. N.

[1] The Application of Lyapuov's Second Method to Interconnected Systems. *SIAM Jour. Control*, 3,(1966), 443-462.

Barbu, V.

[1] *Nonlinear Semigroups and Differential Equation in Banach Spaces.* The Netherlands: Noordhoff International Publishing, (1976).

Butz, R.

[1] Higher Order Derivatives of Lyapunov Functions. *I.E.E.E. Trans. Auto. Control*, 14, (1969), 111-112.

Bellman, R.

[1] Vector Lyapunov Functions. *SIAM Jour. Control*, 1, (1962) 32-34.

Bernfeld, S. R., Ladde, G. S.,and Lakshmikantham, V.

[1] Existence of Solutions of Two Point Boundary Value Problems for Nonlinear Systems. *J. Diff. Eq.*, 18,(1975), 103-110.

Bernfeld, S. R., and Lakshmikantham, V.

[1] *An Introduction to Nonlinear Boundary Value Problems.* New York, Academic Press, (1974).

Carl,S., Heikkila, S., Lakshmikantham, V.,

[1] Fixed Point Theorems in Ordered Banach Spaces via Quasilinearization, *Nonlinear Analysis*, 71, (2009), 3448-3458.

Chandra, J., Lakshmikantham, V., and Leela, S.,

[1] A Monotone Method for Infinite Systems of Nonlinear Boundary Value Problems. *Arch. Rat. Mech. and Anal.*, 68, (1978) 179-190.

Chandra, J., Lakshmikantham, V., and Mitchell, A. R.,

[1] Existence of Solutions of Boundary Value Problems for Nonlinear Second Order Systems in a Banach Space *Nonlinear Analysis*, 2, (1978) 157-168.

Cramer,E. J., Lakshmikantham, V., and Mitchell, A. R.,

[1] On the Existence of Weak Solutions of Differential Equations in Nonreflexive Banach Spaces, *Nonlinear Analysis*, 2, (1978) 169-177.

De Blasi, F. S.,

[1] On a Property of the Unit Sphere in a Banach Space *Bull. Mat.Soc.Sci Math. R.S. Roumanie*, 21, (1977) 259-262.

[2] On the Differentiability of Multifunctions *Pacific. J Math* (1976), 67-81.

Deimling, K., and Lakshmikantham, V.

[1] Existence and Comparison Theorems for Differential Equations in Banach Spaces. *Nonlinear Analysis*, 3, (1979), 569-575.

[2] On Existence of Extremal Solutions of Differential Equations in Banach Spaces. *Nonlinear Analysis*, 3, (1979) 563-568.

Dieudonne, J.

[1] *Foundations of Modern Analysis.* New York, Academic Press, (1971).

Deo, S. G.,

[1] On Vector Lyapunov Functions, *Proc. Amer. Math Soc*, 29, (1971) 575-580.

Drici, Z.

[1] New Directions in the Method of Vector Lyapunov Functions: Application to Large Scale Systems, *J. Math. Analy. Appl.*, (1994), V 184, 317-325.

Dunford, N., and Schwartz, J. T.

[1] *Linear Operators, Part I.* New York, Interscience Publishers Inc, (1958).

Eisenfeld, J., and Lakshmikantham, V.

[1] Existence and comparison theorems for differential equations in Banach spaces. *J. Math. Analy. Appl.*, 49, (1975), 504-511.

[2] On a Measure of Nonconvexity and Applications. *Yokahama Math J.*, 24, (1976) 133-140.

[3] On the Existence of Solutions of Differential Equations in a Banach Space, *Revue. Roum. Math. Pures et Appl.*, 22, (1977), 1215-1221.

Godunov, A. V.

[1] Peano's Theorem in Banach Spaces. *Func Anal. and Appl.*, 9 (1975), 53-55.

Grujic, L. T., Martynyuk, A. A., and Ribbens-Pavella, M.

[1] *Large Scale Systems, Stability Under Structural and Singular Perturbations.* Berlin: Springer-Verlag, (1987).

Hille, E., and Philips, R. S.,

[1] *Functional Analysis and Semigroups.* Providence, R.I., Amer. Math. Soc. Coll. Publ Vol 31, (1957).

Ikeda. M., and Siljak, D. D.,

[1] Generalized Decomposition of Dynamic Systems and Vector Lyapunov Functions, *I.E.E.E Trans. Autom. Control* 26, (1981), 1118-1125.

Karatueva, N. A., and Matrosov, V. M.,

[1] Vector Lyapunov Functions Method and its Application to Immunology, *Mathematical Modeling in Immunology and Medicinc*, North Holland, (1983), 175-186.

Kozlov, R. I., and Matrosov, V. M.,

[1] The Method of Vector Lyapunov Functions in the Qualitative Theory of Differential Equations in Banach Spaces, *Colloquia Mathematica Societatis Janos Bolyai, 30, Qualitative Theory of Differential Equations*, Szeged, Hungary, (1979), 685-707.

Krasnosel'skii, M. A.,

[1] *Positive Solutions of Operator Equations.* Noordhoff International Publishing, The Netherlands, (1964).

Kuratowskii, K.,

[1] *Topology Vol II* New York, Academic Press, (1966).

Ladas G. E., and Lakshmikantham, V.

[1] *Differential Equations in Abstract Spaces.* New York, Academic Press, (1972).

Ladde G. S., and Lakshmikantham, V.,

[1] On Flow Invariant Sets. *Pacific J. Math.*, 51, (1974), 215-220.

Ladde G. S., Lakshmikantham, V., and Leela S.,

[1] A New Technique in Perturbation Theory. *Rocky Mountain Jour,* 6, (1977), 133-140.

Ladde G. S., Lakshmikantham, V., and Vatsala A. S.,

[1] *Monotone Iterative Technique for Nonlinear Differential Equations.* Boston, Pitman Publishing, (1985).

Lakshmikantham, V.,

[1] Vector Lyapunov Functions and Conditional Stability. Journal. Math. Anal. Appl. (1965), 368-377.

[2] Several Lyapunov Functions. *Proc* 5th *Inter. Conf. on Nonlinear Oscillations,* Kiev, USSR, (1969), 268-275.

[3] Differential Equations in Banach Spaces and Extension of Lyapunov's Method. *Proc. Camb. Phi. Soc.*, 59, (1963), 373-381.

[4] Stability and Asymptotic Behavior of Solutions of Differential Equations in a Banach Space. *Lecture Notes, C.I.M.E. Italy,* (1974), 39-98.

[5] Existence and Comparison Results for Differential Equations in a Banach Space. *Proc. Int. Conf. on Differential Equations,* New York, Academic Press, (1975), 459-473.

[6] The Current State of Abstract Cauchy Problem. *Proc. Int. Conf. on Nonlinear Systems and Applications,* New York, (1977).

Lakshmikantham, V., and Leela S.,

[1] *Nonlinear Differential Equations in Abstract Spaces.* Oxford, Pergamon, (1981).

[2] Cone Valued Lyapunov Functions. *Nonlinear Anal.*, 1, (1977), 215-222.

[3] On the Method of Upper and Lower Solutions in Abstract Cones. *Ann. Polon. Math,* 42, (1983), 159-164.

[4] Method of Quasi-Upper and Lower Solutions in Abstract Cones. *Nonlinear Anal.*, 6, (1982), 833-838.

[5] Differential and Integral Inequalities, Vol I, and Vol II. New York, Academic Press, (1969).

[6] On Perturbing Lyapunov Functions. *Math Sys. Theory,* 10, (1976), 85-90.

[7] A Technique in Stability Theory for Delay Differential Equations. *Nonlinear Anal.*, 3, (1979), 317-323.

[8] On the Existence of Zeros of Lyapunov Monotone Operators. *Appl. Math. and Comp.*, 4, (1978), 107-119.

Lakshmikantham, V., Leela S., and Martynyuk, A. A.,

[1] *Stability Analysis of Nonlinear Systems.* New York, Marcel Dekker, Inc. (1989).

[2] *Practical Stability of Nonlinear Systems.* Singapore, World Scientific, (1990).

Lakshmikantham, V., Leela S., and Oguztoreli, M. N.,

[1] Quasisolutions, Vector Lyapunov Functions, and Monotone Method. *IEEE Trans. Autom. Control,* 26, (1981), 1149-1153.

Lakshmikantham, V., Leela S., and Ram Mohan Rao, M.,

[1] New Directions in the Method of Vector Lyapunov Functions, *Nonlinear Anal.,* V 16, 3, (1991), 255-262.

Lakshmikantham, V.,and Papageorgiou, N. S.,

[1] Cone Valued Lyapunov Functions and Stability Theory, *Nonlinear Anal.,* V 22, (1994), 381-398.

Lakshmikantham, V.,and Guo, D.,

[1] *Nonlinear Problems is Abstract Spaces* New York, Academic Press, (1988).

Lakshmikantham, V., Matrosov, V. M., and Sivasundaram S.,

[1] *Vector Lyapunov Functions and Stability Analysis of Nonlinear Systems* Dordrecht, Kluwer Academic Publishers, (1991).

Lakshmikantham, V., Bainov, D. D., and Simenov, P. S.,,,

[1] *Theory of Impulsive Differential Equations* Singapore, World Scientific, (1989).

Lakshmikantham, V.,and Ram Mohan Rao, M.,

[1] Integro Differential Equations and Extension of Lyapunov Method, *J. Math. Anal. Appl.,* 30, (1970), 435-447.

Lakshmikantham, V., and Vatsala, A. S.,

[1] *Generalized Quasilinearization for Nonlinear Problems,* Kluwer Academic Publishers (1998).

Lakshmikantham, V., Leela, S., and Vasundhara Devi, J.,

[1] *Theory of Fractional Dynamic Systems*, Cambridge Scienific Publishers, (2009).

Lumer, G.,

[1] Semi-inner Product Spaces, *Trans, Amer. Math. Soc.*,100, (1961), 29-43.

Matrosov, V. M.,

[1] The Comparison Principle with a Vector Lyapunov Function, I. *Differencial'nye Uravneniya,* 4, (1968), 1374-1386.

[2] Comparison Principles together with Vector Lyapunov Functions II. *Differential Equations* 4, (1968), 1739-1752.

[3] Comparison Principles together with Vector Lyapunov Functions III. *Differential Equations* 5, (1969), 1171-1185.

[4] Comparison Principles together with Vector Lyapunov Functions IV.*Differential Equations* 5, (1969), 2129-2143.

[5] The Method of Comparison in the Dynamic of Systems, I. *Differential Equations* 10, (1974), 1547-1559.

[6] The Method of Comparison in the Dynamic of Systems, II. *Differential Equations* 11, (1975), 403-417.

[7] Vector Lyapunov Functions in the Analysis of Nonlinear Interconnected Systems, *Symp. Math. Roma.,* 6, *Mechanica Nonlinear Stabilita,* (1971), 209-242.

Martin, R. H. Jr.

[1] *Nonlinear Operators and Differential Equations in Banach Spaces.* New York, John Wiley and Sons, (1976).

Mazur, S

[1] Uber Konvexe mengen in linearen normierten Raumen. *Studia Math,* 4, (1933), 70-84.

Mcleod, R.,

[1] Mean Value Theorems for Vector Valued Functions, *Proc. Ed. Math. Soc.,* 14, (1964), 197-209.

Murakami, H.

[1] On Nonlinear Ordinary and Evolution Equations. *Funkcialaj Ekvacioj,* 9, (1966), 151-162.

Nagumo, M.,

[1] Uber die Lage der Integralkurven Gewohnlicker Differential Gleichungen. *Proc. Phys.-Math. Soc. Japan,* 24, (1942), 551-559.

Rama Mohan Rao, M., Sivasundaram, S.,

[1] Perturbing Lyapunov Functions and Lagrange Stability in Terms of Two Measures, *Appl. Math. and Computaion,* 28, (1988), 329-366.

Silajak, D. D.,

[1] *Large Scale Dynamic Systems.* New York, North Holland, (1978).

[2] Competitive Economic Systems: Stability, Decomposition and Aggregation. *IEEE Transactions,* Ae-21, (1976), 149-160.

Szufla, S.,

[1] Some Remarks on Ordinary Differential Equations in Banach Spaces, Bull. Acad. Polon. Sci. Ser. Sci. Math. Astron. Phys., 16, (1968), 795-800.

[2] On the Existence of Solutions of an Ordinary Differential Equation in the Case of Banach Space. Bull. Acad. Polon. Sci. Ser. Sci. Math. Astron. Phys., 16, (1968), 311-315.

[3] Measure of Noncompactness and Ordinary Differential Equations in Banach Spaces.Bull. Acad. Polon. Sci. Ser. Sci. Math. Astron. Phys., 19, (1971), 831-835.

Volkman, P.

[1] Gewonliche Differentialungleichungen mit Quasimonoton Wachsenden Functionen in Topologischen Vector Raumen. *Math. Z.,* 127, (1972), 157-164.

[2] Uber die Invarianz Konvexer mengen und Differential Ungleichungen in Normeriten Raumen. *Math. Ann.,* 203, (1973), 201-210.

[3] Gewonliche Differentialungleichungen mit Quasimonton Wachsenden Functionen in Banach Raumen. *Proc. Conf. on Differential Equations.,* New York:Dundee, Springer-Verlag, Lecture Notes, Vol 415, (1974), 439-443.

[4] Ausdehnung eines Satzes von Max Muller auf Unendliche Systems von Gewohnlichen Differentialgleichungen. *Funkcialaj Ekvacioj.,*21, (1978), 81-96.

Voronov, A. A.,and Matrosov, V. M.,

[1] *Method of Vector Lyapunov Functions in the Theory of Stability.* Nauka, Physics-Math Literature, (1987).

Index

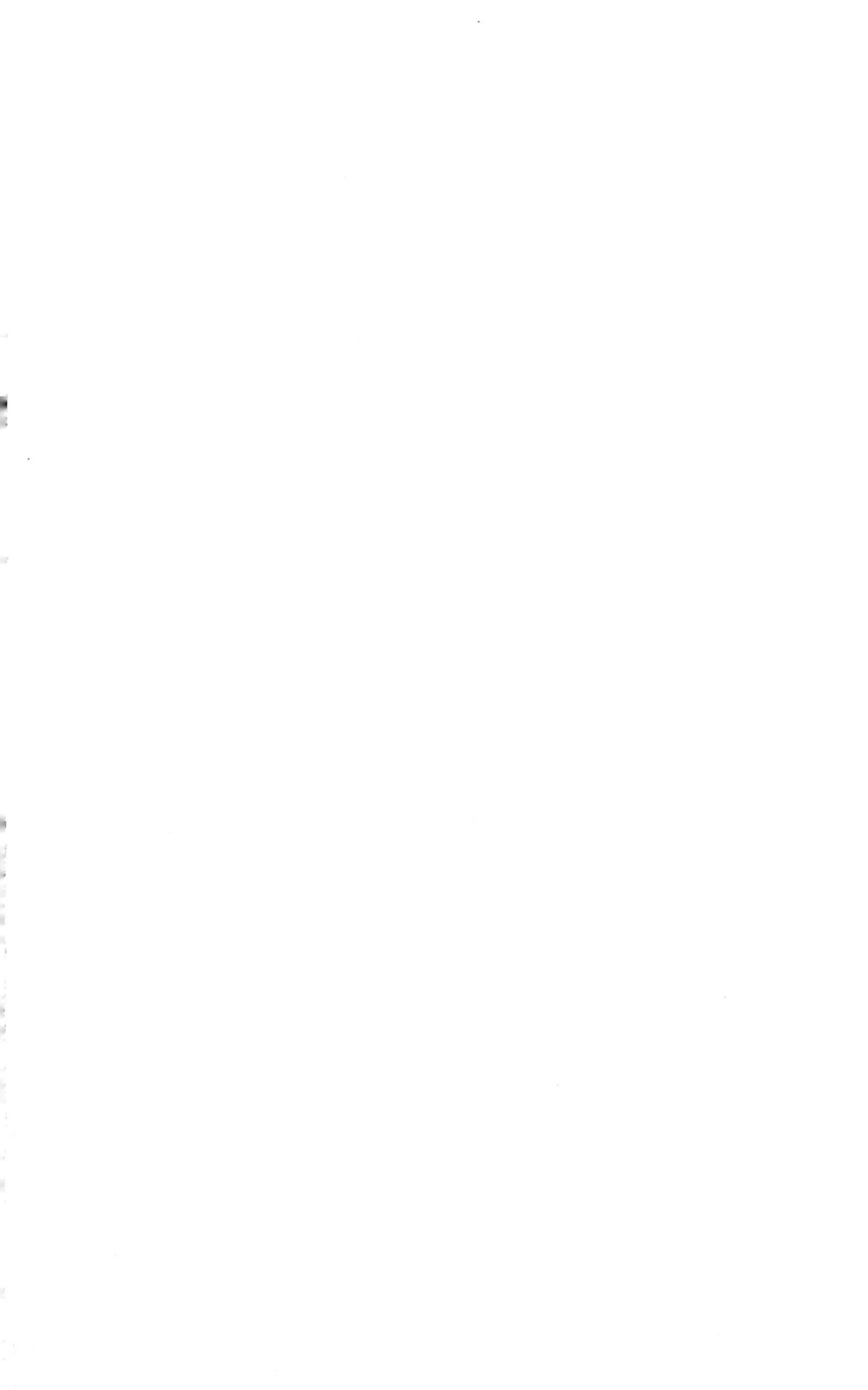